Surety Bonds for Construction Contracts

JEFFREY S. RUSSELL
The University of Wisconsin–Madison

ASCE PRESS
American Society of Civil Engineers
1801 Alexander Bell Drive
Reston, Virginia 20191–4400

Abstract: This book aims to demystify the complex field of suretyship for construction contracts by describing and analyzing each element of the surety bonding process in basic terms for contractors, owners, design professionals, construction accountants and attorneys, and other construction industry personnel. It is the only comprehensive book on surety bonding as it relates to construction contracts written from the perspective of contractors, sureties, and owners. The book outlines the basic players in the surety relationship, what each can expect at every stage of the bonding process, how to properly structure key documents, and what a contractor should look for in a surety and an independent surety agent. In addition to exploring the numerous facets of the bonding relationship, the book moves through many of the subtitles of this complex endeavor, including the surety's role in prequalifying contractors and why an owner might consider requiring, or even waiving, surety bonds. To help elucidate the interworkings of suretyship, numerous examples and extended case studies drawn from the experience of professionals working in the industry, as well as sample bond forms and surety documents, have been included. Finally, this work covers such issues as international guaranty, reinsurance, surety and Design/Build, and trends emerging in the surety market.

Library of Congress Cataloging-in-Publication Data

Russell, Jeffrey S.
Surety bonds for construction contracts / Jeffrey S. Russell.
p. cm.
Includes bibliographical references and index.
ISBN 0-7844-0426-7
1. Insurance, Surety and fidelity—United States. 2. Contractors—Bonding—United States.
I. Title.
KF1228 .R87 1999
346.73'08684—dc21 99-049594

Library of Congress Catalog Card No: 99-049594
ISBN 0-7844-0426-7
Manufactured in the United States of America.

CONTENTS

PART IV THE UNDERWRITING PROCESS

PART V SURETY BOND CLAIMS

PART VI PRACTICAL CONSIDERATIONS REGARDING THE SURETY INDUSTRY

Now faith is being sure of what we hope for and certain of what we do not see.

Hebrews 11:1

ACKNOWLEDGMENTS

I want to express my gratitude to the University of Wisconsin-Madison College of Engineering, Department of Civil and Environmental Engineering for the freedom and flexibility to pursue my research interests. It is in great part due to the academic and administrative support I have received at this institution that the following work was written.

The material presented is based, in-part, on research funded by the National Science Foundation (NSF) and on papers published in various American Society of Civil Engineers (ASCE) journals. I greatly appreciate the opportunity and support from both NSF and ASCE.

Many thanks go to Mr. Brewer Stouffer, who played a critical role in pulling all of the material together and editing the manuscript. Without his continued persistence and dedication, this endeavor could not have been accomplished. I am also deeply grateful to Mr. Dennis Wine, an industry professional who provided extremely thorough and insightful critique of the project. I also acknowledge Mrs. Susan Brunsell, who provided continuous help and technical support.

A special acknowledgment goes to Mr. Clem Winter for providing invaluable information and assistance on Chapter 7: Reinsurance and Co-surety and throughout Part 5. Surety Bond Claims. Mr. Winter has hands-on surety claims experience and knowledge through almost 40 years spent in the field and offices of a primary surety and a major surety reinsurer. He now serves as a surety claims consultant and expert witness on surety claims cases and is President of C. Winter Enterprises, Inc. and Surety Financial Services, LLC of Wichita, Kansas. In addition to providing information, Mr. Winter put the author in touch with numerous industry experts who provided further critique and insight in the complex surety relationship.

My thanks goes out to these and many others too numerous to mention for their help and commitment in making the completion of this project possible. To my family, especially, who invested their time and energy and encouraged me to pursue my dreams—my father, Mr. Ronald F. Russell; my mother, Mrs. Georgia A. Holmes; my uncle, James C. Russell; and my wife, Vicki, and our children, Nicole, Jacob, Matthew, and Rachel—I thank you deeply for the unconditional love and support you continue to show.

PREFACE

This book aims to demystify the complex field of suretyship for construction contracts by describing and analyzing each element of the surety bonding process in basic terms for contractors, owners, design professionals, and other construction industry personnel. While seemingly a straightforward task, the realities of today's complex construction and surety industries make it difficult to linearly describe what surety bonds are and how they work. In addition, since the readership will bring with them varying levels of understanding about suretyship and construction processes, it is necessary to provide adequate background information for many levels of understanding. Therefore, in addition to a comprehensive glossary, great pains have been taken to structure the book for maximum breadth.

As a way to begin understanding the surety relationship, think of an unassembled six-piece puzzle. To construct the larger puzzle from the smaller units, the pieces must be connected together. In a similar fashion, understanding contract surety requires a working knowledge of the components of suretyship. Figure P-1 helps demonstrate this comparison.

Let each puzzle piece represent one of the six sections of the book. To arrive at a working understanding of how surety is used in today's construction industry, which is to fit the pieces together, the reader must not only be aware of the elements of the surety relationship, he or she must also grasp how the different people and parts work together throughout the bonding process. To understand the whole it is necessary to understand each piece, and to understand each piece it is simultaneously necessary to comprehend the whole. If any piece is missing, one's understanding will be more than incomplete; it may be incorrect.

To ultimately describe how suretyship works in today's half-a-trillion dollar construction industry, which is to say to build a working puzzle, an understanding of what suretyship is, how surety bonds are used, who is involved in issuing them, and what they typically guaranty is crucial. Accordingly, the first three sections of the book explore suretyship as a concept, the specific types of surety bonds available for construction contracts, and the basic tenants of the surety relationship including who is involved, in what capacity, and how the various parties interact. How a contractor can go about securing bonding and why the private owner should consider surety bonding to guarantee projects is discussed as well. Therefore, Parts 1 through 3 outline the nature of the surety relationship, providing the background necessary to grasp how suretyship works.

Upon this base of understanding, the author's next task is to outline the significant considerations in the surety underwriting decision process, which comprises Part 4. As Figure P-2 demonstrates, a surety company is subdivided into two main branches: underwriters and claim personnel. The former decides which contractors to write bonds for, while the latter resolves any problems that may arise from a bonded contractor's failure to complete a bonded contract for any reason. These roles are discussed further in Part 2 and in increasing depth in Parts 4 and 5.

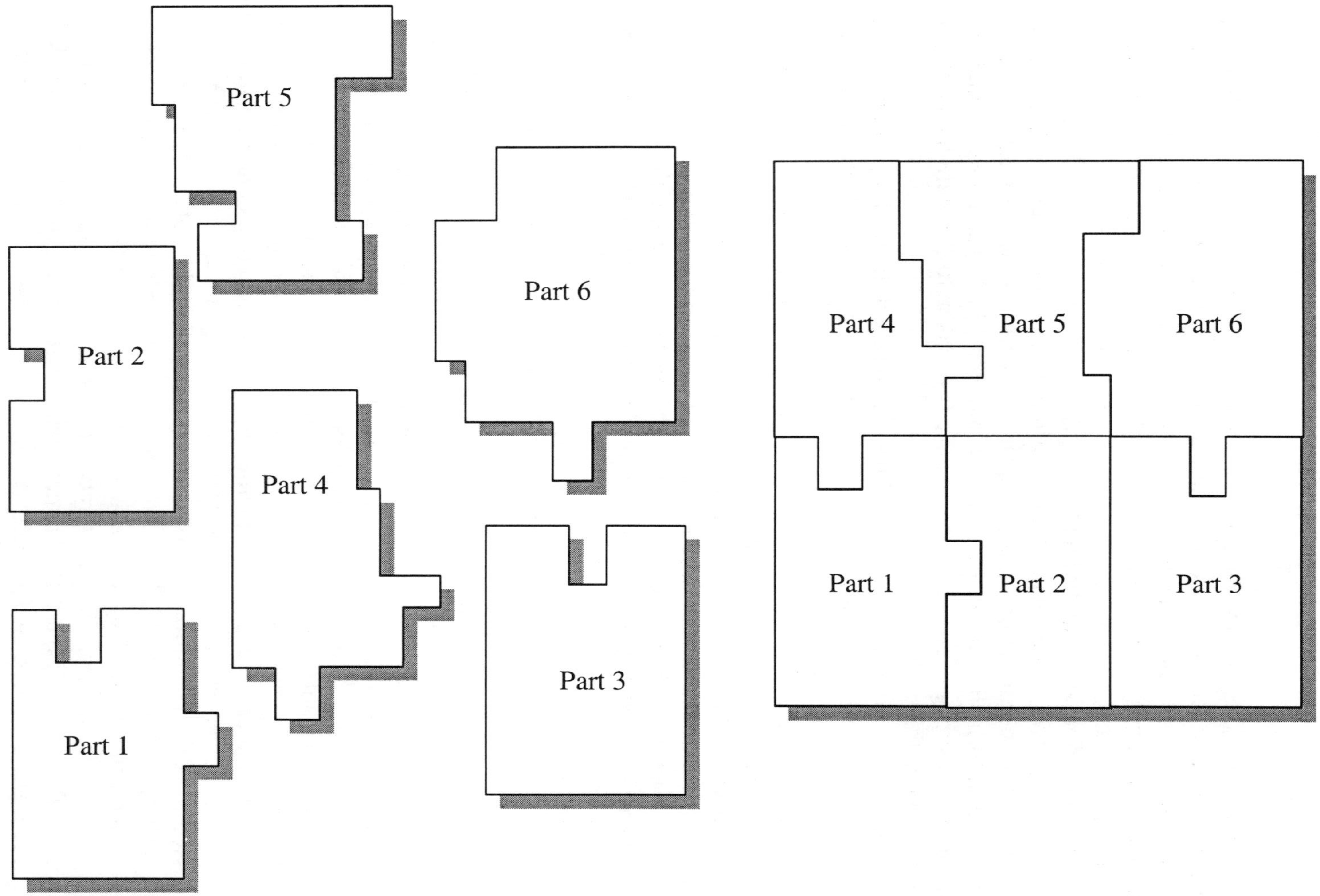

Figure P-1 Constructing the Surety Puzzle

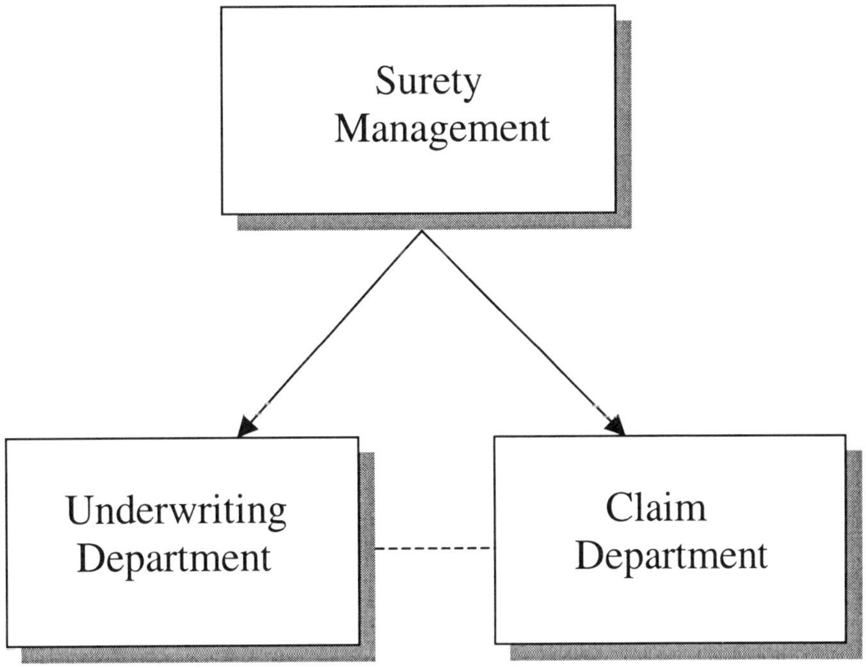

Figure P-2 The Surety Division: Separate but Equal

Fundamentally, evaluation by the underwriter on the basis of the following four major factors sustains the underwriting process of the contractor: character, capacity, capital, and continuity. Surety companies expend a large amount of their evaluation effort analyzing a contractor's financial condition, which is to say its available capital. However, the remaining factors are extremely important, and the contractor that fails to meet any of the expected standards may find it impossible to obtain surety bonds. In addition, communication with insurance agents, banks, attorneys, and the surety, as well as among members of the contracting firm itself, is important and will therefore be discussed in Part 4 as well.

This book provides an extensive description of the financial analysis underwriters perform, including several example analyses of contractor data. Such information should help contractors understand what financial information the underwriters will need and help them prepare it, while also providing owners with some insight as to how they might benefit from this financial analysis. In addition, the author intends to describe how an underwriter's needs and interests are consistent with those of the project owner from evaluation process undertaken during the underwriting process to the resolution of any complications that may arise on a project. Such understanding is critical in reducing the possibility of contractor failure, a goal that everyone in the construction industry, including owners, sureties, and contractors themselves, strives to achieve.

Through gaining a solid understanding of the surety underwriting process, including what the surety requires, prime contractors and subcontractors can substantially reduce the frustration they might otherwise experience in trying to obtain bonds; concurrently, contractors can also improve their chances for establishing a successful bonding relationship. Some contractors believe surety bonding has become easier to obtain during the 1990s. To the extent that sureties have become more aggressive in their marketing as competition has heightened and relaxed in their underwriting requirements, this observation is "on target." However, even though the current prosperous climate has allowed many contractors on the cusp to find a surety willing to underwrite its ventures, with an ever-increasing emphasis on capital, many undercapitalized contractors have found and will continue to find it difficult to locate a surety willing to underwrite bonds.

Individual states have legislated special allowances and conditions to aid contractors who cannot obtain bonds. In New York for example, the Commissioner can require a 20% retainage in lieu of the performance bond. Obtaining surety bonds is complicated and consumes contractors' time and resources. However, one benefit of going through the process is that a strong bond rating is an important tool from a marketing standpoint when competing against other companies for projects. As in all commercial endeavors, a healthy credit rating is important. For contractors that may have experienced difficulties acquiring bonding and for contractors unfamiliar or new to surety bonding, in addition to discussing alternatives to surety products, this book may serve as a guide to help evolve toward a bondable enterprise.

To facilitate understanding between "surety companies" and the independent "agents" or "brokers" whom contractors retain to place their surety bonds, this book will henceforth refer to surety companies as "sureties" and "underwriters," and the agents or brokers as "independent surety agents" or "producers." This book will describe the regulation of underwriters and producers, and will also discuss factors that influence bond premiums. In addition, the reader will find some of the advisory rate tables that The Surety Association of America (SAA) promulgates, with sample premium computations, average agent commissions, and average costs for financial preparations by Certified Public Accountants (CPAs).

To understand how a surety arrives at the decision to underwrite a contractor, it is necessary to grasp how the entire industry operates. Throughout this book, and especially in the final section, the scope will widen to look at how industry-wide concerns can affect the local owner and contractor. Understanding the general economic climate of the surety industry is especially important, as it is a good general indicator of surety underwriters' attitudes. If business is good, they will be more willing to underwrite a potentially qualified contractor, while if business is bad, the surety will only underwrite those contractors that it is practically certain will not fail.

Sureties suffered significant and unexpected annual losses on their contract bond business from 1984 to 1987. This poor result was at least to some extent brought about by a lack of underwriting discipline in the prior few years. Some underwriters believed that if the contractor was large enough, there was little chance of failure. This attitude seemed quite logical as over a long span of years very few of the largest contractors had failed. Accordingly, even if a large contractor wanted a work program that seemed excessive, the surety would tend to be supportive for fear that the contractor would go to its competition if the surety refused.

When a very large, overextended contractor fails, it can generate large losses for the surety. Between 1984 and 1987, several large contractors failed.

To a considerable extent, their losses were a major cause of poor surety industry results during those years. It is important to recognize that not all of the surety companies suffered losses during these years. Those few sureties that maintained their underwriting discipline went through this period virtually unscathed. Even those sureties that suffered substantial losses usually only had one or two poor years out of four. Nonetheless, the significant losses forced many surety companies to adopt more stringent underwriting criteria and caused others to discontinue writing bonds altogether. Hence, in the late 1980s and early 1990s, some contractors, particularly those that were financially, managerially, or technically weak, found themselves forced to take only unbonded work or discontinue operations altogether.

However, in recent years the surety industry has been profitable. Contractors and producers report that it is now much easier for a contractor to get a bond than it was even in the early 1990s. Since 1988, this string of profitable years has created a strong desire among the sureties to expand, or at least maintain, their market share in view of perceived potential profit. Since sureties compete for business on the basis of the service they offer, the prices they charge, and the underwriting criteria they apply, to attract or retain business, the surety might indicate a willingness to support a larger work program or offer a lower premium rate to the contractor. As a result of this desire by the sureties to maintain or increase their market share, many contractors that were once having difficulty obtaining bonds have found a surety willing to handle their account. However, even in the best of times there will always be contractors lacking some important quality or asset that will be declined bonding. Accordingly, this book discusses possible alternatives to surety bonding, including how to acquire a waiver of bonds, individual sureties, letters of credit, and certified checks or cash.

The end of the 20th century and the beginning of the 21st century may yield different industry results. The surety industry tends to be cyclical, with this current period of growth and profitability going on 11 years. There was a slight loss in 1996, but profits were solid in 1997. The current profitable climate has attracted many new companies, and when sureties are aggressively seeking to write more bonds in a competitive environment, there is a tendency for some companies to loosen underwriting standards. As competition increases, underwriting inexperience can become significant as many personnel lack the discipline or experience to focus their analysis to render sound underwriting decisions. Heightened competition coupled with underqualified personnel may result in some contractors being guaranteed surety bonds who are actually not fully qualified to perform the work associated with a project. This could result in contractor bankruptcy and could force less-successful surety companies out of the market. Recently, many industry experts have pointed to first signs of this occurring.

Given the nature of the construction industry and surety underwriting, experience is essential for long-term success. In the current climate with a dearth of experience, once in a while a surety bond may be issued to a contractor who proves to be unqualified to perform the contract. Consequently, some owners may choose to use other contractor evaluation methods such as owner-contractor prequalification, even if they are requiring surety bonds. This method of evaluation is one of the author's areas of expertise and the subject of another book by him entitled *Constructor Prequalification: Choosing the Best Constructor and Avoiding Constructor Failure* (Russell 1996).

Many construction project owners, especially those in the private sector, will depend on their own contractor evaluations and not require surety bonds. However, if they make a mistake in judging the contractor, they will incur the financial consequences of the contractor failure. When the surety makes a mistake in judgment, regardless of the reason, the construction project owner and those who furnish their labor, materials, or services at least have the potential of recovering some or all of their loss from the surety.

In the event that the contractor fails to perform the contract and is defaulted by the project owner, a surety bond claim usually results. A surety bond claim occurs when the project owner or a payment bond claimant invokes his or her rights in accordance with the terms and conditions set forth in the payment bond, in an attempt to avoid or recoup financial losses through payments from or through performance by the surety. Contractor "default" or "failure" implies that the contractor is unable to fulfill one or more of its contractual obligations. A contractor's default may be caused by a variety of underlying problems such as inefficient job site administration, subcontractor failure, unexpected material shortages, labor disputes, price increases, scheduling conflicts, bad weather, personnel changes, or the insufficient availability of cash to pay laborers, subcontractors, and material suppliers.

In Part 5, this book will discuss considerations in handling surety bond claims and will also outline the steps a surety company may take when contractor default occurs. In the order that a surety would execute them, these steps are: 1) investigation, 2) evaluation, 3) decision-making, 4) implementation, and 5) monitoring. Examples of actual surety bond claims and an extended hypothetical account based on interviews with surety claim personnel and investigations of case law are given as a supplement to the procedural information. In addition, an outline of the actions taken to discharge a claim written by a claim manager with over 30 years of surety and reinsurance experience is found in Appendix XIII.

In the last section of the book, Part 6, practical applications regarding surety bonding, including the U.S. Government Small Business Administration Surety Bond Guarantee Program, special concerns currently confronting the surety industry, the aggregate performance of the surety industry, sources of information that one can access to check the authenticity of a surety bond, typical disputes arising from surety contract bonds, and obligee considerations regarding surety bonds are all addressed in depth. Finally, the author describes and projects surety industry trends, including developments brought about by the increasing use of Design/Build delivery and maintenance bonds.

With an understanding of how the surety industry is structured, how individual sureties make underwriting and claim decisions, and how individual bonds function, contractors, owners, design and construction professionals will benefit. All will make better informed decisions about whether to enter the world of bonding to begin with, and thereafter, how best to negotiate that terrain.

Part I

THE BIG PICTURE

The following section presents the foundation for understanding the complex surety puzzle. The concept of suretyship and how it is used in the construction industry is introduced in general terms, including a description of who will most likely benefit from such an understanding. This section also explores how suretyship and construction have evolved over time, describing the current use of surety bonds in public and private construction.

Chapter 1

SURETY BONDING: WHO BENEFITS

Suretyship is a concept that has existed for thousands of years. Suretyship is simply an agreement in writing, under which one party, the surety, guarantees to a second party, the obligee, to answer for the debt, default or miscarriage of a third party, the principal. Prior to the last decade of the 19th century, the surety was almost always an individual or group of individuals such as Lloyd's of London. During the 1890s the idea of corporate suretyship arose in the United States, and it proved to be far superior to the old individual style. Since the start of the 20th century, corporate suretyship has virtually eliminated the use of individual sureties.

Construction is an important segment of our economy involving about two million enterprises performing approximately $500 billion worth of construction work annually. In the United States, various laws require that surety bonds shall be provided by contractors to protect public entities and suppliers on most federal, state, and local public construction projects. At the discretion of the project owner, surety bonds may also be used to protect privately owned construction projects. Clearly, the administration of any construction contract that requires the contractor to provide surety bonds requires both the contractor and project owner and their representatives (i.e., design engineer or architect) to understand the various types of bonds on construction projects, such as bid, performance, and payment. Private construction project owners can benefit from a clear understanding of surety bonding and the protection various bonds provide, so that they may make an educated decision as to whether to require surety bonds in the first place. This book attempts to provide an explanation of the surety industry and its underwriting procedures, to help owners, prime contractors, subcontractors, design professionals, and others grasp this essential topic. It is hopeful that this analysis will enhance these individuals' abilities to understand and successfully deal with the surety bonding process.

Except for those engaged in a business or profession where surety bonding is customarily required, it is unlikely that the average American will ever encounter the subject of surety bonding. However, it is quite likely that surety bonding will be encountered at some point during your business or professional career if you are a construction contractor or subcontractor, public or private construction project owner, design professional, attorney, accountant, banker, or credit professional. This book intends to both explain how surety bonds are procured and used in construction projects and to be a resource to turn to for specific questions dealing with all stages of the construction process, including underwriting, maintaining a good surety relationship, and claims.

Government contracting officials involved in the construction procurement process will frequently encounter surety bonding since the federal government and most states have laws requiring contractors to provide bid, payment, performance, and sometimes maintenance bonds. Since turnover is common in the public sector, officials are often left with contracts and bonds acquired by the previous administration. Furthermore, state and local elected officials are sometimes forced to rapidly familiarize themselves with bonding procedures, something they may not have professional expertise with. This book should help aid in the process.

Design professionals, whether architects or engineers, need to understand the surety bonding process in order to help their private clients decide whether to require surety bonds and to determine that their public clients are in compliance with any statutory surety and/or bonding requirements. Also, on occasion a construction project owner may require that the design professional provide a surety bond to secure the terms of a contract for the design of a project.

Although attorneys have a general understanding of surety bonding as part of their legal training, they may not have been exposed to the practical side of the surety bonding process. In practice, attorneys may be called upon to provide advice on surety bonding in a variety of situations. For example, when a client is considering a construction project the attorney may be asked for advice on whether to require bid, performance, or payment bonds, and or even all of them together. If the client decides to require surety bonding, he or she may want the attorney to review the bond form and check the executed surety bond when it is presented by the construction contractor to make sure it is properly prepared. The attorney could also have construction contractor clients that need advice in connection with matters relating to surety bonding. Then, there is always the possibility that the attorney may represent a subcontractor or material supplier that needs advice with regard to a claim that is or may be made against the surety bond.

Accountants who understand the surety bonding process are in a much better position to advise their construction contractor clients and help them provide all the necessary information that the surety requires. Construction accounting is somewhat specialized and it is very helpful to the contractor-surety relationship if the accountant has had experience preparing a contractor's financial statements.

Bankers will benefit in two ways by understanding something about surety bonding. First, the banker may come into contact with surety bonding when providing regular banking services to a construction contractor client. The contractor may need a letter from the bank confirming a line of credit or verifying a deposit account to support an application for surety bonding. Second, the bank loan officer may want the protection or the rights involved in a surety bond when the borrower is using the loan proceeds to finance a construction project.

Credit professionals who extend credit to construction contractors for supplies and materials often depend on the surety bond as a source of payment in the event the customer becomes insolvent or does not pay for the materials supplied to the job. The credit manager needs to have a sound knowledge of the terms and conditions found in the payment bond and the underlying statutory provisions that apply to construction credit transactions in order to file notices and claims in an effective and timely manner.

According to industry experts, approximately 83% of all contract bonds are underwritten for public entities. The balance, 17%, are for private owners. Although surety companies underwrite contract bonds predominantly for the public sector, more and more private owners are finding that, especially on large projects, their lenders require surety bonds. As will be discussed in Part 6, many within the surety industry are trying to expand into the profitable private market represented by the large circle in Figure 1-1. Although it is unclear how successfully the surety product will be embraced by owners unfamiliar or misinformed about the product's benefits, there is clearly room for expansion into this sector. However, even though there are a considerable number of occasions when surety bonding may arise, the subject has been generally ignored in the curriculum offered by our institutions of higher learning. Furthermore, there is virtually no objective information available to those who might be interested in educating themselves. All too often the need to learn about surety bonding is not recognized until it becomes necessary to enter into a transaction that requires a surety bond. Clearly, it is preferable to gain this requisite knowledge beforehand.

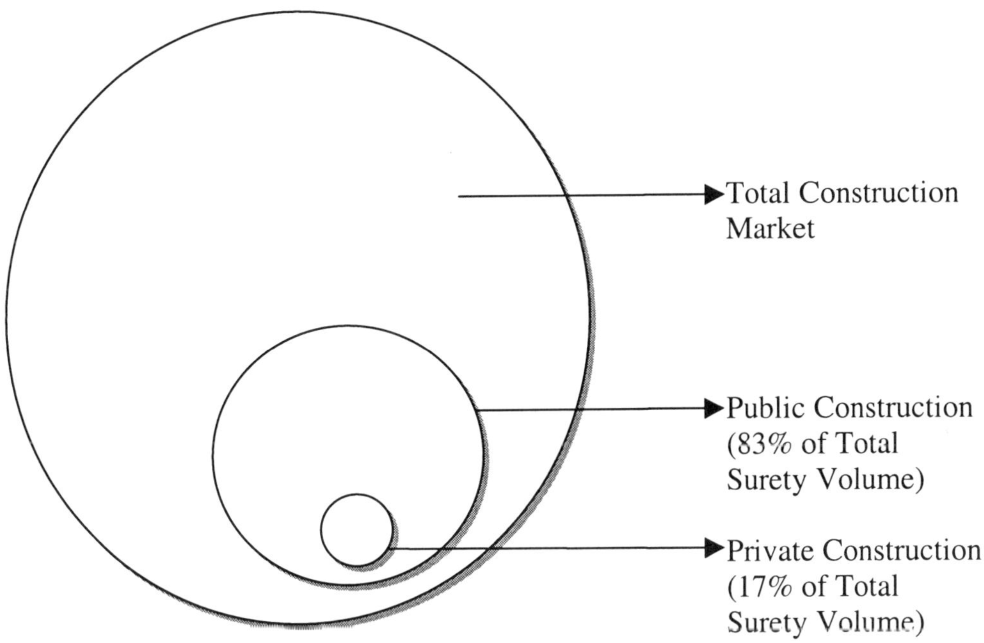

Figure 1-1 Surety Market and Out-of-Box Thinking

Many project owners, contractors, design professionals, and others who will come into contact with surety bonds are unfamiliar with the surety industry and its procedures. This lack of knowledge often leads to misunderstandings and unnecessary disputes, translating into inefficiencies and disappointment with the surety product and process, often resulting in financial complications and loss. With more project owners using a variety of project delivery systems, there is a degree of increased uncertainty and complexity in today's construction industry, and the contract bond underwriting process becomes even more sophisticated. Thus, understanding the surety process takes on even more importance.

Chapter 2

DEFINITION OF SURETYSHIP

The definition of suretyship is the obligation to pay the debt of, or answer for, the default of another (*Glossary Fidelity Surety* 1997). Thus, surety bonding is a three-party relationship, which Figure 2-1 helps illustrate. In this relationship the (first party) guarantor (underwriter or "surety company") guarantees to the (second party) obligee (project owner) that the (third party) principal (contractor) will perform work agreed to in the contract. If the principal fails to perform, the obligee can default the principal and demand that the underwriter accomplish the work according to the plans, specifications, and contract terms, or pay damages up to the total penal sum of the bond. The underwriter's maximum liability is the penal sum of the bond unless they choose to finance a contractor, in which case costs might exceed the original penal sum of the bond. If the prime contractor requires significant specialty contractors to provide bonds, then on each of those bonds, the specialty contractor is the principal, the prime contractor is the obligee, and the surety guarantees the principal's (i.e., specialty contractor's) performance to the contractor.

The surety transaction is in effect an extension of credit for a fee, generally refereed to as the premium. A bond is a credit tool analogous to a bank loan that someone cosigns. The surety guarantees to the project owner that the contractor will perform the contract at the agreed upon price within the time allowed. A contractor's failure to perform these obligations means the obligee may call upon the surety to do so. The surety collects a fee or premium for lending its credit worthiness to the contractor. The premium charged for a particular bond will vary depending on factors such as the type of bond, total project cost, type of construction, stipulated duration of the project, and contractor's financial condition.

The reader should note the difference between surety and other types of insurance, such as health, life, automobile, and property coverage. These differences are discussed more at length in Chapter 6, but a preliminary understanding is needed to grasp the basic tenets of suretyship. Premiums for non-surety insurance are actuarially determined. That is to say, they vary according to data that both documents past losses of an individual person or company and predicts the frequency of future losses among a larger *group* that includes the individual seeking coverage. By contrast, surety underwriters assess their risk by looking at the record and current status of only the *individual* contractor who seeks bonding, and the bulk of the premium goes to administrative costs and profit.

The primary objective of the surety underwriter is to issue bonds on behalf of contractors whom they deem capable to perform the construction contract, rejecting all others. Thus underwriters are in the business of prequalifying contractors. When the underwriter makes a mistake in judgment, it may be able to collect the costs of that error from the principal, otherwise the surety will have to use its own resources and suffers the loss itself. The underwriter is in effect extending credit, usually unsecured by collateral, on the basis of the signature from the contractor and from any indemnitors. While underwriters have historically not required collateral, the practice of requiring collateral for bonds is on the rise. However, in current practice, the collateral is usually in an amount that is a small percentage of the contract price.

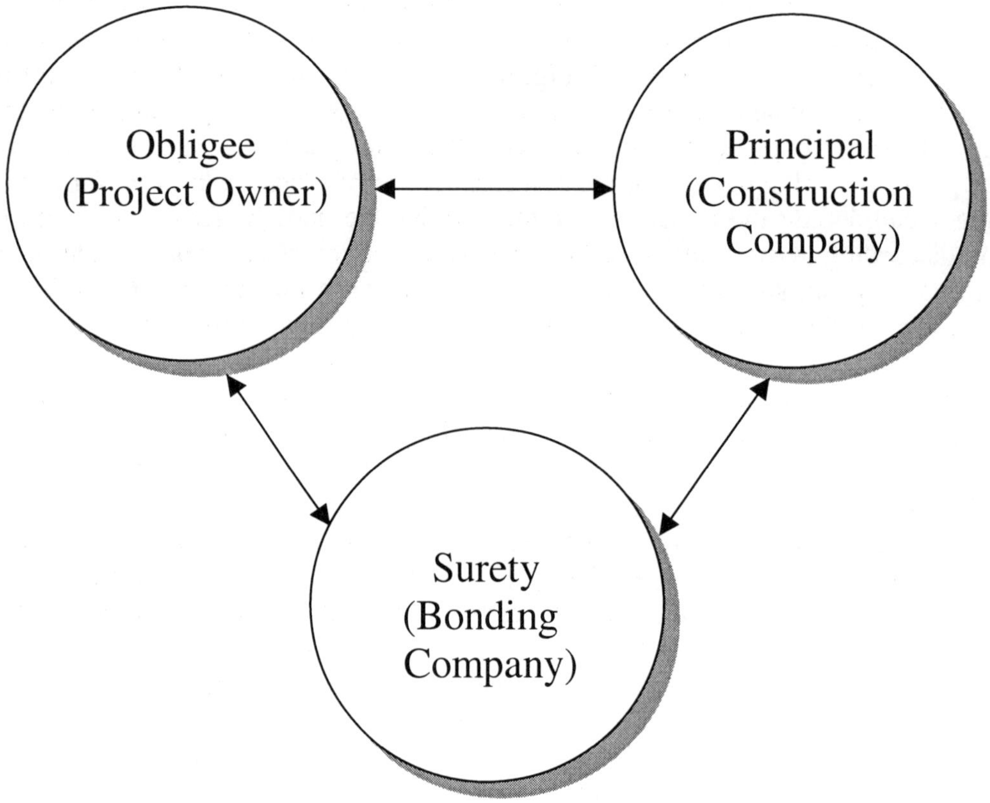

Figure 2-1 Representation of Bonding Relationship
Source: Adapted from D.W. Halpin and R.W. Woodhead

Underwriters operate under the assumption that they will not suffer losses on bonds. (Otherwise, why would they be in the business?) They therefore extend suretyship on the basis that a principal will ultimately be able to reimburse the surety for any loss the surety incurs arising from underwriting the bond. In addition, the underwriter typically demands the indemnification of third parties who have a vested interest in the principal, such as the individual owner of the construction company and his or her spouse or stock holders or financial backers. Appendix I contains a sample indemnification agreement. Typically these indemnity agreements are in writing. The rationale for this request is that if the owner of a construction company and relevant interested parties are not willing to pledge their assets in support of the enterprise, then why should the underwriter do so?

Chapter 3

HISTORY OF SURETYSHIP

Suretyship has a long and rich history. The earliest known record of a contract of suretyship is a Mesopotamian tablet written around the year 2750 B.C. According to the tablet, a farmer drafted into service of the king was unable to tend his fields. Accordingly, he entered into contract with a second farmer, whereby the first farmer allowed the second to farm his fields on the condition that each would receive one-half the proceeds from the first farmer's fields. To ensure that the first farmer would receive his share, a local merchant guaranteed the second farmer's compliance with the promise, fulfilling the essence of a suretyship contract.

The Code of Hammurabi, authored around 1750 B.C. in Babylonia, mentions suretyship contracts, and a Babylonian contract of financial guarantee from 670 B.C. survives to this day. Persia and Assyria were using suretyship as early as 525 B.C., and in 509 B.C., Rome and Carthage used a form of surety to guarantee goods bought and sold between the two empires. As did the ancient Hebrews, the Roman Empire used suretyship to guarantee not only the price of goods but the completion of public works and the exchange of hostages. By 150 A.D., the Roman Empire had developed highly technical laws of suretyship so complete that many of these principles have changed little to the present day.

As time progressed, so did suretyship. In Anglo-Saxon England, suretyship served as a method of criminal law enforcement that required every man to have a *bohr,* or surety, who was responsible for the criminal acts of his principal. Also, in order to buy or sell anything in the marketplace, a person had to have a bond and a witness to the sale. Sir Walter Raleigh and Shakespeare mention suretyship in their works, but throughout the middle ages and the early modern period, suretyship was still performed by individuals.

The year 1720 marks the first known attempt to create a company engaged in the business of bonding the fidelity of persons holding positions of trust, but it wasn't until 1837 that the first corporate surety successfully organized, the Guarantee Society of London. Over the next 20 years, seven more surety companies were created. Many more institutions were involved in the burgeoning fidelity bond market in England between 1837 and 1875, and in 1865 the first U.S. venture opened its doors. The Fidelity Insurance Company became the first American company to underwrite bonds. This venture soon failed, but others stepped in, and between 1883 and 1898, nearly 25 new surety companies came into existence in the United States.

As corporate surety became more recognized and utilized, it soon spread to the construction industry. No record exists of a company guaranteeing the completion of a construction contract before 1875, but in that year, the *Insurance Encyclopedia* matter-of-factly declares that the surety industry had "extended to the guarantee of contractors for public works" (Petro 1987). In 1887, The American Surety Company, organized in 1884, began underwriting contract bonds in this country. Recognizing the benefits of surety, Congress passed the Heard Act in 1894 requiring surety bonds for public projects and further legislation allowing corporate sureties to underwrite these projects.

By the turn of the century, the surety industry was taking off on both sides of the Atlantic, but not without problems. In spite of the growth of corporate suretyship, the general public remained largely unfamiliar with the modern surety concept and regarded underwriters skeptically. The fear of big business stepping into local affairs was pervasive, and mistrust toward large companies spanned the population from farmers to politicians and from lawyers to shopkeepers. Individual suretyship, though unreliable to a large extent because of its wide-ranging variance, was at least familiar, and if suretyship was needed at all, people tended to look to individuals. Accordingly, the growth of the surety industry after the 1870's was rocky. The failure rate of surety companies accurately reflects the public's fears: between 1883 and 1898, 11 of 25 new surety companies (44%) discontinued surety operations or failed completely.

For corporate surety to gain public trust, the burgeoning industry had to distance itself from the unreliability of individual sureties and early corporate failures. Companies had to promote not only the suretyship concept, but the idea that a large conglomerate could readily serve common men and women at a time when monopolistic capitalism unsettled the popular imagination.

CORPORATE AND INDIVIDUAL SURETIES DIVERGE

Originally, as discussed above, suretyship was personal. For example, a Roman gateway construction contract of 106 B.C. required that whoever received the contract award would furnish bondsmen secured by real estate to the satisfaction of the magistrates. Consequently, it was the individual acting as the surety who had to demonstrate sufficient assets available in the event of default and/or loss.

As might be expected, the major problem associated with personal or individual suretyship was collection in the event of a loss. In many cases, the individual sureties failed to possess the resources they claimed, could not be found, or had died, making collection extremely difficult, if not impossible. This problem still persists today among individual sureties and among individual contractors who become "self-sureties" and underwrite their own surety needs. A self-surety is in effect the waiving of a bond when an obligee allows the contractor to be both principal and surety, with or without collateral or a form of indemnification. Recent attempts have been made to require that individual sureties meet the same requirements as a commercial corporate surety and such will be discussed in Part 6 ("Individual Sureties" 1990). Suffice to say, individual sureties claiming resources and otherwise fraudulent bonding is more prevalent than many people realize. Therefore, project owners and contractors should carefully select the surety sources with whom they do business.

In response to the problem of collecting money from the individual sureties in the 19th century, corporate surety emerged as an alternative. Corporate surety eventually thrived for the basic and compelling reason that corporate sureties were more likely to possess sufficient financial resources than a personal guarantor due to greater financial strength, perpetual existence of the enterprise, and better developed technical skills. Today, corporate sureties are typically insurance companies or divisions or subsidiaries of insurance companies, having the financial resources to cover most all unexpected losses.

However, at the turn of the century, the American populace was skeptical of corporate surety. Even though surety bonds were required on federal and many state projects, the industry was not viewed positively by Joe Public. The high failure rate of early corporate ventures spurred a professional interest from within the surety industry to ensure stability, and in 1908, surety industry leaders created the Surety Association of America (SAA) to regulate the volatile, risky surety business, and to uphold professional standards. Since that time, SAA has served as a regulating body tracking industry trends, a lobbying concern for political issues, and a focal point for organizing the industry.

Through SAA efforts and an improving performance record, corporate suretyship eventually found its rightful place in the construction industry. Underwriters gradually gained acceptance in the public sector, especially after the passage of the Heard Act, but as the industry developed, more complicated problems than public image began to arise. Under the Heard Act, the contractor provided a single bond that protected the U.S. government as well as specialty contractors, material suppliers, and laborers. If a specialty contractor, material supplier, or laborer wanted to file a claim against the contractor on a public project, the creditor party had to wait at least six months after the completion of the project before taking action, allowing the U.S. government time to file the first claims against the contractor. If the government did file a claim, specialty contractors, material suppliers, and laborers had to wait until after the settlement to file their claims. The end result was that these parties often experienced undue delays—sometimes several years—before judicial action occurred on their claim. This set the stage for the landmark passage of the Miller Act of 1935, a more comprehensive body of legislation to regulate federal construction projects, which with only minor changes is still in use today.

The Miller Act of 1935 required a contractor to provide both a performance and a payment bond on federally funded projects to protect the U.S. government as well as those individuals providing labor and materials to the project. The performance bond guaranteed that the project would be completed per the terms of the contract, and the payment bond ensured that laborers, suppliers, and subcontractors would be paid by the surety in the case of contractor default. The Federal Acquisition Regulation (FAR) originally required that construction firms obtain payment and performance bonds when contracts exceed $25,000, but that amount was recently raised to $100,000.

Most states have adopted similar legislation for public works projects, and these are known as the "Baby" or "Little Miller Acts." Today, nearly every agency of federal, state, county, and city governments, as well as most political subdivisions, require payment and performance bonds to guarantee the successful completion of construction, supply, and certain miscellaneous contracts. However, the surety industry is still trying to expand into the private sector. Recent efforts have been made to popularize the use of surety bonds for private owners, but to a large extent, the private market has remained steady with only marginal growth (see Figure 1-1). The private-owner market currently accounts for roughly 17% of all surety business.

Today, the general public is still largely unfamiliar with the surety concept for construction contacts. What is surprising is how little those who stand to profit most from such an understanding know about the surety product line. Few courses are on the books at institutions of higher learning, and there are few objective sources exploring how surety is used in today's complex construction market. In response to a perceived incomplete understanding, it is a primary purpose of this book to help familiarize those parties who may benefit from further knowledge about suretyship.

Part II

THE SURETY RELATIONSHIP

The following section explores in detail the individual players in the surety relationship, paying special attention to how the various parties interact and at what stages of the bonding process. After the major participants have been fleshed out, surety and insurance are compared, with an emphasis on how risk is assessed. To this end, how sureties calculate premiums is shown to fundamentally differ from actuarially derived insurance premiums. Finally, the concept of reinsurance is introduced, along with how reinsurance is used by sureties.

Chapter 4

THE KEY PLAYERS

Many industry professionals believe that surety bonds are the foundation for a competitive, properly functioning bidding system. In competitively bid contracts where the owner awards the contract to the lowest responsible bidder, surety bonds help ensure that all bids are in earnest and the selected contractor will perform to the specifications outlined in contract documents. Surety bonds serve as evidence of the contractor's competence and capability. They also guarantee that suppliers of labor and material have some recourse against financial loss that stems from contractors who fail to fulfill their obligations. On projects requiring bonds, suppliers and specialty contractors (subcontractors) will generally remove any allowances for extra funds they may have factored into their quotes to cover the risk of not receiving payment for materials or services. Another significant benefit is the presence of an impartial third party who extensively investigates and evaluates the qualifications of a contractor prior to contract award. Owners should keep in mind that the *surety underwriter has access to financial data that contractors rarely release to project owners*. Therefore, as a practical matter, surety bonding can be extremely beneficial to the owner. The underwriting evaluation process receives further attention later in this book.

The bonded construction enterprise itself is founded on the relationships between the key players. Along with contractors and owners, the main players in the surety relationship are surety companies and independent surety agents. While it is relatively easy to describe how each player functions, as with any complex arrangement, there are many local nuances in the way each surety agent or surety company does business. In addition, bonding relationships include reinsurers, architects, engineers, bankers, creditors, subcontractors, and material suppliers who must all understand their roles, obligations, and rights in the bonded construction cycle. Limning the contours of the surety relationship is no easy task, as with every project roles deviate and change according to specific and unique requirements of the project. What follows is an exploration of the primary roles and functions of the key players, along with an emphasis on how various parties interact in the course of the bonding process.

OWNERS

Owners initiate the construction cycle by conceiving and funding construction projects. By their very nature, owners can vary a great deal. However, owners can be subdivided into two general categories, public and private, with public owners further divided into federal, state, and local governments, and private owners into individuals, companies, or corporations. If the private owner desires the protection provided by surety bonds for its investment or if the owner is a public entity (most all public projects of any substance require contract bonds), then the owner will make it known in contract documents that it wants a bonded contractor. In that case, only contractors able to secure bonding will be considered for the job, and interested contractors will guarantee their intent to provide the required contract bonds by putting up a bid bond or some other form of guarantee with their bid.

CONTRACTORS

The contractor is the party that will perform the building or construction function. Likewise, the subcontractor or specialty contractor will provide a portion of the work on a given project (e.g., concrete construction, electrical, or landscaping). The contractor can be an individual, company, or corporation. To bid on construction projects with owners demanding surety bonds, otherwise known as bonded work, the contractor must secure a relationship with a willing surety prior to bidding for such work. Generally this process is facilitated by an independent surety agent. Once a contractor has found a surety willing to underwrite his or her enterprise, the contractor is then said to be a bonded contractor.

SURETIES

Surety underwriters are generally insurance companies that offer surety contract bonds as a product line. However, independent surety companies, as well as individual sureties exist. American surety companies typically have licenses to write contract bonds in one or more states, and their rates are regulated by state insurance departments. A surety company is divided into two major divisions: underwriting and claims (see Figure P-2). The former makes the decision whether to accept certain contractors as appropriate risks, while the latter handles any contractors who, during the course of a project bonded by the surety company, default (i.e., fail to complete the contract for whatever reason). Surety underwriting personnel, the people who make the decision whether to accept individual contractors as risk, possess sound analytical skills and generally hold a degree in business or finance with an emphasis in accounting courses. Claim personnel have similar qualifications, although they might have an emphasis in forensic accounting. In addition, an emphasis in negotiation skills is generally required as claim situations present many difficult and complex options requiring timely resolution.

Underwriters carefully select contractors to bond, but due to the vagaries of the construction cycle, it is impossible to determine which contractors will fail. Claim personnel are experts in rapidly assessing the situation of particular defaults and in mitigating as much loss as is possible. An old adage in the surety industry says that the underwriters are the eternal optimists while the claim personnel are the eternal pessimists. Notwithstanding the idiosyncrasies of individual persons, it is easy to see how such a belief has been fostered. A large portion of this book is devoted to underwriting and claims, and in Parts 4 and 5, the roles of each part of the surety will be explored in much more detail.

INDEPENDENT SURETY AGENTS

One party serving more than one function in bonded construction is the independent surety agent. Surety agents, or "producers," are essentially "middle people" operating independently from the owner, surety, and contractor. Surety agents are typically organized in agencies, although it is common for one agent to be the primary contact on a particular contractor's account. The agent's principal responsibilities are to evaluate the contractor's ability to do bonded work and facilitate communication between the contractor and the surety underwriting the contract bonds. The agent is also generally authorized to sign bonds on behalf of the surety for projects that fall within acceptable ranges established by the surety for each contractor. In many cases, but not all, the agent will also meet the insurance requirements of the contractor (see Figure 4-1).

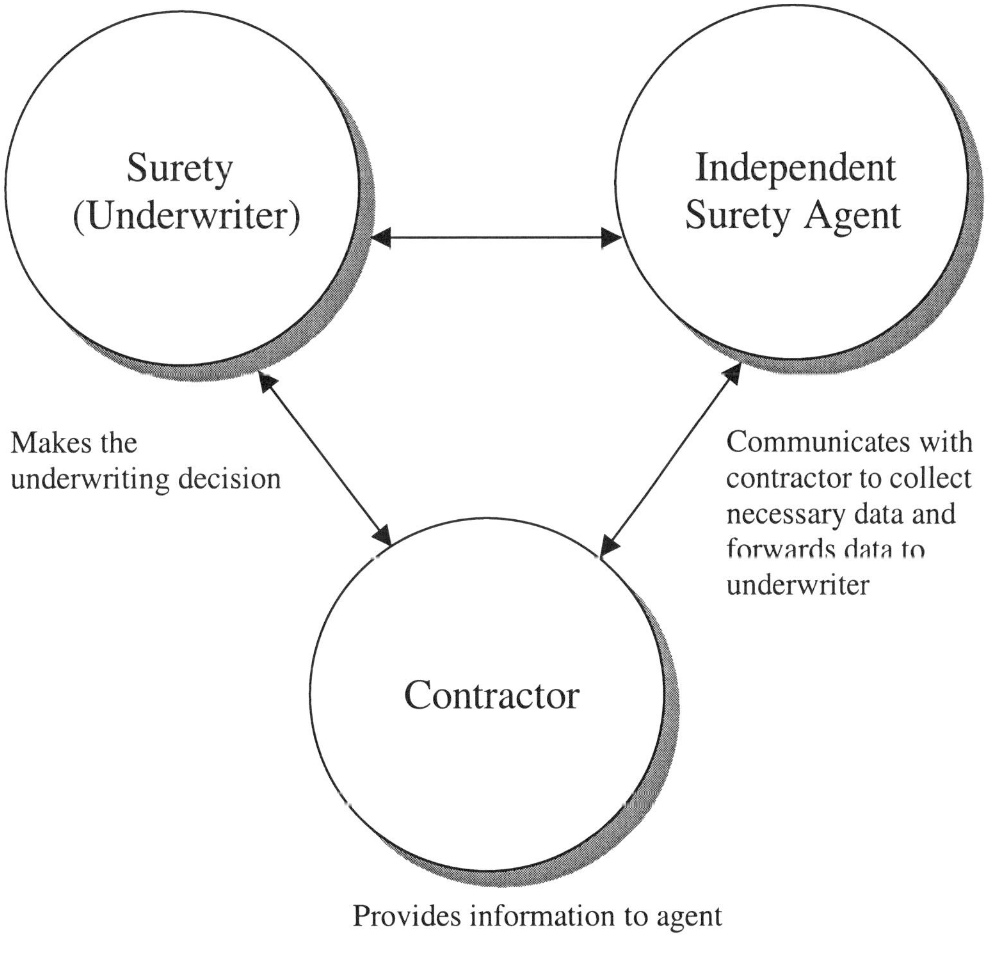

Figure 4-1 Relationship Between Surety Company, Independent Surety Agent, and Contractor

Operating within a given region or territory, independent surety agents develop familiarity with the construction and surety industries there, cultivating personal relationships with contractors, subcontractors, and surety personnel.

Agents develop a client base of contractors to match with sureties operating in the region willing to underwrite them. It is the agent's job to contact these potential sureties and arrange meetings as appropriate. To facilitate this process, the agent performs what is known as the production function as well as the initial underwriting function.

Through interviews and visits to the contractor's office(s) and job sites, the agent compiles information on questionnaires about the contractor's financial and professional history. While doing so, the agent evaluates the contractor's technical and financial ability to perform bonded work. If in the agent's opinion the contractor is capable of performing such work satisfactorily, the contractor data is clarified and organized to create a comprehensive profile. This information is then submitted to a potential surety and becomes the basis of the surety's decision whether to underwrite the contractor. If the contractor is deemed an unacceptable risk by either the surety or agent, the data may be submitted to another surety and/or the agent would no doubt advise the contractor how he or she could develop to perform bonded work.

Once a contractor has been accepted by a surety, meaning that after the surety's comprehensive underwriting analysis the surety accepts the contractor and establishes a maximum work-program (see Part 4: The Underwriting Process), the agent then facilitates the bonding process in several ways. The agent reviews potential job specifications, approves and drafts contract bonds, arranges meetings between contractors and sureties, and continually updates contractor information with visits to offices and job sites.

The agent reviews all potential job specifications provided by owners and determines whether projects fall within the contractor's normal construction practices, geographic area program, and credit line. If they do, then the agent will prepare and sign the bid bond form on behalf of the surety. An agent signs an agreement with a surety known as a line of authority that grants the agent the power to sign for the surety on bid and contract bonds falling within a contractor's "normal" parameters. These parameters are established when a surety agrees to underwrite a contractor and are updated annually to reflect changes in the contractor's capital and experience. If a project deviates in any way from that provided for in the line of authority, the agent must get prior consent from the surety before signing a bond. Some factors requiring the surety's consent might include multiple owners, maintenance bonds for periods longer than two years, or relet projects.

The surety company always makes the final decision to issue or decline to underwrite any contract bond. The independent agent (akin to an insurance broker) services the contractor's account and may initiate writing a bond form, but the surety agent usually *does not* have the intention nor the resources to underwrite bonds. Based on the agent's recommendation, the surety will consider issuing the bond, perform its own analysis, and decide whether or not to assume the risk of contractor default, i.e., underwrite the bond. In addition to assessing the contractor, the surety company underwriting the bonds is interested in: 1) the identity of the agency and the agent servicing the contractor account, 2) the length of time the agent has been handling the account, 3) the experience and successful history of the agent, and 4) whether or not the agent is professionally licensed. All of these factor into the underwriting decision, and this is discussed at greater length in the Part 4.

Surety agents typically work with many surety companies and many contracting firms. There are a few sureties that perform production functions in-house, meaning that they contact contractors directly, but these are by far the exception. Sureties in general rely heavily on independent agents to provide bondable enterprises. The agent's reputation is based on his or her ability to do so, and therefore, agents are meticulous about presenting accurate and up-to-date contractor information to the surety. Likewise, since the agent represents the contractor's interests as well, his or her reputation depends on how forthright and honest he or she is in dealing with contractors.

For simplicity, but also due to the highly subjective nature of bonding, contractors generally try to use one surety for all their bonding needs. A contractor works to cultivate a relationship with its surety agent and a surety company in the hopes that as the contractor grows or faces challenges, both parties will be there in a supportive capacity. Alternatively, when a bond program is established through careful underwriting, the surety fully expects the contractor to stay within the established limits. Sureties generally resent when a contractor tries to expand its work program beyond the aggregate job limit, i.e., the preapproved amount up to which the surety will underwrite bonds for the contractor, by obtaining another surety or by using alternatives to surety bonding such as self-surety, because this almost always increases the chance of default. For these reasons, it is normally impractical for a contractor to use more than one underwriter at the same time. Even so, a contractor will often have a confidential "backup" underwriter in case the primary underwriter is unwilling or unable to provide a bond on a specific project.

The contractor should be aware at all times of its aggregate job limit, constantly keeping its independent agent apprised of plans for expansion or entry into a new specialty field. However, even with thorough planning and good communication, it is conceivable that a situation might arise where a surety refuses to underwrite a project that the contractor wants to perform. This might occur if 1) the project exceeds the single-job or aggregate limit set by the surety, 2) the contract requires the contractor to use techniques for the first time, 3) the contractor is growing too large for the surety, 4) the surety adopts more conservative underwriting criteria, or 5) the underwriter discontinues its surety line altogether.

If for whatever reason the contractor's primary surety is unwilling to underwrite a project, the contractor is faced with a difficult proposition. To bid on the project, the contractor may have to find a new surety, possibly in a very short amount of time. Since cultivating a surety relationship takes time, if a contractor has no backup underwriter, it may fail to acquire the necessary bonds quickly enough. However, even if the contractor manages to secure backing from another surety source, the contractor risks alienating its primary surety, while if it doesn't bid on the project at all, the contractor may lose the opportunity to make a profit and/or gain new skills and techniques. Furthermore, it is also possible that a contractor might wish to leave his existing primary surety for personal or other business-related reasons, such as a rise in premiums. Therefore, establishing contact with a backup surety may be in the contractor's best interest.

If the agent does not immediately provide a backup or secondary surety, a contractor might want to inquire about the possibility of acquiring one. Since this process can be complex with severe consequences if handled improperly, accordingly, the contractor's agent will generally facilitate securing the backup underwriter. From the outset, the agent will declare to the potential secondary surety that the relationship is of a stand-by nature. The agent will present the contractor's financial information to the surety in much the same way as is done with the primary surety. If after the surety's underwriting analysis is performed the surety agrees to write bonds for the contractor, it will offer a line of credit to be updated annually.

The more the agent knows about the contractor's plans, whether for an isolated job or in the hopes of forming a new primary surety relationship, the better are the chances for a smooth transition from one surety to another. On the other hand, it is paramount that a contractor be aware of its surety agent's expertise in construction contract bonds and seek an agent who will take an active role in the development of the construction firm. The agent usually serves as a consulting member of the contractor's professional advising team, providing critical input about the contractor's operations. Generally the contractor's team also includes at least one accountant, attorney, and banker. An agent's level of knowledge and expertise will greatly affect his or her ability to provide constructive input and direction to the management of the contractor. Membership in the National Association of Surety Bond Producers (NASBP), which certifies that the underwriter has met high professional standards applicable to this line of business, speaks highly of an agent, but many qualified agents are not members. For starters, an agent or agency seeking admission to NASBP must be nominated by other agents in the region and must have produced a minimum of $200,000 in premiums in the previous year. Since an agent receives only a portion of the premium for the surety bond, representing from 1% to 5% of the total contract amount, members of NASBP must have high volumes of sales and/or deal with large contractors. Table 4.1 provides typical agent commissions on Class B construction contracts. Therefore, many smaller agents who do not qualify for membership are still worthy of consideration by the contractor.

Table 4.1 Typical Surety Agent Commissions for Contract Bonds
Sources: SAA Standard 1986 Rates; Agent Rates Compiled by Author

Amount of contract in $1000s (1)		SAA average rates for Class B Construction per $1,000 ($) (2)	Agent commission (%) (3)	Agent Commission amount ($) (4)
First	100	25.00	30.00	750.00
Next	400	15.00	30.00	1,800.00
Next	2,000	10.00	30.00	6,000.00
Next	2,500	7.50	15.00	2,812.50
Next	2,500	7.00	12.50	2,187.50
Next	2,500	6.50	5.00	812.50
Total	10,000	—	—	14,362.50

Before selecting an agent, it is customary for a contractor to inquire what other contractors the agent represents. This may help a contractor find an agent with expertise in the particular areas of construction expertise held by the contractor. Figure 4-2 provides a concise checklist for contractors and subcontractors seeking an agent. While the list is not exhaustive, it does provide some of the most important considerations for this important decision. However, simply satisfying the listed requirements will not ensure a good agent-contractor relationship. Even a well-qualified agent may not be right for a certain contractor. The contractor must feel comfortable with the agent and confident in his or her ability to represent the contractor's interest to the surety. Therefore, a contractor should spend time getting to know an agent and should not be timid about being forthright. If the agent fails to provide satisfactory answers to the contractor's questions, then he or she might not be a good fit.

Experience as surety agent

Licensed in region with appropriate state and local regulations

Familiarity with region and market

Good reputation with contractors and construction industry in area

Openness to advise and serve as a consultant for firm

Membership in National Association of Surety Bond Producers or equivalent state group

Expertise with particular types of construction

If subcontractor, experience representing subcontractors

Figure 4-2 What a Contractor Should Look for in an Independent Surety Agent

THE RELATIONSHIPS BETWEEN UNDERWRITERS, CLAIM PERSONNEL AND INDEPENDENT SURETY AGENTS

Open communication is crucial to the bonding process. Underwriters, claim personnel, and agents should have a business relationship founded upon trust and open, unhampered communication. Especially when it appears that a contractor is near contract default, open communication can help prevent large losses to the underwriter and owner by providing an opportunity to respond quickly and preemptively to the conditions precipitating the default. In addition, timely assistance by a surety can help a contractor avoid defaulting or even worse: bankruptcy. Open communication will help the surety develop an accurate loss projection, crucial to completing a troubled project. This reduces expenses incurred by retaining outside consultants such as accountants, engineers, and attorneys to assist in the claim investigation and helps prevent excessive losses from the actual completion of the project and/or paying any obligations under the contract. Thus, open communication may help avoid much undue expense.

Underwriters, bond claim personnel, and surety agents must communicate with each other if they sense a problem; it is in their own economic interest to do so (see Chapters 19 and 20). All three parties are links in the chain reaction of a surety bonding claim. Therefore, they must keep each other informed of events because no one wins if a contractor enters into default on a project. Surety companies and owners both stand to lose money, bond underwriters' reputations can suffer, and surety agents can lose money due to lost revenue from the failed account and resultant damage to their reputation, which is crucial to their business. For the surety industry to thrive, these parties must keep their lines of communication open.

Chapter 5

SURETY AS RISK MANAGEMENT

Construction is an industry rife with risks for all participants. Variable costs, unexpected material shortages, labor disputes, increased costs of non-performing subcontractors, long project duration, guarantee of a fixed price before all costs are known, high cost of management and overhead expenses, poor quality plans, and inconsistent specifications exemplify the types of problems and risks that contractors and owners routinely experience. These multiple risks compound the difficulties of projects that are already complex by nature due to specifications, geography, or the presence of hazardous waste. Couple the high risk with contractors themselves, many of whom are entrepreneurial risk takers by nature, and economic factors such as inflation, recession, and fluctuating interest rates, and the risk factor in the construction industry becomes ever more complicated.

Project owners must navigate through the abundant risks inherent in the practice of frequently relying on the performance of contractors with whom they may never have worked. In addition to all the risks involved in successfully completing a project, commercial project owners must face the real risk that even though the contractor builds the facility to the owner's specifications, the project may be a failure anyway. For instance, the economic climate may change and adversely affect the project's future, or perhaps the public simply does not respond to a facility as the owner anticipated. On the other hand, contractors must constantly balance a work schedule that includes projects of varying levels of physical completion with bidding for new work.

Contractor failure, which takes many forms and levels of loss, occurs on average between 1% and 5% of all construction projects. Reasons for failure range from contractors running over budget or beyond schedule, to total default (non-completion or lack of performance) on a project (see Russell 1996). Based on the contractor information packet compiled by the contractor's independent surety agent and the agent's recommendation, an underwriter will assess the qualifications of the contractor. The professional surety agent will assist the contractor by advising on what information needs to be submitted to the surety and by seeking to place the contractor with the surety that will best serve the contractor's bonding needs.

Should the surety company underwriter decide to accept the inherent risks presented by the contractor, that is to say, authorize bonds for the contractor, the contractor pays a premium for this service that covers the original underwriting analysis and subsequent risk assumption by the surety company providing a particular bond. This bond will then guarantee to the obligee that the contractor will perform, otherwise the surety will complete the project or compensate the owner up to the penal sum of the bond. Hence, contract surety bonds provide a mechanism through which a construction project owner can buy financial protection against the direct financial losses that might occur in the event of a contractor's failure to complete the project satisfactorily (or at all). Surety bonds are also a mechanism to pay certain bills that are incurred on the project by guaranteeing payment for goods and services supplied to and incorporated in the construction project.

However, surety contract bonding is only one of several alternative methods owners use in attempting to avoid contractor failure. Some owners conduct their own contractor prequalification program and some use project progress monitoring processes to decrease the probability of contractor failure. Contractor prequalification implies that the owner evaluates contractors' qualifications prior to receiving bids. See *Constructor Prequalification: Choosing the Best Constructor and Avoiding Constructor Failure* for detailed information on that procedure (Russell 1996). Project progress monitoring processes include periodic cost monitoring, which includes, among other things, routine investigations into percentage of cost to complete projects. The owner may also require Critical Path Management evaluations, which are computer-generated analyses of the status of the project at various times. A strong grasp of each of these methods for avoiding contractor failure will help owners decide which tools to employ and when on any given project.

Three specific signs related to jeopardy on contractor performance are: 1) apparent low bidder refuses to enter into contract with the owner at the price bid and/or fails to provide the specified bonds (usually payment and performance); 2) contractor fails to move the project in accordance with the reasonable time-frame in the plans and specifications as called for in contract documents; and 3) contractor fails to pay labor and material bills associated with project, resulting in mechanic's liens threatened against the project owner's property. Mechanic's liens are liens on the real property equal to the value of any labor, materials, or professional services the owner agreed to pay or did not pay or make settlement for. No one can obtain a clear title to the property until the owner clears up these outstanding claims. In order to cover these risks, companies offer surety contract bonds. Contract bonds for contractors consist of four primary types: 1) bid, 2) performance, 3) payment, and 4) maintenance. See below for discussions of supply, dual obligee, completion, and other types of bonds, which may be required on occasion.

Chapter 6

INSURANCE AND SURETYSHIP

Surety, like other types of insurance, helps the purchaser avoid losses. However, although insurance companies underwrite surety credit and surety is regulated as a line of insurance, there are important distinctions between the two. Insurance is a loss *funding* mechanism designed to compensate the insured against unforeseen adverse events. Insurance companies forecast the probability of the occurrence of events such as accidents, fire, and windstorms, among others, over a large population. Individuals seeking financial protection from such occurrences pay prorated shares of anticipated group (or population) losses in the form of insurance premiums. Insurers pool these funds and use them to pay for individual losses as they occur. Individuals transfer their risk of loss to the group and, in turn, to the insurance company. Losses are inevitable and thus *anticipated*, providing the basis for risk pooling. For an outline of the types of insurance policies that contractors typically require, consult Russell (1991a).

In contrast with insurance, suretyship is a loss *avoidance* mechanism designed to prequalify individuals based on their *credit strength and construction expertise.* Construction company principals retain the economic risk of contract default by signing an indemnity agreement that, in essence, protects the underwriter from incurring losses. While suretyship and insurance differ, both are risk transfer mechanisms. The project owner transfers to the surety the economic risk that might result from contractor failure. The surety guarantee protects the owner from losses due to default by the contractor, but unlike insurance, all parties enter into the agreement assuming no one will incur losses. Table 6.1 further compares surety bonds and insurance. More detailed information about the basic differences between surety credit and insurance follow below.

Chapter 2 alluded to the important difference between casualty and property insurance and surety bonding with respect to premiums. With casualty and property insurance, calculated loss expectations determine the premium, whereas with suretyship, there is little expectation of loss on an individual bond. When and if the contractor fails, rather than the premium providing the funds the surety uses to fulfill its obligation, the principal's funds, guaranteed to the surety by the indemnification agreement, are supposed to be used instead. Whereas this is theoretically true, a surety tends to lose money from contractors who become insolvent, and for this reason, a portion of every surety premium dollar goes to pay for loss.

To date no surety has discovered a way to underwrite with such perfection as to avoid all losses. Even with careful, thorough underwriting, discussed at length in Part 4, bad weather, labor strikes, unforeseen site conditions, and economic and political cycles can work to topple even the well-capitalized builder. A close look at surety industry performance statistics reveals that in every year sureties use from 10% to 50% of their revenues to pay for loss.

Table 6.1 Comparison Between Suretyship and Insurance

Surety bonds (1)	Insurance (2)
Guaranty agreement (three parties)	Indemnity agreement (two parties)
No losses assumed	Losses taken for granted
Premium–fee for extension of credit	Premium–based on actuarial likelihood of loss
Principal retains risk	Insured risk of loss transferred to insurer
Noncancelable	Cancelable
Agreement must be in writing	Agreement may or may not be in writing
Nonpayment of premium–doesn't affect surety liability of bond	Nonpayment of premium–grounds for cancellation
If bond is fraudulently obtained, surety is still liable	Policies often contain warranties as condition precedent to recovery

Assuming 60% of premiums earned goes to operating costs (the standard amount for SAA calculations), from 1984 to 1987 the surety industry lost an average of 40% per year. This means that effectively 80% of premiums earned went to pay for losses (see Table 23.1). Since then the industry has been marginally to highly profitable, but even in good years, many individual companies suffer losses and leave the market (see Table 23.7). In 1992, for example, 12,300 contractors failed with losses totaling over \$5 billion—\$374 million of which the surety industry absorbed. Despite earning a 7.3% profit from contract bonds that year, this example demonstrates that income from defaulted contractors does not completely offset surety expenses, as 92.7% of the premium went to cover expenses and loss and not the 60% assumed in standard calculations to cover operating costs alone.

In short, even with careful prequalification and underwriting, the nature of the building cycle is such that a surety will inevitably be responsible for defaulted and insolvent contractors. Therefore, a surety must provide some amount for loss, and this comes from the premiums it charges contractors. Accordingly, an insurance or surety premium has three fundamental components: expenses, loss cost, and profit factor. The expense component is intended to cover all operating expenses such as salaries, commissions, rent, utilities, postage, communications, consultants, supplies, taxes, etc. The loss component is intended to cover costs associated with investigating and settling losses, and what remains is profit. With property or casualty insurance, anticipated loss is actuarially determined and in good part determines the premium.

With surety premiums, however, loss is avoided through the careful underwriting of solvent contractors deemed capable of performing the construction contract. Compared to other types of insurance, surety is a high expense line because so much effort is devoted to the underwriting process. The surety must attempt to properly prequalify the contractor before the bond is approved, as once the bond is executed and delivered to the obligee, the bond is not subject to recall or cancellation as are other types of insurance. Inevitable losses will arise, and from 10% to 40% of the surety premium can be viewed as accounting for unexpected contingencies and contractor failures.

Traditionally, surety bond premiums have been judgment-based whereas premiums for other types of insurance are actuarially derived. In both cases the product price is established before the actual costs are known, making it seem that the actuarial approach would work for surety as well as it does for other lines of insurance. To understand why the judgment approach has been applied to the surety bond premiums it is useful to compare the extension of surety credit with bank lending.

Both the bank loan officer and surety underwriter want to avoid losses by prequalifying their applicants. Before making a loan, the loan officer will want to be satisfied that the borrower will repay the loan. The interest rate charged by the bank will vary to some extent depending on the financial strength of the borrower just as the financial strength of the contractor may have an effect on the premium charged by the surety. In both cases, those with the strongest financial statements will likely enjoy the lowest interest charges or premium rates. But if the loan officer concludes that the loan applicant is unlikely to repay the loan, charging an exorbitant amount of interest will not make the loan any more attractive. Likewise, charging a higher premium to those contractors with minimal financial resources who are unlikely to perform their contracts will not make the case more attractive to the surety underwriter.

In either situation, the loss sustained will far exceed the potential interest or premium, which may not even be collected. Surety premiums thus serve primarily to cover the expenses of underwriting and day-to-day business. Bond premiums generally do not exceed 5% of the contract price. The majority of contractors probably pay between 1% and 2% of the contract price for bonds. Those contractors that are exceptionally strong financially may be able to obtain bonds at premium charges below 1% of the contract price on their larger projects. Surety rate structures will be discussed further in Chapter 17.

Virtually all other insurance products are aimed at protecting against accidental or calamitous events and involve the assumption that there will be losses. The premiums charged are derived actuarially based on a variety of factors using the law of large numbers. A portion of the premium dollar is allocated to pay for the expected losses. In the event the actual losses are greater than the expected losses, it is quite likely that the insurance company will seek a rate increase, assuming it does not have any way of reducing the frequency or severity of its future expected losses.

Turning back to the surety situation, let us assume that sound underwriting judgments will produce a relatively low level of losses. Remember, the intent of the surety underwriting process is to prequalify the contractor to avoid providing bonds to those contractors likely to fail. Because of the nature of the objective of the surety risk, it is difficult in view of the enormous possibility of loss involved with a defaulted contractor to expect the comparatively minuscule premium to offset average losses. If prior losses were to be used as the predominant factor in developing the premium level for surety bonds, the effect would be to reward the worst underwriter with the highest rates and to drive potential customers away because of the excessive premium charges. A deterioration in underwriting quality would likely ensue and simply tend to drive up the premium rate level rather than force the surety to improve the quality of its underwriting decisions. Although increasing the premium rate might allow the surety to recover some or all of the losses that were incurred because of its poor underwriting, in the long run, this approach would tend to increase the cost of surety bonds and drive away potential customers.

Chapter 7

REINSURANCE AND CO-SURETY

Although owners, contractors, and agents will have little or no direct contact with a reinsurer or co-surety, reinsurance and co-suretyship are crucial components of the bonding cycle. Moreover, it is common practice for a surety to reduce its maximum exposure on any given account or bond by ceding or transferring some of its risk by entering into a reinsurance transaction or co-surety agreement. Such arrangements allow sureties to increase bonding capacities while decreasing individual risk. Reinsurers and co-sureties are sometimes associated with large insurance companies but are also sometimes independent entities. With a co-surety agreement, multiple sureties sign the bond form and are liable for a specific percentage or dollar amount of the risk but jointly liable.

With reinsurance, a surety finds a reinsurer to accept a set portion or a percentage of an individual contractor or on its entire account. When an surety transfers some or all of the exposure it has assumed or the amount above some certain minimum to one or more reinsurers, the surety is known as the ceding surety and the transaction is referred to as a cession of reinsurance. An insurer that provides reinsurance is called a reinsurer and is said to assume or accept the reinsurance that was offered by the ceding insurer or surety. The portion of an exposure that is not reinsured by the ceding insurer is called the ceding insurer's retention. It is common practice for a reinsurer to further divide its share of risk with other reinsurers or retrocessionaires, another type of specialized financial entity that buys and sells risk but deals primarily with reinsurers.

Many sureties use reinsurance, but this does not suggest the reinsurer will necessarily become directly involved with underwriting or investigating bond claims. Neither the principal nor the obligee on a surety bond need to be concerned about what portion of the exposure is reinsured, except where statutory requirements mandate that risk over a certain amount be prohibited by reinsurance or co-surety. Ceding reinsurance does not release the surety from its primary obligations to obligees or to suppliers, and usually neither obligees nor other potential claimants have any direct right of action against the reinsurer unless the reinsurer has also signed the bond (in which case it becomes a co-surety).

When a claim arises, the lead surety (usually the first surety named on the bond) controls the handling and investigation of any claims, the payment of proper claims to the owner and materialmen, and thereafter collects indemnity from the contractor and individual indemnitors, just as it would in the absence of any reinsurance. A reinsurer may be involved in what option a surety takes, but any negotiations, suggestions, or directions discussed with owners and contractors would be solely by the lead surety. There is unlikely to be any direct communication between a reinsurer and the principal or obligee.

By ceding reinsurance or by creating a co-surety agreement, the surety simply reduces its maximum exposure in the event of loss, since it has acquired the right to reimbursement for the reinsured portion of its loss from the reinsurer or co-surety. The primary role of the reinsurer is to stabilize earnings of the ceding surety by reducing the impact of exceptionally large losses. Of course reinsurers may also serve as a way to provide capacity to write bonds that are larger than would otherwise be allowed. If the reinsurer signs the bond, then it becomes a co-surety on that bond. However, reinsurance can indirectly affect a contractor who is seeking a surety bond and there are situations involving reinsurance that may be of concern to an obligee, including some claim situations and bonds that approach a surety's Federal Treasury Limit or the risk limitations that are set forth in the various state statures.

THE U.S. TREASURY AND SINGLE BOND LIMITS

The federal government and many states have laws or rules that prohibit a surety from providing a bond in an amount that would exceed 10% of its capital and surplus unless the excess is protected by reinsurance or co-surety. The U.S. Treasury annually provides a list of all sureties and their "underwriting limitation." In effect, a surety cannot write a bond over the amount of this underwriting limitation unless it gets a co-surety or reinsurer to protect the amount of the bond in excess of the lead surety's federally mandated underwriting limit.

For example, if a surety has $100 million of allowed capital and surplus it cannot individually underwrite a bond exceeding $10 million unless the excess amount over $10 million is protected. If a surety wants to underwrite a bond for $15 million, they would have to find a co-surety or reinsurer to guarantee the excess $5 million.

Thus, if a contractor wants a surety to provide a bond that exceeds 10% of the surety's capital and surplus, the surety must obtain reinsurance or provide co-surety to protect the excess amount. In the event the surety is unable to cede the excess portion of the risk or locate a co-surety, it will have no choice but to decline to provide the bond for the contractor. This same problem could arise even if the bond being requested is in an amount far less than the 10% underwriting limitation, as it is not unusual for surety management to make a policy decision to keep their exposure on a single contract well below 10% of the insurer's capital and surplus. Usually, the surety will not provide a bond unless it is able to reinsure that portion of the exposure that exceeds the single risk limitation decided on and imposed by management, whatever that limitation may be.

If the bond amount is below the maximum permitted by statute or managerial policy, depending on the surety, there is really no reason for the obligee or contractor to be concerned about reinsurance or co-surety. In the event the bond amount exceeds the statutory limitation, most obligees will want some evidence that the required reinsurance is in place or check the bond to see that it is properly executed by the required co-sureties. If the surety agrees to write a bond that exceeds its statutory limit, the surety usually provides the reinsurance information in a report after the bond is submitted to the obligee, if requested.

When selecting a surety company, the contractor and its agent should consider whether the surety has the capacity to handle the contractor's current and future bonding needs and anticipated work program. After the contractor has established a relationship with a particular surety and then discovers that the surety does not have the capacity to handle a particular bond request, the contractor will have to make a decision. Time permitting, the contractor may be able to locate a surety that has the capacity to provide the bond. Otherwise, the contractor may have to forego the project unless the contractor is able to persuade the owner to either reduce or waive the bond requirement. Often, however, if the contractor is a valuable client, the surety will find a way to accommodate an expanding work program through use of reinsurers and/or co-sureties.

Occasionally, there may be very large construction projects that might approach or exceed the available capacity of the entire surety industry, making it difficult or impossible for the contractor to obtain a bond in an amount equal to the contract price. The capacity of the surety industry to provide a bond for a single project may move either up or down over a period of time. When the surety industry is profitable, the single job capacity tends to expand as new insurers and reinsurers enter the market and existing insurers and reinsures are willing to assume larger exposures.

The surety industry currently appears to have the capacity to handle single projects of up to about $800 million. However, the ability to assemble the capacity to provide a bond exceeding $500 million depends on the market condition at the time the bond is requested and the reputation of the contractor. The contract terms must be reasonable and the contractor must be exceptionally well-financed and qualified to perform the project. If it can be demonstrated that a single bond equal to 100% of the contract price is beyond surety industry capacity or that competition would be severely restricted with such bonding requirements, many project owners including government bodies will reduce the amount of the required bond. For federal bonds, the agent or agency in charge of obtaining the mandated bonds has the right to lower the percentage covered by the performance bond. For all federal or state contracts exceeding $5 million, the payment bond is capped at $2.5 million.

REINSURANCE TREATIES

For extremely large jobs it is obvious how reinsurance or co-surety is used: to allow a surety to decrease its exposure and increase the amount it can legally and responsibly bond. In actuality, most sureties have an established on-going relationship in place with a reinsurer(s) called a treaty to cover portions of the bonds that they write. There are two primary types of treaty agreements between sureties and reinsurers: 1) a pro rata treaty and 2) an excess treaty. The primary difference between the two is how risk is allocated. When a reinsurer takes a percentage of the risk on a particular bond or contractor, it is called pro-rata reinsurance. When a reinsurer agrees to assume the loss of a surety over a certain amount and up to an established, capped figure, it is called standard excess of loss reinsurance, or excess reinsurance. These differences will be discussed below, followed by a discussion of how sureties and reinsurers operate.

Pro Rata Treaty

Under a pro rata agreement, a surety signs a treaty with a reinsurer to the effect that for all bonds the surety underwrites over a certain amount, the reinsurer automatically assumes a certain percentage of the risk. The surety is then responsible for all bonds up to the established amount and a percentage of every bond underwritten over that amount. If a co-surety is added, the reinsurer is still responsible for the same amount of risk. On a very large bond that approaches or exceeds the surety's U.S. Treasury limit, some arrangement would most probably be in place with the reinsurer to limit the reinsurer's ultimate exposure.

For its assumption of risk, the pro rata reinsurer receives the same percentage of the premium as the percentage of the risk it assumes. If the reinsurer assumes 20% of the risk, it then receives 20% of the premium paid to the surety by the contractor. Since the surety underwrites and handles the individual accounts and contracts, it incurs administrative costs; therefore, a percentage of the premium is paid back to the primary surety as commission. Whereas usually the surety will earn the premium over the period of the contract, this arrangement allows the surety to translate the commission into immediate income. Reinsurance not only controls the lead surety's exposure, but pro rata reinsurance helps the surety to cover its underwriting and miscellaneous costs associated with the surety business.

For example, Action Bonding signs a pro rata treaty with Rose Reinsurance to the effect that Action is responsible for all bonds up to $2 million. For any bond written over that amount, Rose Reinsurance will be responsible for 75% of the bond and Action for 25%. If Action writes a bond for $10 million, Action would be responsible for 25%, or $2.5 million, and Rose Reinsurance would be responsible for 75% of $10 million, or $7.5 million. If Action's federal underwriting limit is $10 million and it wants to underwrite a bond for $45 million, then as mandated by statute it would have to obtain a co-surety or cede further reinsurance to protect the excess, as any bond over $40 million would make Action responsible for more than its legal single-bond limit (25% of $40 million equals $10 million, i.e., the single-bond limit). Rose Reinsurance might be the co-surety or reinsurer for this amount, but another reinsurer or surety could be used as well.

Excess Treaty

Under an excess treaty, the lead surety sets its maximum exposure and one or more reinsurers provide coverage on all losses over this set amount, which could be on a contractor's work program or on the entire surety company, called the surety's "book." In addition, the surety and reinsurer agree to a limit for the largest loss or amount the reinsurer will cover. For this reinsurance, a lump sum premium is paid annually by the surety.

In the case of a loss, the entire amount in excess of the surety's retention becomes reinsured up to the agreed-upon amount. For amounts greater than this value, the surety is responsible unless they have what is called second-level reinsurance. Second-level reinsurance essentially works the same way as first-level reinsurance, but it sets the surety's and reinsurer's caps much higher. Additionally, third- and fourth-level reinsurance is used by many large sureties, with a progressively higher cap at each level.

For example, Action Bonding signs an excess treaty with Rose Reinsurance on a contractor. Action's limit is $2 million dollars, and Rose's limit is $13 million. This is called a 13 x 2 excess treaty. If the contractor defaults causing a $10 million loss, Action would assume $2 million and Rose, $8 million. Likewise, if the contractor caused a loss of $15 million, Action would still pay $2 million, but Rose would pay $13 million, its maximum exposure. If the contractor had losses totaling $20 million, Rose would again be responsible for $13 million. The remaining $5 million, in addition to Action's $2 million retention, would fall to Action. If they have a second-level excess treaty with Rose or another reinsurer, say a 40 x 15 excess treaty, then the $5 million would be paid by the second-level reinsurer. If Action does not have second-level reinsurance, then it would have to pay out $7 million on a $20 million dollar loss with a 13 x 2 first-level excess treaty—$2 million retention and $5 million over excess coverage.

How Treaties Work

It is important to point out that all treaties between sureties and reinsurers are negotiated individually, meaning that there might be any number of stipulations between the surety and the reinsurer. It is also common for a surety and a reinsurer to form a hybrid between the two types of treaties, signing pro-ratas for bonds and excesses for contractor accounts or the entire book. These two categories of treaties should be seen as generalized formats only. In actuality, the only people who know the details of an individual treaty are the primary surety and the reinsurer who signed it. In addition, it is common for a reinsurer to parcel out its assumed exposure to other reinsurers as well. If this happens, the original reinsurer becomes the lead reinsurer and would be the primary contact between the reinsurers and the surety.

It is also significant that there is limited government regulation of reinsurers. Whereas the primary sureties are subject to regulation by the various state insurance departments, reinsurers are generally not regulated by such departments. However, if they are going to provide bonds for the Federal Government, both the primary sureties and their reinsurers must comply with the Treasury Department regulations. Treaties between a surety and a reinsurer virtually always have legally-binding arbitration clauses and may exclude reinsurers and co-sureties from punitive damages. Any disputes between a surety and its reinsurer will be settled by arbitration, without court action. If an owner or claimant files a lawsuit against a surety, the surety will cover the legal expenses and then share the same with any co-sureties and reinsurers as set out in agreements and treaties, possibly including maximum exposure limits for each party. This may become important when dealing with large losses.

If a surety takes numerous or severe "hits" or losses, its reinsurer will undoubtedly lose money as well. Occasionally, the reinsurer re-negotiates with the surety to recoup its losses through a "loss clause" or "payback clause." This arrangement is similar to the indemnification the contractor signs for the surety. In the case of a financial loss, the agreement would entitle the reinsurer to receive a portion of its loss from the surety. To keep the surety from "divorcing" its reinsurer after a large hit or a series of hits, an "exit hook" is also generally included in the treaty. Essentially, this clause insures that the surety has to re-negotiate or compensate the reinsurer in the case of a loss and cannot leave the reinsurer without having done so first.

In a claim situation, the lead surety holds the folder, meaning that it decides how to contend with its obligations under the bond. If the surety decides to relet or buy the bond back, for example, it does not need explicit consent from its reinsurer(s). The only occasion when a reinsurer would normally become involved in a claim decision is when a surety is deciding whether to finance a contractor. Such a maneuver can prove highly costly for several reasons, including that costs to complete are very difficult to estimate and advances to a contractor may not reduce the obligation under the penal sum of the bond. Because this option is highly risky, before making a decision to grant an extension of credit to a defaulted contractor, the surety usually must obtain the consent of all pro-rata reinsurers.

As will be discussed more in the final section of the book, reinsurance has worked well for the surety industry and has allowed it to expand its capacity in the last few years. However, if the construction industry were to experience an economic downturn, more defaults would likely occur and the surety industry would become less profitable or even unprofitable. There would be some erosion in the number of reinsurers, but the major providers would probably still continue to sign reinsurance treaties with the well-managed primary sureties. It is also possible that the remaining reinsurers might charge higher premium rates, which could eventually increase the cost of bonding for the contractor, and ultimately, the cost of building for the owner. Understanding sound underwriting procedures and following them can only aid the construction industry in the long run.

Part III

TYPES OF SURETY BONDS

While most bonded contractors and owners requiring bonds will encounter bid, performance, and payment bonds, it is important to understand what other products the surety offers, such as maintenance and completion bonds. The following defines each category of bond and offers insight into how these surety products are used with construction contracts—information crucial to understanding the underwriting and default processes.

Chapter 8

BID BOND

The purpose of the bid bond is to prevent frivolous bidders by ensuring that the contractor will enter into the contract at the bid price if awarded. When providing a bid bond, the surety company promises that if the owner accepts the constructor's bid, the contractor will enter into the contract as well as provide any other required bonds within the time period allowed by the project specifications. If the contractor fails to enter into the contract without a valid legal excuse, then the contractor and his or her surety are typically liable for the difference between the low bid and the cost of awarding the project to the next qualified bidder, or the penal sum of the bid bond, whichever is less. Although this "bid spread" type of bid bond is the most common, there are two other ways to structure the bid bond application.

Some bid bonds are forfeiture in nature, calling for the forfeiture of the entire penal sum of the bid bond if the contractor fails to enter into the contract and/or fails to post any other required bonds. These forfeiture bid bonds are somewhat more hazardous and contractors should be aware of the exposure they are assuming. If the forfeiture bid bond is in an amount of 5% of the bid amount, or at most 10% of the bond amount, most sureties will go along provided the contractor has enough liquidity to cover a claim by the obligee. Lastly, there is the "actual damages" type of bid bond when the contractor agrees to reimburse the obligee for all costs and damages that result as a consequence of the contractor's failure to enter into the contract and post any other required bonds. This type of bid bond is not very popular because of the disputes that often arise in the event a claim is made against the surety. Such disputes usually involve the calculation of "actual damages."

There are no "standard" bid bond forms, but some professional trade associations promulgate bond forms and make them available for purchase by any obligee that desires to use them. The typical bid bond form identifies the project, delineates the parties and defines their obligations, and stipulates the amount of the bond. Appendix II presents a sample bid bond form presented by The American Institute of Architects referred to as AIA Document A310. While some project owners prefer to use forms promulgated by others, many owners, both public and private will develop their own bond forms. It is important that the contractor read many surety bond forms, bid bond or otherwise, as such forms can include a wide variety of terms and conditions.

Bid bonds are usually for 5% or 10% of the bid amount, although they may be considerably higher. However, they are seldom more than 50% of the bid amount. The Federal Government requires a bid bond totaling 20% of the bid amount. This requirement reflects the prevalent opinion that the taxpayer deserves protection from the loss that might otherwise be incurred when the contractor with the low bid fails to enter a contract with the government entity. In other words, the legislature has mandated that public project owners must take steps to avoid unnecessary expenditure of public moneys due to contractors who submit frivolous bids.

When executing a bid bond, care must be taken to make certain that the bond is properly and fully completed. Errors in the bid bond may cause the contractor's bid to be rejected or lead to expensive litigation. Special care should be taken to make certain that the date of the bid bond is either on or before the date of the bid. If the date of the bid bond is after the date of the bid, the owner may reject the bid because the bond was not in effect on the day the bids were opened.

The completed bid bond form usually accompanies the contract documents and bears the principal's signature. The bond will also carry the signature of the representative of the surety, usually an Attorney-In-Fact or a senior officer of the surety. Accompanying the bid bond should be a form stating that the surety's Attorney-In-Fact has the legal authority to sign the bid bond on behalf of the surety company.

Chapter 9

PERFORMANCE BOND

In a performance bond, the underwriter guarantees the contractor's performance of the terms of a written contract (i.e., performance of the work according to the plans and specifications, within the time allowed, at the agreed-upon price). If the contractor fails to perform the contract, the obligee may declare the contractor to be in default and call upon the surety to perform its obligation under the bond. Upon receiving such a demand, the surety will usually conduct its own investigation of the situation and then decide upon a course of action that is appropriate and acceptable to the obligee and fair to the contractor. (For a full discussion of the claim process, see Part 5.)

Most construction project owners require both a performance bond and a payment bond, each in an amount equal to 100% of the contract price. A few owners will require only a performance bond, but they will usually include wording in the performance bond to cover certain laborers, material suppliers, and subcontractors in the event they are not paid for their services. The disadvantage of this combined approach is that the bond amount may be inadequate to cover both the additional cost of performance and the unpaid bills in the event the contractor defaults. Also, if the bond amount should prove to be inadequate, the question arises as to which of the potential claimants will be shortchanged. As will be discussed later in the book, there is no financial incentive for requiring only a performance bond; the two bonds in tandem cost the same as a performance bond by itself but collectively provide more coverage.

The properly executed performance bond should be delivered to the obligee after the contractor's bid has been accepted and before the commencement of any work on the project. Obtaining a performance bond after work has commenced can be quite complicated, as the surety will be concerned that something has already gone wrong or caused the obligee to fear that the contractor might default.

Just as with a bid bond, there is no "standard" or "universal" performance bond form. The following choices are generally available to obligees who are in need of a form, assuming that a statute, rule, or regulation does not mandate the bond form that must be used:

1. The obligee may elect to use one of the bond forms promulgated by organizations such as the American Institute of Architects (AIA), the Associated General Contractors of America (AGCA), or the Engineers Joint Contract Documents Committee. Most, if not all, of these organizations will sell the forms they promulgate to both members and non-members.
2. The obligee may develop a bond form of its own. If this option is selected it is advisable to seek the assistance of legal counsel

in drafting the bond form language and in reviewing the final form.

3. The obligee does not have to select or provide any bond form. Instead, some obligees will simply go along with a bond form that is provided by the surety of the low-bid contractor. Most sureties maintain a stock of bond forms for use whenever a specific bond form is not provided or requested by the obligee.

Regardless of where it originates, prudent obligees will want to read and understand any bond form that they are considering accepting. Most performance bond forms are rather brief, but others may consume several pages. However, longer is not necessarily better. Every bond form should at a minimum clearly delineate the parties to the bond, indicate the penal amount of the bond, state the obligation of the bond, include the construction contract and specifications by reference, provide a brief description of the project, and give the location of the project. Once the bond has been executed by the surety and the contractor and turned over to the obligee, it should be carefully reviewed to make certain that all of the necessary data has been correctly entered and that the bond is properly signed and dated. Appendix III presents a sample of the performance bond form that is promulgated by the AIA commonly referred to as AIA Document A312.

Chapter 10

PAYMENT BOND

A payment bond (sometimes referred to as a labor and material bond) is given by the contractor to guarantee payment to laborers, material suppliers, and subcontractors for the labor or service that is incorporated in the project that they perform and the material that they furnish in performing the construction contract. Most construction project owners, both public and private, will require a separate payment bond in addition to the performance bond.

When a payment bond is provided on a private project, it also has the effect of protecting the owner against the possible financial loss caused by mechanic's liens, which can jeopardize the owner's interest in the property, by serving as an alternative remedy. Mechanic's liens are legal procedures that protect laborers, material suppliers, and specialty contractors against their inability to collect funds the contractor legitimately owes to them. Without payment bonds, mechanic's liens would be the sole remedy for such unpaid persons. Private owners must recognize that even though the contractor has been paid on schedule, the contractor may use the money for other purposes or to pay bills on another project. Payment to the contractor does not preclude unpaid persons or subcontractors who furnished labor or material incorporated in the construction from filing mechanic's liens against the owner's property. Although public projects are not usually subject to mechanic's liens, making a claim against the payment bond on public projects may be the only remedy available to laborers, material suppliers, and subcontractors if they are not paid for their services or materials.

Once the labor has been performed or the material has been installed on the project, neither can be retrieved in the event of non-payment. This reality created the need for mechanic's lien laws that apply to private construction projects. Accordingly, the various states enacted their mechanic's lien laws to provide a remedy for those laborers, material suppliers, and subcontractors who are unable to collect funds legitimately owed to them. When the private project owner has the protection of a payment bond in place, potential lien claimants may elect to seek payment from the surety and thus avoid the need to file a mechanic's lien, which effectively reduces the equity of the owner on the property.

It is important for all of those who furnish labor or materials to recognize that the detailed provisions of the mechanic's lien laws tend to differ somewhat from one state to another. Everyone who is doing work or performing a service on a construction project should make the effort to understand the mechanic's lien laws and notice requirements of the state where the work is being performed, as many payment bonds are designed to follow the provisions of the mechanic's lien law of the state where the work is being performed. Those who are depending on the protection provided by a payment bond should first make certain they are in fact covered, and then make certain that they are in compliance with any notice requirements that might apply. Failure to give a required notice may make it impossible to collect from the surety or even to file a mechanic's lien. Care should be taken to understand and comply with any time limitations that may apply to filing notices and/or law suits, as there are almost always restrictions.

Public project owners, whether federal, state, or local, generally require that the contractor provide a payment bond to protect those who perform labor or furnish materials to the project. The public policy desire to provide some remedy for those who are not able to collect for the labor or material they furnish on public projects has led most public bodies to mandate the use of payment bonds by law, rule, or regulation. Without the protection of a payment bond, those who are unable to otherwise collect what the contractor legitimately owes them would be without any remedy since mechanic's liens cannot be filed against public property. In some jurisdictions, an unpaid party can file a lien against project funds that are still in possession of the project owner. However, this remedy often proves to be of little value because the funds may have been alienated before the claim is made or the remaining funds may be inadequate to cover all of the claims that are filed.

Contractors may find that subcontractors and material suppliers are more willing to extend credit when a payment bond protects the project. Since even some public projects may not require a payment bond (especially projects involving very small dollar amounts), the contractor may find that an increasing number of material suppliers and subcontractors will want to know if the project is bonded. If the project is bonded, such persons may ask for a copy of the payment bond before entering into a contract.

As there are no "standard" or "universal" payment bond forms (just as there are no standard or universal performance bond forms), each obligee is confronted with the same bond form choices that are outlined in Chapter 10. There are, however, some additional points to consider with regard to the language content of the payment bond.

In general, care should be taken to make sure the bond form is not broader than the mechanic's lien law that applies to the project. With the payment bond, it is also desirable to specify a time limitation for filing suit and to spell out exactly which potential claimants are covered. The latter may be accomplished by referencing the appropriate statutory provisions or by including language in the bond form that clearly states which laborers, material suppliers, and subcontractors are protected. For example, the Federal Statute (40 U.S.C. 270b) extends the payment bond protection to every person who has furnished labor or material in the prosecution of the work and who has not been paid in full before the expiration of 90 days after the day on which the last of the labor was done or performed, or material was furnished or supplied. Those that have a direct contractual relationship with a subcontractor but no contractual relationship with the contractor furnishing the payment bond must give a written notice to the contractor in accordance with the terms of the statute or they lose the protection afforded by the payment bond.

Thus, the federal payment bond covers those persons who are in privity with the contractor that furnished the payment bond and those who are in privity with a subcontractor, provided they give proper notice. Coverage stops at the second tier; i.e., a sub-subcontractor or a supplier to a subcontractor is protected. Third-tier subcontractors and suppliers to a supplier are not protected. If the subcontractor does not pay a sub-subcontractor or a supplier to the subcontractor for work properly performed or materials supplied, the contractor may be obligated to pay them even though the subcontractor was already paid in full.

The logic of extending protection to the second tier, but not beyond, hinges on the prerogative of the contractor to require performance and payment bonds from the subcontractors. There is no way that the contractor can protect against non-payment of those persons that are beyond the second tier or to even know who they may be. Accordingly, it has been considered improper to burden the contractor with a risk over which there is no way it can exercise any reasonable control or obtain any protection.

Payment bonds are usually provided in conjunction with the performance bond but they may also be issued as a stand-alone product. When issued in conjunction with a performance bond, the payment bond premium is included in the premium for the performance bond. The premium for a stand-alone payment bond will be somewhat less than premium charged when both bonds are issued together. However, the savings is not as much as what might be expected, since, according to industry experts, claims against payment bonds exceed those against performance bonds by a significant margin.

Consult Appendix IV for a sample of the payment bond form promulgated by the American Institute of Architects, commonly known as AIA Document A312. (Since the AIA performance and payment bond forms are packaged together and sold as one item, both bond forms have the same AIA Document number.)

Both the performance and payment bonds should carry a date that is either the same or subsequent to the date of the contract that is referenced in the bond. These bonds must never predate the date of the underlying contract. A power of attorney or a jurat (notarial acknowledgment) should accompany the bonds, and in some parts of the country, the surety may attach an abbreviated copy of its most recent financial statement. What must be attached to a particular bond can vary somewhat depending on the jurisdiction where the bond is executed or where the construction project is located.

Chapter 11

MAINTENANCE BOND

Maintenance bonds are guarantees that oblige contractors to return to completed projects to perform maintenance or corrective work during a specified guarantee period. Work can include any number of duties defined in contract documents, but is usually limited to repairing any substandard construction or installation, whether due to workmanship or materials. However, it can also include such things as upkeep.

Whereas no precise numbers exist, maintenance bonds are estimated to account for less than 5% of the total annual bonding volume. Although relatively uncommon now, maintenance bonds are required in many areas of public construction such as highways and co-generation projects and will be increasingly used in the future. Furthermore, even a few years after construction has been completed, much can happen to a facility or a contractor, and it is often extremely difficult to determine what or who is at fault for a lack of performance. As such, maintenance bond claims are some of the most costly for the surety to investigate and often result in lengthy litigation. It is important for owners and contractors to understand what coverage they can supply and how best to structure any maintenance bond agreements.

Maintenance bonds are usually issued in tandem with performance and payment bonds and are commonly included on the same bond form. When a performance bond protects the construction project, the performance bond automatically covers any maintenance covenants included in the contract documents. However, some project owners may require a separate maintenance bond form, and this is advisable to help clarify the nature of any maintenance agreement. If a warranty is stipulated in the contract document, regardless of the surety's actual awareness of such a guarantee, the contractor and surety are bound by law to the contractual agreement. Even if a contractor is not forthcoming about such a contractual guarantee, the surety must honor any stipulations of guarantee regardless of any premiums received. Since contract documents are often highly technical and detailed, it is uncommon for a surety to review or even receive a complete copy of the contract. Therefore, requiring a separate maintenance bond allows the surety to better respond to owners' needs should a claim situation arise, because the surety's and contractor's obligations will be known to all three parties. Appendix V presents a sample of a separate maintenance bond form.

Whether covered under the performance bond or by a separate bond, a maintenance agreement covering defective materials and workmanship for a period of one year will generally not incur additional premium charge. However, a premium charge will generally be made if the term of the maintenance agreement exceeds one year or if the scope of the maintenance agreement goes beyond defective materials and workmanship.

Of course, when a separate maintenance agreement is required, the owner will need to specify the amount of the bond. Many project owners will make a judgment estimate as to the dollar amount of the maintenance exposure and then specify a bond amount that is believed to be adequate. Usually a maintenance bond in the amount of 5% to 15% of the contract price will provide adequate protection, but some warranties are for up to 100% of the contract. If a maintenance bond is issued for a contract that is not protected by a performance bond, or if the duration exceeds whatever period is included in the performance bond, a premium will be charged.

When seeking a performance bond or when asking the surety to consider a separate maintenance bond, the contractor should be ready to answer questions posed by the surety about the scope and duration of any maintenance covenants in the contract. In order to determine exactly what the maintenance bond obligation involves, it is necessary to carefully read the contract documents. The risk being assumed in conjunction with a maintenance obligation can range from minimal to significant depending on the project delivery system being used, the terms of the agreement, the type and technical scope of the project, and the duration of the obligation. The surety agent will want to know all of these factors when determining whether to underwrite any maintenance agreement.

Sometimes a contractor in good standing with its surety will receive a second year of coverage at no additional charge, but whenever the duration of the maintenance obligation exceeds two years, the contractor should immediately make the surety aware of the requirement when applying for the bid bond. While the surety may be willing to consider covering a maintenance obligation exceeding two years, the contractor should not assume the surety will automatically approve the case. In addition to ascertaining that the surety will approve the long-duration maintenance obligation, the contractor should request what, if any, additional premium charge would be made so the amount can be included in the bid calculation. These premiums, while charged at a lower rate than performance bonds, can become significant over an extended time.

After as little as two to three years, it is often very difficult to determine whether workmanship, materials, unintended or over-use, or design flaws account for any undesirable performance. As such, due to increases in exposure—both physical and capital—premiums become significant. Furthermore, as mentioned above, it is in general difficult to predict the performance of a construction product beyond a two or three-year period. Therefore, it is problematic to identify and quantify the performance risk, both for an owner and for the surety. It is thus advisable for an owner to seriously consider how they will be served by such a guarantee, since in the end, they absorb the premium cost.

When the term of the maintenance covenant extends beyond one or two years, the contractor is in reality guaranteeing the quality of the work. Even if the work was done in conformance with all of the plans and specifications, there is still a chance that over a period of several years something will fail. For example, a roof might start to leak because a building settled or the supporting materials expanded or contracted. Also, it is important to recognize that in effect, the surety is guaranteeing that the contractor will remain in business for the duration of any maintenance obligation. While it may be possible to predict with some accuracy that a particular contractor will still be in business one year in the future, it is virtually impossible to predict that a particular contractor will be in business several years in the future. Thus, the contractor must recognize that a maintenance obligation exceeding two years may cause the surety to decline the bond that would otherwise be accepted. Project owners should keep in mind that maintenance obligations of a long duration might limit competition, and thus indirectly increase the cost of the project.

When a maintenance obligation is included in a contract where the contractor is responsible for some or all of the project design, the risk being assumed by the contractor and the surety is substantially expanded. In effect, the contractor and the surety are guaranteeing that the product selected and installed by the constructor will perform adequately for the allotted amount of time. For example, assume a state highway department provides the contractor with a performance specification for pavement markings and leaves it up to the contractor to select the material and apply it to the pavement. The project owner wants the markings to last for at least four years. Accordingly, under the terms of the contract, the contractor is required to post a maintenance bond guaranteeing it will "maintain" the pavement markings for a period of four years. In the event the pavement marking fails within the four years, the cost of the required repair or replacement could be substantial. For instance, if the contractor must replace the markings, it might have to pay the state a lane rental charge for any traffic lanes that need to be closed. Moreover, there is a possibility that the pavement markings may need to be repaired or replaced more than once during the four-year time span.

A performance specification may occasionally be included in a contract where the basic project design is provided by a design professional retained by the project owner. For example, the design professional of a high-rise building may describe how the exterior of the building is to look and leave the actual design of the curtain wall to the contractor. Then, the contractor is responsible if the curtain wall cracks, leaks, fades, or if the glazing falls out. If a maintenance bond is required, the contractor and the surety have guaranteed that the products selected and installed by the contractor will perform as specified.

Even more hazardous are contracts requiring a maintenance bond where the constructor is required to design, build, operate and maintain the project for a period of years, as they are in a Design/Build contract. For example, the contract for a light rail system could be awarded under which the contractor would design and build the system, then operate and maintain it for a specified number of years. Design/Build is rapidly changing the way construction is performed, and is discussed at length in the final section of the book. Suffice it to say that with detailed or extended maintenance bonds, the contractor's and surety's liabilities can become significant.

Contractors must watch for contract language that makes them responsible for warranties or guaranties that are provided by product manufacturers. If such warranties are simply passed through to the project owner, the contractor is not at risk. For example, suppose a manufacturer of roofing material offers a 20-year warranty. In the event the material fails, the manufacturer will replace it, perhaps at a prorated price. However, if the contractor is made responsible for performing the labor needed to remove and replace the failed product, the risk to the contractor is considerable. Contractors should require warranties from suppliers for particularly critical and technical materials such as roofs, concrete, electrical systems, and HVAC systems, and thoroughly examine them to make sure that the coverage is clearly articulated.

Chapter 12

OTHER SURETY BONDS

SUPPLY BOND

Supply bonds are similar in function to performance bonds; however, supply bonds differ in that sureties issue them for contracts to supply materials, such as boilers and machinery, at a specified time and place. For sample clauses for inclusion in the contract specifications for bid, performance, payment, and supply bonds, see Appendix VI.

Supply bonds are used to guarantee the quality, quantity, or price of goods and delivery at a future date. Public entities can ensure that materials needed at specific times will be available to the exact specifications required. For example, many municipalities in the northern regions of the country require salt to de-ice roads in the winter. Storage space for adequate supplies may not be available, and the supply bond guarantees that salt will be available to the city when it needs it, either from the contracted supplier or from another supplier if the first one defaults.

Supply bonds also ensure that materials will be of a quality designated in contract documents. Appliances for apartments, concrete, lumber, and gravel can all be guaranteed with a supply bond to be of an appropriate grade or class. If inferior materials are supplied to a project, the maintenance bond also generally guarantees the removal and replacement and any costs associated with the inferior product.

COMPLETION BOND

Underwriters issue completion bonds in the private and public sectors. The motivation behind this bond is to protect the lending institution that has advanced to the owner the necessary funds for construction or the municipality authorizing construction. The completion bond enables the lender, who has a mortgage on the facility, to hold the first lien against the property. In theory, the bond acts as collateral for the mortgage loan for the construction period regardless of whether the owner or the contractor performs the work (Remmen 1977).

One situation that would require a completion bond would be when the owner and contractor are the same entity. For example, a city or lending institution may require a real estate developer to obtain a completion bond if that developer plans to build a subdivision. In such a case where the real estate developer is both owner and contractor, the city or lending institution must protect its interests in the event the real estate developer does not complete construction of the subdivision. Leaving a partly developed landscape would of course do nobody any good, especially not the municipality or lender. In this case, the bond might also be referred to as a subdivision bond.

A standard performance bond protects the owner against the risk that the contractor may not complete the facility. In addition, the contractor receives protection against "owner failure." Under such a performance bond, should the owner default, the contractor may cease work without liability, since the owner can no longer pay for the completed facility. On the other hand, a completion bond guarantees completion of the project independent of the owner's obligation. Hence, the underwriter guarantees the completion of the project even in cases where the lender or the owner does not provide the financial resources at the agreed-upon time.

On certain federal projects, the government prints bond forms that cover these situations. Federal projects that fall under this category are housing projects under the U.S. Department of Housing and Urban Development (HUD) and projects involving the U.S. Postal Service. For HUD projects, such as the development of low- or middle-income housing, the government requires real estate operators to obtain a completion bond to guarantee the completion of these projects. The policy of the U.S. Postal Service is to lease its post offices upon completion of the project rather than actually owning the facility. This system guarantees that the project will be complete before the owners (investors such as life insurance companies and other mortgage outlets) receive any money.

When a private owner desires a completion bond for a project, the performance and payment bond guarantee should include the following contract language ("Bonds & Insurance for Contractors" 1989):

The right of the mortgagee hereunder is conditioned upon:

1. The payment by or on behalf of the owner of the payments to which the principal is entitled under the terms of said written contract with owner and the performance of all its obligation under the contract by the owner.
2. The mortgagee withholding its consent to any change or alteration that increases the contract price of the improvements until sufficient money has been deposited with the mortgagee to cover any such increase.
3. Recovery hereunder by mortgagee being reduced by the loss, if any, resulting from failure of owner's representative to obtain Builders All Risk insurance as designated in the latest edition of the AIA Document 201.

Completion bonds represent more risk than traditional bid, performance, and payment bonds. For these reasons, underwriters are usually reluctant to underwrite a completion bond, and premiums for such coverage is more expensive than for other types of construction contract bonds. The principle reason for this is that they are guaranteeing items, such as financial funds to pay the contractor, that are beyond the control of the owner.

DUAL OBLIGEE BOND

A dual obligee bond is simply a surety bond with more than one obligee named on the bond. Generally, the obligees are the project owner and the financial institution financing the project. The lender desires to have security for its mortgage loan in the event of owner default. In the absence of a dual obligee bond, the lender might have a difficult time gaining control of a defaulted project. However, if the owner defaults under a dual obligee bond, the co-obligee can then protect its mortgage by taking over the project.

As an example, if the owner does not pay the contractor, the lender (as a co-obligee) can pay the contractor, and then the contractor legally must complete the job. Without such an agreement, the contractor could cease to work on the project and neither the contractor nor the surety would be liable for its completion. After all, a half-completed project is of no use to a lending institution. For example, if a landlord plans to pay off a mortgage for a new apartment complex with rent from the new tenants, and the project is not complete, the lender will most likely default on its mortgage. Thus the bank will not be able to collect the money it loaned. With a dual obligee bond, the lender has all the rights of the owner and thus the assurance that it can "step in" and complete the project so that it can receive a return on its investment.

NON-CONTRACT BONDS

Other surety bonds exist besides contract bonds. They include fidelity bonds, public official bonds, fiduciary bonds, judicial bonds, union bonds, license and permits bonds, federal bonds, and miscellaneous bonds. A fidelity bond protects employers against dishonest and corrupt employees who can cause an employer to lose money. Statutes generally require public official bonds for public officials who handle money (e.g., tax collectors, probate officers, and treasurers). These bonds protect the public from loss through embezzlement of funds. A fiduciary bond ensures that the administrator, executor, guardian, trustee, or court-appointed receiver of property perform their functions honestly and within the stipulated laws. Judicial bonds provide security to the opposing litigant in the event the final decision is adverse to the principal. Additionally, unions may require a bond from the contractor to guarantee the payment of union dues, pension-fund payments, or other funds due the union and deducted from the workers' wages. While they are officially classified as miscellaneous surety bonds, these bonds are commonly referred to as union bonds within the industry.

License and permit bonds ensure that the principal engaged in a specific activity will meet all the requirements and regulations associated with that activity; examples include remittance of social security taxes or sales taxes. Federal bonds consist of three types: 1) custom, 2) excise, and 3) immigrant. A custom bond guarantees that an importer or exporter pays relevant duties. An excise bond requires the following to exist: adequacy of premises and compliance with regulations that are subject to supervisory controls over the manufacturer, handling, storage, and transportation of a given product, as well as the payment of relevant taxes due. An immigrant bond guarantees that aliens temporarily entering the country will leave upon completion of their authorized stay. Of all of these bonds, fidelity bonds, public official bonds, union bonds, and judicial bonds are the most relevant to this book because these bonds cover people likely to be involved in the construction industry.

Part IV

THE UNDERWRITING PROCESS

Part 4 is a comprehensive exploration of the nuts and bolts of the underwriting task of the contractor written primarily from the perspective of the surety. This section identifies what information the surety typically requires to complete the contractor's review, including the character, capacity, continuity, and capital or financial portions. The section describes what form such contractor information generally takes and discusses why such information is important and how it is analyzed. Contractors and owners will learn the methods underwriters employ to make their underwriting decisions and understand how surety products are priced and regulated by state insurance commissions. To bring it all to life, an extensive case study looks at the underwriting of a fictional firm, complete with examples of most all the documents sureties require.

Chapter 13

OVERVIEW OF UNDERWRITING PROCESS

The surety underwriting decision is extremely complex. Underwriting personnel must consider numerous uncertainties and elements of risk, many of which are unquantifiable. Contractors face myriad concerns, such as how to obtain an adequate volume of construction work and the availability of labor, materials, and equipment at fixed or knowable costs. They must concern themselves with such disparate issues as whether contract and bond form language is fair, and if weather conditions are favorable. In addition, the qualifications, reasonableness, and ethics of owners, construction managers, sureties, independent surety agents, architects, engineers, and inspectors can all affect a contractor's organization.

Surety bonding has grown significantly in the 1990s, and in the current profitable cycle, many contractors are experiencing low rates and superior service as sureties vie to underwrite healthy contractors. However, as will always be the case, undercapitalized contractors or contractors lacking experience or managerial strength will have a hard time securing bonding. Even though the industry has been profitable since 1988, the substantial losses in the mid-1980s stimulated a re-evaluation of previous underwriting rules and the development of stricter evaluative criteria in an attempt to counteract the risks and uncertainty involved in surety bonding. While many have noted a gradual loosening of the strict underwriting that made the industry profitable again, small- to medium-sized companies with sales range between $1 million and $10 million trying to obtain bonded work are most likely to fall short of even lax underwriting standards.

Because many contractors will not be able to secure bonding, they must seek alternatives such as self-surety or bid unbonded work. In extreme cases, they may be forced to discontinue business. A major reason for bonding difficulty is that smaller applicants often have lower working capital and net worth than their larger counterparts. Larger contractors generally have greater resources of capital to draw from than smaller companies, and this is particularly attractive to a surety considering accepting the risk on a bond. If the larger builder were to suffer setbacks due to weather, labor strikes, or a downturn in the economic cycle, chances are that they would be able to surmount these problems. The builder with a smaller capital base, however, is particularly vulnerable to the normal risks associated with the construction business and may be unable to find a surety willing to underwrite their bonds. In the current climate, however, where approximately 600 surety companies are competing when in 1980 there were less than 200, a small company with a sound record may indeed secure bonding. For those that cannot, this book discusses alternatives to surety bonding and ways in which a contractor might evolve toward bonded work.

Chapter 14

SEQUENCE OF EVENTS: A CONTRACTOR'S PERSPECTIVE

Contractors cannot establish bonding credit overnight. It may take anywhere from a few weeks to several months to establish a surety relationship, and thus *well before* any bonding deadlines for specific projects, a contractor should establish a business relationship with a surety agent. This will provide sufficient time for the agent to become familiar with the contractor's account, submit the account to a surety company for review, and obtain a response, hopefully positive. To initiate the process, the contractor typically submits data in the form of a questionnaire for analysis by the surety agent and surety underwriter. Appendix VII presents a blank sample of such a questionnaire. An additional contractor information form used in the underwriting case study of the Prometheus Construction Company can be found in Appendix XII. Other methods agents and underwriters use to obtain information regarding a contractor include: 1) visits to the contractor's business premises, 2) interviews, 3) credit rating services (e.g., Dun and Bradstreet and Robert Morris Associates), and 4) references from bankers, accountants, suppliers, specialty contractors, architects, engineers, former clients, and insurance companies, among others.

Table 14.1 illustrates data that surety underwriters typically require to open a new account. Although each underwriter may require different information, independent surety agents generally know what information the various underwriting companies require. Even after an account is established, surety agents will need updated financial statements as well as a continual flow of job performance information to stay abreast of the contractor's status. Thus, the contractor should keep all financial and project information current and have it ready to submit to the agent as appropriate.

After establishing the surety relationship, a contractor can initiate the bonding process by contacting his or her agent to request a bid bond for a specific project. At this time, the contractor describes to the agent the project characteristics including bid deadline, start and completion date, type, size, location, and, if stipulated, amount of per diem liquidated damages, among others. If successful, the low-bidder then requests any bonds the owner requires, possibly including performance, payment, and maintenance.

After establishing a surety relationship, a contractor may secure an underwriting decision for a specific project within a few hours or days. The turnaround time depends on the completeness and accuracy of the data the contractor submits, the agent's familiarity with the contractor's account, and the project characteristics. Figure 14-1 presents a flow diagram representing the underwriting process.

Table 14.1 Data Typically Requested by a Surety Company
Source: The Travelers, Cincinnati, Ohio

	Information needed (1)		Explanation (2)
1.	Audited financial reports for the last three (3) years; a review with tax statements is acceptable.	1.	We track their past financial record to help make decisions about their current and future financial strength.
2.	Current schedule of work on hand along with work on hand schedules for the last eight quarters.	2.	This information allows us to track profits and billings to see what contractors have done historically.
3.	Current personal financial statements on owners of the company.	3.	This gives us an indication as to the personnel financial strength of all owners involved for personal indemnity purposes.
4.	Name and address of banker along with their contact information.	4.	We try to develop a relationship with their banker so that we can work together to help the contractor.
5.	Name and address of ten (10) suppliers of materials, along with their phone number.		Historical information for tracking their construction record.
6.	List of ten (10) largest projects completed in the past. Please include contract amount, final gross profit, year completed, name and address of owner, name of architect and/or engineer, and phone numbers for all involved.	6.	Historical information for tracking their construction record.
7.	Organizational chart of company, including all key employees. This chart should go down through field superintendents. The educational and experience resumes of all key employees should be provided.	7.	This information gives us an indication of their potential capabilities with their present personnel.
8.	Summary of stock ownership and details of company's continuity program.	8.	This information is used to judge what kind of planning has been done to protect the company's and our interest in the event of a death of key personnel.
9.	Contemplated program; largest single job and work backlog.	9.	This information helps us plan for their future needs. We want to be aware of their goals so we may plan for them together.
10.	Past program; largest single job and work backlog	10.	This information gives us an indication as to their past capabilities to help estimate their current capabilities.

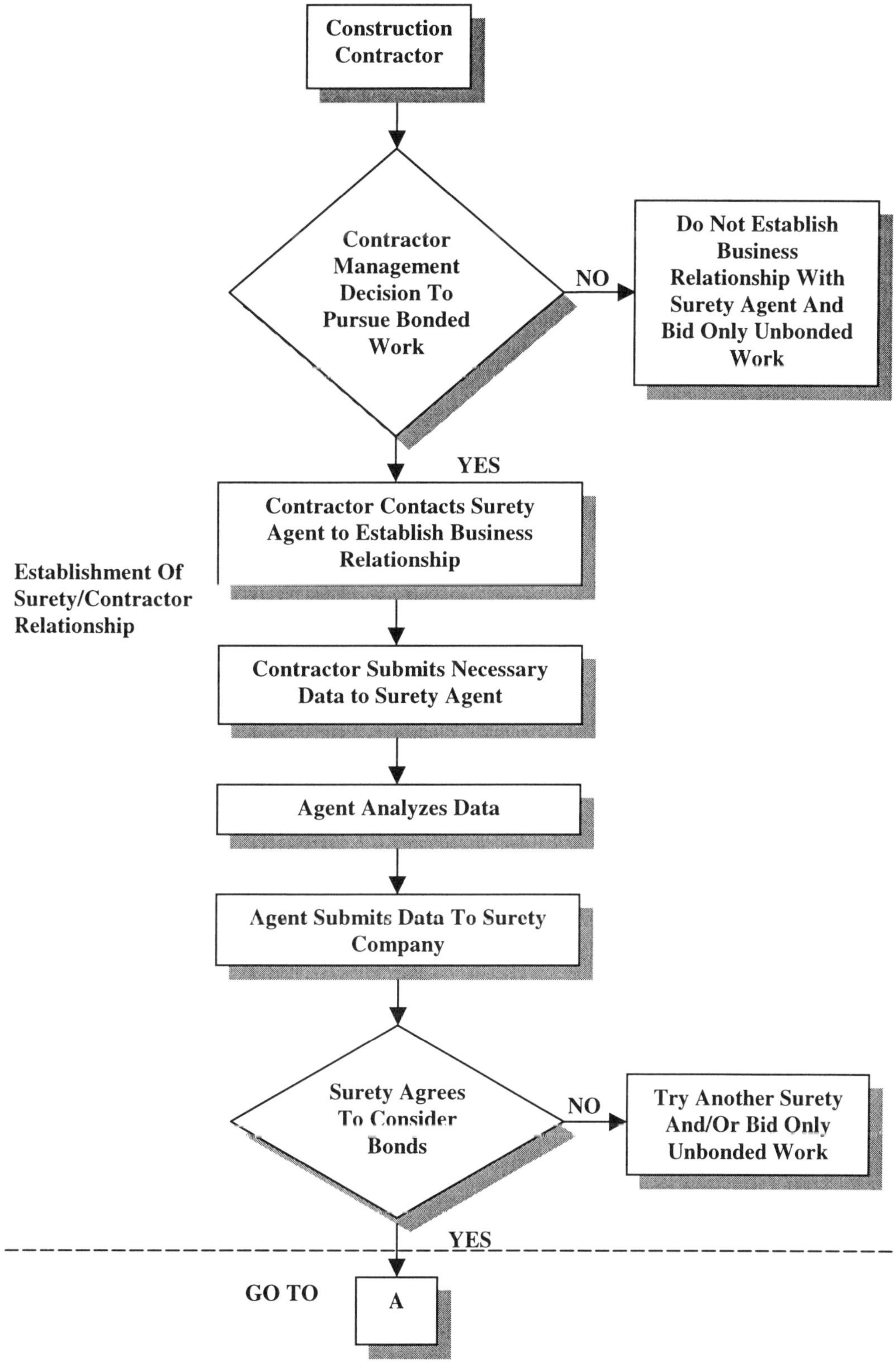

Figure 14-1 Flow Diagram of Surety/Contractor Relationship and Underwriting Process

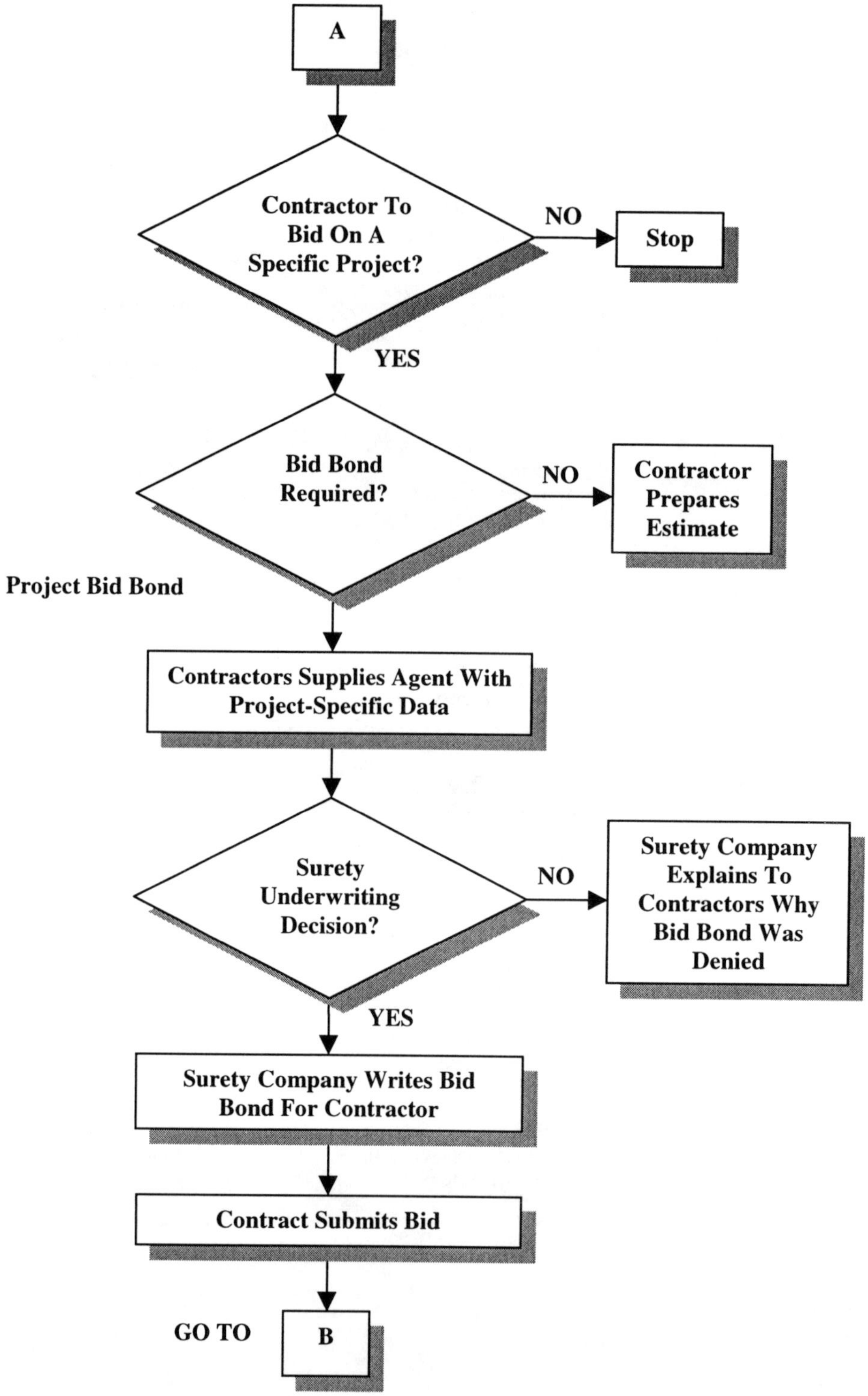

Figure 14-1 Flow Diagram of Surety/Contractor Relationship and Underwriting Process (Continued)

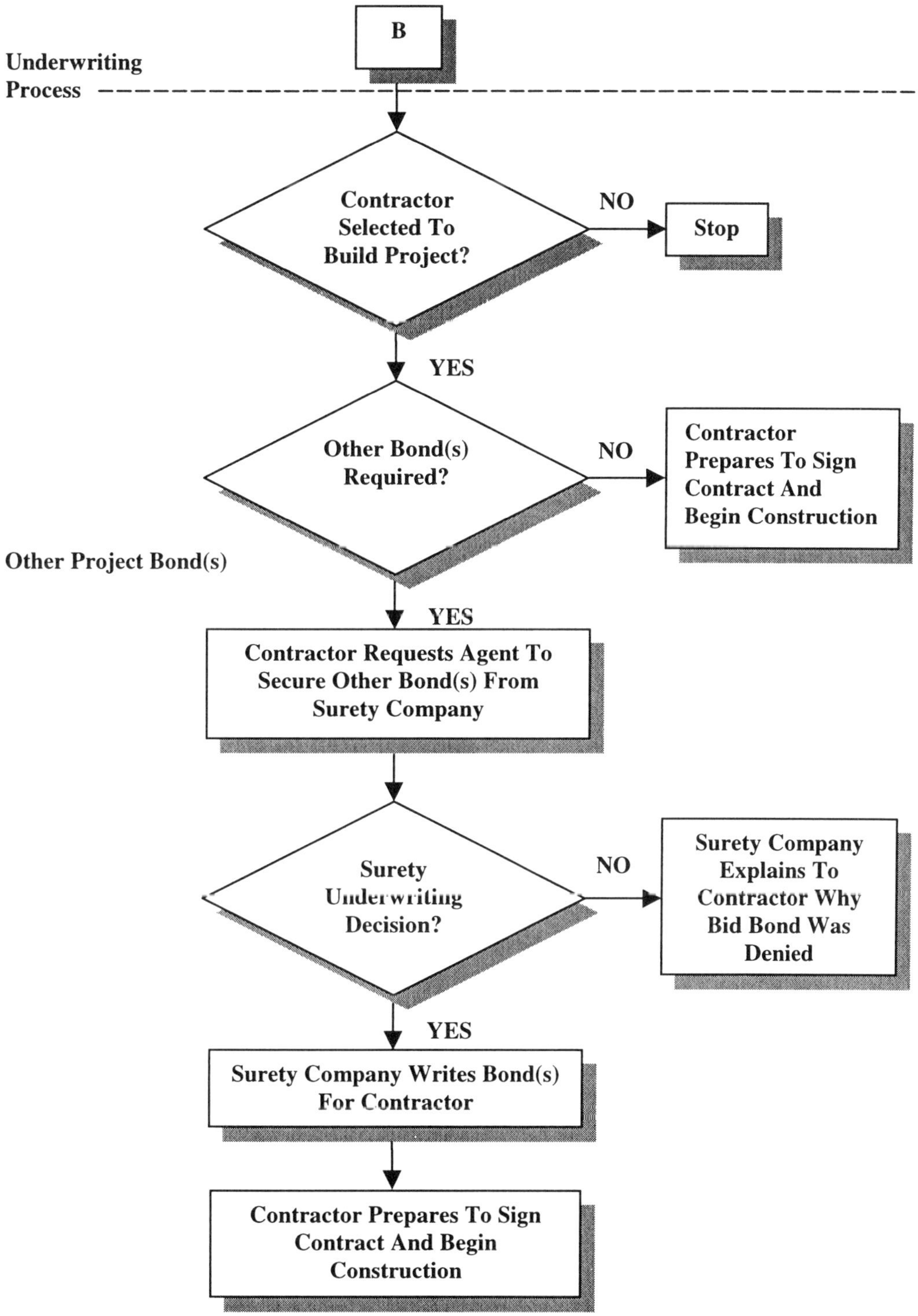

Figure 14-1 Flow Diagram of Surety/Contractor Relationship and Underwriting Process (Continued)

It is often the case that a surety underwriter will not examine the contract documents in their entirety. The underwriter is concerned with the contractor's account and performance in the past, and whether any changes might have occurred to warrant a serious examination of the proposed bid. In addition, the underwriter is trained to look for any anomalies such as a contractor bidding a much larger contract than is normal, bidding too many jobs at once, or entering into a new or particularly risky kind of construction.

For these reasons, a surety generally sets a cap on the maximum single bond and aggregate work program that it will underwrite, meaning that the surety will most likely refuse to underwrite a bond for an amount over the single-job limit or a bond that might push the contractor over his or her aggregate limit. For example, a contractor has a single bond limit of $5 million and an aggregate limit of $20 million. If the contractor has $15 million of on-going work and wishes to bid a $6 million job, the surety underwriter would have to seriously consider whether to underwrite the bond, as the new job, if the bid is accepted, would push the contractor over both his limits.

A surety that underwrites a bid bond ordinarily expects to write the performance and payment bonds if the owner awards the contract to the contractor. However, the underwriter may refuse to underwrite these bonds if: 1) the contractor's bid is not appropriate (e.g., 10% or more below other bidders), 2) the project owner has not secured the funds necessary to build the project, 3) the key contractor personnel or principals are unavailable, 4) the contractor is insolvent, or 5) the contractor has suffered severe financial setbacks. Thus, the underwriter does have the option of not issuing the subsequent performance and payment bonds. The only situation that requires a surety to underwrite the additional bonds is one where the agent has committed to the owner and contractor that the surety company will provide, or underwrite, these bonds. Despite the fact that the surety agent is an independent entity, his or her word may be deemed an agent of the underwriting surety company, for which reason the agent must be careful not to misrepresent the intentions of the underwriter. Courts have interpreted such indications from even an independent agent as implied contracts, and have required the surety company to underwrite the bonds.

However, the fact that a surety approves and underwrites one type of bond does not guarantee it will provide other types of surety bonds for the same construction project or future project. The risk on each bond is always assessed independently. For example, the Federal District Court for the State of Colorado has held that an underwriter that issues a bid bond is under no obligation to issue performance and payment bonds for the same project—L&M Enterprises Inc. v. Hartford Architect and Indemnity Co., 700 F. Supp. 517 (D. Colorado 1988). Although it is customary for the surety to provide the necessary bonds once a bid has been accepted, the contractor should be prepared for the occasional refusal by the surety company.

Chapter 15

SURETY UNDERWRITING CRITERIA

The previous sections aimed to help readers grasp the context in which contractors establish a relationship with surety agents and underwriters and operate within the surety bonding process. This next section describes the details associated with the underwriting decision from the surety underwriter's perspective. Since the beginning of corporate suretyship, underwriters have arrived at underwriting decisions using the three "Cs"—1) character, 2) capacity, and 3) capital. Evaluating character involves considering a contractor's past performance and how well the organization fulfills obligations. Measuring capacity requires evaluating the technical resources the contractor needs to complete projects or particular work programs successfully. To quantify capital, the underwriter analyzes the financial strength and performance of the contractor's company. Due to the complexities of the modern construction industry, another "C" has entered the underwriting equation—continuity. Continuity addresses issues that relate to a company's longevity. It can also be argued that "communication" is crucial as well to the bonding endeavor, and therefore the contractor who has good communication skills will benefit. It is important, however, to point out that the underwriting process *does vary* from underwriter to underwriter, and thus what follows is more a template and not a script.

In underwriting a specific bond request, the surety underwriter needs to be aware of the terms and conditions of the contract their firm will be guaranteeing. This ensures that the underwriter has all of the information he or she needs to make an informed decision. Additional issues of concern to an underwriter are: 1) time for completion, 2) liquidated damages clause, 3) guarantees and warranties, 4) retainage clause, 5) indemnification clause, 6) whether the bond includes details regarding project design, 7) specific performance warranty, 8) project financing, 9) turnkey agreements, and 10) other payment clauses.

CHARACTER

Character pertains to the evaluation of a contractor's past performance. Of particular interest is the reputation of the company and its owner(s). Items underwriters use to evaluate character include determining whether or not any governmental agencies have debarred the contractor from bidding projects. Also at issue is determining whether the contractor has colluded with other bidders, engaged in fraudulent activities (e.g., false representation of financial information), failed in a previous business, failed to complete a project, or performed poorly on previous projects (e.g., finishing projects behind schedule or producing poor quality work). Underwriters also research whether the contractor is prone toward litigation that may yield bad faith and punitive damages.

Other character issues include whether or not the contractor keeps his or her promises, conducts business transactions fairly, and performs obligations in a timely manner. Surety companies can obtain this information by contacting references such as prior clients, architects, engineers, specialty contractors, or material suppliers, and through meetings between the agent, surety company, and contractor.

CAPACITY

Surety companies measure a contractor's capacity by evaluating the technical resources necessary for the contractor to complete a particular job successfully. The amount and quality of resources are what the surety underwriter tries to determine. Resources can include education of available management, equipment, labor, material, and financial reserves. Evaluators must also analyze resources from the viewpoint of total dollar amount, current work load, and dollar amount available, which are further analyzed in the capital evaluation. To obtain information regarding these items, the surety underwriters contact relevant references such as architects and engineers, owners, specialty contractors, and suppliers.

Management Considerations

Construction companies maintain two types of management: 1) company and 2) project. While each type has its own set of peculiarities, both affect the company's future productivity and profitability. Therefore, the contractor's management ability at the project and office levels is a concern for an owner or underwriter who may select that particular contractor.

Company Management

When evaluating company management, surety agents and underwriters consider the experience of the contractor's personnel and the work the organization previously completed. Underwriting personnel consider the company's ability to administer activities related to its operation and to properly estimate and control field operations at the current and anticipated levels of activity. To establish the quality of a contractor's company management, evaluators consider the efficiency in processing paperwork (including pay requisitions) and change orders. Underwriters also seek to confirm the existence and quality of cost accounting records, as well as comprehending financial matters including cash flow, importance of cost documentation, and business planning.

Several key items in company management that surety underwriters look for are:

1. Does the company have leadership with vision and the ability to lead?
2. How well do they communicate this vision to the rest of the company?
3. At what level and to what degree are the leaders engaged in the company?
4. Do the leaders have a formal or informal business plan?
5. What is their bidding philosophy?

6. Does a plan exist to reduce overhead when insufficient work is available?

Company philosophies and procedures affect the level of efficiency that a firm achieves. It is management's job to make sure company expectations are understood. A crucial element to this process is estimating procedures and job cost monitoring systems. Without the ability to accurately predict how much the firm stands to make on a certain job and overall, a contractor is much more susceptible to problems and becomes a more risky enterprise to underwrite. To determine if a contractor's estimating procedures are acceptable, the underwriter's evaluators pose questions related to the estimation process, including who is involved, their backgrounds, previous success in pricing product, who establishes project profit margins, the existence of checks and balances in the processes, and typical spreads on project bids. Questions pertaining to job cost monitoring include determining who in the field is responsible for collection of cost data and how the home office learns about uses and cost information and project status. Communication channels must be identified and proven secure and effective at every level of operation.

Another important aspect is the company's position on subcontracting. Monitoring and control of subcontractors reveals to underwriters how a contractor manages people, projects, and resources. Questions an underwriter poses to a contractor regarding subcontracting practices include the amount of work the organization generally subcontracts and procedures the contractor uses to ensure the subcontractor's qualifications. Specifically, they will want to know if bonds are required on subcontractors and, if so, under what conditions.

Business planning is also revealing of a contractor's company management. The planning process identifies the type and amount of future work desired, the construction type and maximum size of any single project (measured in contract amount) the contractor can take on, total (aggregate) work program, span of operations, and duration of on-going projects. An underwriter also investigates issues related to expansion of operations such as geographic location, union relations, labor pool, project type and size, and bank line of credit, including both single and aggregate amounts.

Surety underwriters are interested in the organizational structure of a construction company. They often request and review a company's organizational chart that depicts management positions, titles, responsibilities, and relationships with other positions. Underwriters evaluate each individual's qualifications to assess skills and experience and how they fit into the framework of the company. The underwriter is particularly concerned with whether or not the *right person* is in charge of the *right functions*. If not, the company may encounter future difficulties, and underwriting a mismanaged firm is inherently more risky than a well-managed one.

Underwriters should investigate a company's history of ownership transfer. Most construction companies are family-owned businesses. Few such as Bechtel and Fluor Daniel are publicly traded. In the surety industry it is said that the first generation who starts the business are good tradesman, the second generation are better businessmen and managers by virtue of the opportunity for education, and the third generation is so far removed from what it took to build the business they have a higher rate of failure. Thus, succession is of concern to the surety. What provisions has the company made to facilitate any future or recent transference of ownership? Are future successors qualified? Will any transition of ownership or management be smooth? These issues are assessed in tandem with the parallel continuity investigation discussed below, and let it be pointed out here that this is not uncommon. All categories of the underwriting investigation impinge on all others, and thus any problems will usually be manifest in several guises. This level of thoroughness helps sureties increase their underwriting effectiveness, as it helps pinpoint potential concerns and risks.

Management items related to the operation of the company such as amount of overhead, profit, and anticipated changes reveal to the underwriter the contractor's ability to remain commercially viable. According to expert opinion from industry, a contractor's ability to control the amount of overhead during a depressed economic cycle affects his or her ability to maintain an ongoing business concern. Related to this concern is the contractor's current workload. Today, surety underwriters spend much more time than in previous eras reviewing both in-progress and completed project schedules to determine overall success (or failure) of the contractor.

This analysis includes a project-by-project evaluation of anticipated and actual profit and of overbillings and underbillings. An underbilling or overbilling is the amount that a contractor has failed to bill or billed in excess of the payment that corresponds to the actual amount of work completed to date or as outlined in contract documents. In the early 1980s, many underwriters improperly analyzed the underbillings accounts contractors presented on the balance sheet. Contractors were either unaware of or hid large losses by claiming they were underbillings when in fact they were *unrecognized losses*, and thus of concern to the surety. A similar concern exists in regard to the age and collectability of accounts receivable and anticipated compensation dates. Assuming no substantial changes in the scope of the project, contractors do not make a profit on projects not completed on time, and accounts receivable more than 90 to 120 days overdue are generally considered uncollectable or severely discounted by the underwriter. In addition, such long-overdue receivables may reveal the owner's lack of satisfaction with the quality of the work billed, which would undoubtedly concern the surety. Again, this analysis dovetails with the capital analysis, demonstrating the great importance of a healthy financial record in the underwriting evaluation.

Capacity also reflects what the contractor has done historically. To this end, the underwriter considers the contractor's track record. This evaluation takes into account items such as three to five years of completed job history, types of projects, field personnel, project locations, owner name and reputation, project duration, contract amount (bid and final), and gross profit including bid and final amounts. The underwriter analyzes all employees by determining their responsibility level and previous experience. This information helps the underwriter judge whether or not the contractor has the adequate technical expertise in company management, estimating, and field management necessary to complete with success the current portfolio of projects plus the potential project(s) under consideration. By determining a contractor's total technical resource base and identifying what resources are committed to ongoing projects, an underwriter can determine the resources available for other projects. Contractors seeking a sustainable bondable enterprise can use this information, combined with a contemplated work program, to create a strategy for future operations.

Project Management

While company management keeps the whole organism in tact, the key to remaining viable and profitable is performing work in the field efficiently and effectively. Figure 15-1 provides the basic model for successful contracting, and to accomplish this objective, field staff must be competent in the uses of construction methods and techniques, planning and scheduling, and quality, safety, and cost control. Additionally, a contractor must have a sufficient number of managers to execute necessary tasks and monitor the status of the progress and quality of work performed. These project managers and superintendents must have adequate experience and support staff to perform the work efficiently. Project control procedures (e.g., cost, schedule, quality, and safety) by field staff are vital management tools for assessing and subsequently enhancing field performance. Through this part of the investigation, surety underwriters determine the contractor's ability to maintain a profitable job site.

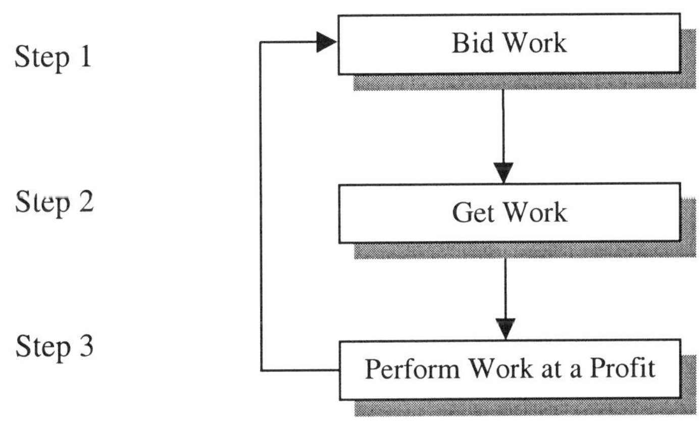

Figure 15-1 Model for Success in Contracting

Equipment

The underwriter reviews the amount and efficiency of the construction equipment and any plans for maintaining and acquiring equipment as needed for particular jobs and in general. How the contractor goes about deciding to purchase new equipment or rent equipment is of interest because such expenditures can negatively or positively affect the financial strength, future profitability, and sales of the contractor. The underwriter looks for signs of "yellow fever," or the urge to buy equipment when it isn't needed. Another issue is whether the contractor has access to the specialized equipment necessary for particular kinds of construction the contractor claims to perform, especially when that type of equipment is difficult or expensive to rent. Maintenance and repair capabilities can also help a surety anticipate the contractor's responsiveness in the event of mechanical difficulties and overall management of resources. Knowledgeable surety underwriters also review equipment cost control procedures including how a contractor computes the rates for using any necessary equipment such as the cost of wear and tear, and that the contractor is in fact including these costs in its bids.

Labor

Construction work is labor-intensive. Depending upon the type of construction, labor can represent between 10% and 40% of the total project cost. Thus, availability and quality of labor in the geographic area of a potential project merits investigation. The *type* of labor a project requires is equally important. Therefore, the status of union contracts and when they will be negotiated is of concern to the surety underwriter. Each of these items has the potential to restrict the contractor's ability to complete ongoing projects successfully.

Material

A contractor must have access to the appropriate building materials necessary to construct a facility. Often projects require special materials or equipment crucial to completing the project within the scheduled amount of time. Unavailability of these materials or equipment can result in project delays and subsequent losses through liquidated damages. Price stability of the materials is also of concern. Frequently, material costs are tied to the price of energy or weather (e.g., asphalt, a petroleum-based product) and require analysis. Is there risk of price escalation after the contract has been awarded? Are there allowances for such an increase in contract documents? If at all possible, contractors should account for the risk of price escalation in their cost estimation process. If particularly expensive or rare materials are needed at a later stage in a project, their price or quantity might be secured with a supply bond.

CONTINUITY

Due to the intricacies of the construction industry, another "C"—continuity—has become part of the underwriting equation. Construction firms must have plans that ensure continuity by identifying who carries on the business activities in the event of disability or death of key employees. The surety has four objectives that generally should be consistent with those of management: 1) transition of ownership/leadership, 2) preservation of capital, 3) maintenance of surety credit, and 4) continuation of the business.

Sureties concern themselves with the company's longevity and want assurance of its continuance, at least through the duration of ongoing work. Such assurance protects all individuals and businesses associated with the contractor's operations. Typically the surety requests a copy of the buy/sell agreement as well as the declaration page of the contractor's life insurance policy for owners and key employees that the surety can review.

In the construction industry, approximately 98% of all companies are closely held family businesses (i.e., not publicly traded on a stock exchange). Hence, management transition from one family generation to another is a common occurrence. In such family-owned businesses, the succession from one generation to another can result in transferring company leadership to an individual not well-suited for the position. Therefore, the experience levels of planned successors are of interest to the surety.

In industry, for example, it is common to see succession where the first generation is skilled in the construction business (i.e., good tradesman), the second generation is usually better at the administration of construction (i.e., management), and the third generation is typically *too far* removed from the original founder's ideals and principles to succeed. Sureties are particularly concerned with making sure that the principal(s) directing the company are well-qualified.

A construction business can be organized in one of several ways. These ownership types include (Mc Kenna 1972):

1. *Proprietorship*—also known as sole or single proprietorship or individual ownership, is the simplest and most prevalent form of business entity. In a majority of cases, the operation of the business and all policy decisions reside with one individual. This is the easiest form of business venture to start and depending on the state, may require only a license to commence operation.
2. *General Partnership*—a venture in which two or more individuals combine their resources to engage in a business, usually on a permanent basis. A partnership is formed with oral or written agreement by all parties for formation of partnership and nature of business. Each partner is responsible for the business actions and for obligation incurred by the venture. A partnership is simple to start and needs only to comply with licensing laws of the state.
3. *Limited Partnership* a venture that permits limiting the liability of one or more partners. A limited partner does not have any input into how the company is operated but must have an actual investment by means of cash or property. Services rendered by the limited partner that contribute to the operation of the company are not acceptable as equity. At least one general partner, however, must be designated. Limited partnerships are organized under state law and federal regulations under the Uniform Limited Partnership Act. Written agreement must be filed with proper state officials and must comply with statutory requirements.
4. *Corporation*—a legal entity created by law upon the filing of articles of incorporation and compliance with laws of the state where incorporated. This entity exists separate and apart from its individual stockholders. Hence, it represents a legal personality that can act in many respects as an individual with its powers and duties expressed in the corporation charter. Two principal types of corporations exist under tax laws: 1) C corporation and 2) S corporation. The difference between these two lies in the means by which they are taxed. The C corporation is a taxable entity paying income tax on earnings of the corporation. With an S corporation, the individual stockholders of the business are taxed personally on the earnings of the corporation based on their individual equity in the venture. The S corporation is of concern to sureties in that the sureties have no control when and how much money is withdrawn from the company by the owners. Large amounts of funds withdrawn from a contractor's operation can severely affect its financial position.

5. *Subsidiary*—a corporation in which a majority percentage is owned by another entity called a parent company. When 100% of stock is owned by another corporation, it is called a "wholly owned subsidiary." Each corporation is a separate legal entity and is not responsible for the debts of the other unless one of the corporations has formally agreed to guarantee a specific obligation. Under certain circumstances, the courts can "pierce the corporate veil" and treat the corporation(s) as one entity.
6. *Affiliate*—partial interest of this company is owned by another corporation or by the same owners as another corporation.
7. *Division*—does not represent a separate legal entity. This is the creation of a separate unit of a corporation for a specific purpose. There is no legal separation of the assets of the division from the assets of the corporation as a whole. Hence, corporations are fully responsible for the debts of their divisions.
8. *Joint Venture*—usually a temporarily established entity that consists of joining two or more individuals or corporations for the purpose of combining certain assets, expertise, experience, and/or personnel to complete a well-defined task. It is a means of sharing risks on large projects and is widely used in the construction industry. They are treated similar to partnerships in regard to how profits or losses are distributed to each of the participants. In some states, joint ventures are subjected to licensing laws similar to that of a partnership.
9. *Limited Liability Company/LLC*—is a new form of business that has been created by statute in several states including Wyoming, Florida, Kansas, Colorado, and Virginia. LLCs combine both corporate and partnership characteristics (Weitz 1992). More specifically, an LLC combines the limited liability of its member owners with the tax "flow through" benefits that accrue to a partnership. The LLC is not a taxable entity, and each member owner, however, is subject to tax on his or her share of the earnings from the LLC, just like shareholders of subchapter S Corporations and partners of limited partnerships. Unlike subchapter S Corporations, there is no restriction on the number of shareholders in the LLC, and the latter is more flexible regarding distributions and special allocations of funds. Unlike limited partnerships where only the limited partners have limited liability, all members of the LLC have limited liability. The LLC is not restricted in terms of types of activities it may engage in; however, the LLC terminates upon the death, resignation, incapacity, or bankruptcy of a member. A word of caution is in order: because the statutes creating this type of business organization have only been drafted in the past few years, there has been little case law on this topic. Therefore, it remains to be seen whether this development will succeed in the long run or spread to other states. As with all legal matters, the format for the entity to be used should be made after consultation with an

attorney to determine whether this format is suitable for your company.

For a given account, sureties are interested in the ownership structure of the company. Each ownership type introduces risks and special concerns that a surety will normally investigate. In particular, the S corporation is of concern to sureties because a significant part of the underwriting is based upon the contractor's commitment to growth through retention of earnings. With an S corporation, the surety, in theory, has no control over when and how much money moves from the company to the owner(s). In practice, the surety generally executes an agreement with the owner(s) not to take more money in dividends out of the company than is necessary to pay their personal tax liability on corporate earnings. In some cases, the surety requires the S corporation to agree to retain a predetermined percentage of profit or, at a minimum, maintain equity at a specified level.

As previously mentioned, the surety frequently requests the company's organizational chart. This provides sureties with a clear understanding of the contractor's organizational structure. Sureties should look for a contingency plan that minimizes the initial impact of a loss (i.e., death or disability) of owners and key employees.

The proper term for such a contingency plan is a "continuity plan." A continuity plan is a documented plan detailing the continuation of an entity in the event of a change of ownership for whatever reason. Surety companies value such plans because they provide for financial and performance viability if a traumatic event such as death or disability should occur. These plans are also important to those people whom the event touches most, such as the employees of the company and the spouse and children of the owners.

Generally speaking, three types of plans together form a solid continuity plan: 1) succession plans of key personnel within the company, 2) buy-sell agreements for the transfer of company stock, and 3) life insurance covering owners. The beneficiary for life insurance on key employees and owners should be the company to the extent that it is needed, and thereafter, would normally be the spouse or children of the owner or key employee. Family survivors demanding payment for the deceased or disabled key person's investment in the business can severely impair the financial strength and longevity of the company. Therefore, key person insurance may be an integral part of a contingency plan in the event of the death or disability of an owner or key individual. The prudent business person should create a continuity plan that uses aspects of the three options, thereby: 1) providing operational continuity in the event of a important event, 2) providing financial continuity that may benefit a spouse and/or children of key individuals in the event of death or disability, and 3) indicating to a surety company a commitment to business viability and performance.

A surety must review the contractor's existing insurance program to determine the types of coverages and their per incident and aggregate benefits. The underwriter will want to know if the insurance program under review compares favorably with the insurance carried by similar contractors, as well as the quality and solvency of the insurance carrier. For a more complete description of the insurance industry and standard policies as they relate to the construction industry, see Russell (1991a).

Often the primary motive for incorporating ownership of a business is to limit personal liability to creditors. Ironically, most sureties still use personal assets as an additional guarantee by requiring the owner of the contractor to pledge personal indemnity. This means the contractor or another interested party may pledge his or her house, car, or other assets as collateral for the surety bond. Consequently, an owner's death may leave his or her family penniless. One solution may be life insurance that is irrevocably in trust for the family, which creditors are not able to access. Usually, however, through indemnity agreements, the surety will have access to these funds. Clearly, estate planning should be of concern to contractors. As the typical contractor becomes personally liable by providing personal indemnity (including spouse's assets) to the surety in order to obtain bonding, proper estate planning is important to the surety as well as the other parties normally considered. In community property states such as Wisconsin the spouse may have to approve using joint property as security.

Business longevity is essential to adequate estate planning (Dockray 1979). The question contractors must ask themselves is: "Do I want to keep the business in the family, sell it to my employees, or sell it to an unknown buyer upon my death?" Realistically, very few contractors will be able to perpetuate their businesses successfully, and if they can, will probably need assistance. Contractors must consider the following issues:

1. Who is going to operate the business and where is the supporting management staff?
2. Where is the source of money for estate taxes?
3. How will the family receive income? If the contractor receives a salary, the dependents will only receive a taxed dividend—not earned income.
4. In regard to situations involving children, how will assets of the company be distributed?

Contractors who determine that they want the business sold upon death must consider a buy/sell agreement or stock redemption plan. Potential issues include:

1. Who will purchase the business?
2. Who is competent to purchase the business form a continuance standpoint?
3. What is a reasonable sale price?
4. What sources of funds exist for the purchaser?

These are fluid, complex concerns with tax consequences, and today's solution may be inappropriate tomorrow. It is important to point out that while sureties do *not* authorize bonds solely on the basis of a continuity plan, it may be a deciding factor in a small number of cases. As the above hopefully demonstrates, it is also almost always a good business strategy.

Chapter 16

CAPITAL AND FINANCIAL REVIEW

Since the surety essentially sells credit to the contractor, it has a vested interest in the contractor's finances. The analysis of a construction company's financial condition typically involves an in-depth review of the contractor's financial statement, including balance sheet and income statement. Surety companies frequently require at least three years of financial data. The analysis consists of two parts: 1) review of the quality of data the contractor presents and 2) an analysis of the entries and numbers the statement contains. Each of these parts is analyzed below.

FINANCIAL DATA CONSIDERATIONS

Prior to analyzing financial information, sureties must examine issues related to the presentation of contractor data. Of particular importance is the 1) certified public account's (CPA) opinion, 2) quality of the financial statement, and 3) type of accounting method used to recognize revenues earned. These issues are significant because they weigh heavily in conclusions drawn by the surety in the underwriting decision process and receive further description below.

CPA Opinion

The certified public account (CPA) or auditor expresses his or her opinion in a cover letter accompanying the financial statement. It will be one of the following types (American Institute of Certified Public Accountants 1981):

1. *Unqualified opinion*—pronounces the entire examination and statement to have been prepared and presented in conformity with general auditing and accounting principles.

2. *Qualified opinion*—rendered in cases where generally accepted auditing and accounting practices have been followed except for one or more qualifications. These qualifications may result from a deviation in the scope of the audit or statement and may reflect uncertainty about certain assets. The uncertainties causing the qualification should be clearly stated in the opinion letter.

3. *Disclaimer of opinion*—covers those statements where there have been limitations on the scope of examination such that the presentation may or may not be fairly representative.

4. *Adverse opinion*—rendered when the auditor knows that the statement has not presented the contractor's financial position accurately. This should be clearly detailed in the opinion letter.

The type of opinion influences the usefulness and amount of credibility underwriters ascribe to the reported data as they make decisions regarding underwriting bonds. Surety underwriters prefer an unqualified financial statement and typically disqualify any contractor whose statement contains a disclaimer or adverse opinion.

Quality of Financial Statement

The financial presentations fall into one of three categories (American Institute of Certified Public Accountants 1981):

1. *Compilation*—accountant provides no assurance regarding the numbers presented in client's financial statement. Hence, a compilation comprehends no detailed examination of the figures in the statement (i.e., no normal audit procedures were performed to verify the numbers) but is merely a presentation of the figures generated by the company in conformity with usual presentations for the type of statement issued. Compilations are performed infrequently, and then usually only for small contractors.

2. *Review*—somewhat more comprehensive than a compilation, accountant has given important figures more than routine examination because of their relevance to the whole picture. The accountant performs enough investigation such that a reasonable degree of assurance exists that there are no material misrepresentations in the reported financial information.

3. *Audit*—accountant applies industry-accepted auditing procedures including verification of relevant items in the statements with both internal and external investigations of their truthfulness and accuracy. Accountant certifies that the financial statement is presented in accordance with generally accepted accounting principles.

When bonding programs are substantial, sureties will generally require audited statements to determine the financial status of the company. For the initial underwriting process of a contractor, as opposed to annual underwriting updates, sureties will also typically require an audited financial statement. With this form of preparation, the CPA has applied more extensive procedures to verify the presented data and the surety can rest assured that the audit fairly represents the contractor's financial situation at the time the audit was performed. Audits lend more credibility to financial statements because during audits the contractor's finances are subject to rigorous accounting methods. Only when a statement is prepared under the audit does a surety consider the evaluation of the contractor's internal financial control structure to be complete. Audited statements provide third party verifications on items such as amount of cash and certificates of deposits, accounts receivables, and accounts payable, among others. As Table 16.1 shows, an audit costs anywhere from $25,000 to $70,000 or more depending on the auditor, the size of the construction company, and the market in which both the contractor and CPA work. A mid-size contractor with an annual volume of $50 million of work can expect to pay around $35,000 for an audit in all but the most expensive markets.

Table 16.1 Estimated Average Cost of Financial Preparation by Size of Contractor
Source: Rates Compiled by Author

Type of financial preparation (1)	Cost (2)
(a) Audit	
Small contractor ($5 to $30 million)	$25,000 to $40,000
Mid-size contractor ($40 to $60 million)	$30,000 to $45,000
Large contractor ($80 million plus)	$40,000 to $60,000+
(b) Review	
Small contractor ($5 to $30 million)	40% to 60% of audit (as low as $5,000)
Mid-size contractor ($40 to $60 million)	40% to 60% of audit (as low as $10,000)
Large contractor ($80 million plus)	40% to 60% of audit (as low as $15,000)
(c) Compilation	
For all contractors	30% to 40% of audit

In some cases, an underwriter may accept a non-audited review for the initial underwriting and subsequent annual updates. An example would be if the bond requirement is less than a specified minimum amount or when the personal net worth of the contracting firm owner(s) is large. Additionally, a review may be accepted when the contractor is small in size but has a strong capital base and a track record of successful performance. With small contractors who do not have extensive financial reserves and who may only be working on a few jobs at a time, a review is sometimes allowed by a surety because an audit may not be cost-effective, as it can be very expensive for smaller contractors. If a review is accepted, this means the surety is confident that it will be able to ascertain the relative financial strength of the contractor without an audit; if there is any question, the surety will require an audit. As a contractor grows, requiring more credit and holding higher leverage, an audit at least every other year becomes more and more essential.

CPAs report that in the current profitable climate, sureties are less adamant about audits than they were even six years ago. Perhaps this has something to do with the competition to underwrite contractors and the surety's willingness to hang onto accounts. A review costs from 40% to 60% of an audit, and for a small healthy company, an audit may be overly burdensome. Even for a large company, the cost may be excessive if internal auditing systems are deemed accurate and extensive by the surety and financial data is kept up-to-date. Some sureties allow their best accounts to alternate yearly between reviews and audits, saving contractors of all sizes money.

Whenever a review is accepted by a surety, it must take the proper format. This generally includes a schedule of completed contracts and a schedule of contracts-in-progress. As with a compilation, an underwriter may use a multiplier to weight the assets and liabilities listed in the review to account for the unknown accuracy of the financial preparation, and such a procedure is often referred to as "discounting" (see Table 16.2). For example, if a contractor has $200,000 in cash-on-hand and the financial statements were reviewed and not audited, the surety may consider that equal to $132,000 ($200,000 * .66).

Table 16.2 Financial Preparation Multipliers

Type of financial preparation (1)	Multiplier for assets (2)	Multiplier for liabilities (3)
Audit	1.00	1.00
Review	0.66	1.50
Compilation	0.50	2.00

Additionally, sureties tend to favor certain accountants' pronouncements and will therefore allow reviews from one accountant when insisting on audits from another. A contractor should inquire through its agent as to which CPAs do business most often with which sureties. Alternatively, a contractor happy with its CPA should inquire through its accountant which sureties the CPA firm regularly does business with, especially those sureties with which it enjoys good rapport.

Sureties do not generally permit compilations, as this form of financial preparation does little more than list the assets and liabilities of a contractor with little or no verification or analysis. Furthermore, the cost of a compilation for small contractors is almost that of a review, which provides a much more accurate representation of the contractor's finances. Banks, primarily interested in holdings and collateral, sometimes allow compilations for loans, but sureties prefer an audit or a review. However, if the contractor is extremely small and represents minimal exposure, the surety may allow a compilation.

Type of Accounting Method

The method the accountant uses to prepare a contractor's financial statement is significant to underwriters because it indicates how the accountant has recognized the contractor's income. Income recognition occurs through one of the following methods (American Institute of Certified Public Accountants 1981):

1. *Cash Basis*—income is taken into account when it is received and costs and expenses are taken into account when they are paid. Rarely used now.
2. *Straight Accrual*—income is taken into account when it is earned regardless of when it is received and costs and expenses are taken into account when they are incurred regardless of when they are paid. IRS requires modifications.
3. *Percentage of Completion*—income is recognized as the work progresses. Income is determined by estimating the percentage completion of each project. This method is now required by the IRS.
4. *Completed Contract*—income is only reported in the year that the contract is actually (or substantially) completed and accepted by the owner. The contract revenue and costs are accumulated for each project and the difference between the two is taken into account as profit or loss when the job is completed. No longer allowed by the IRS.

Underwriters prefer not to see cash basis or straight accrual statements. The percentage of completion method gives the most accurate picture of a contractor's financial position because accountants recognize income in proportion to cost incurred. The American Institute of Certified Public Accountants (AICPA) recommends that contractors using the percentage of completion method determine income in one of the following three ways:

1. The ratio of costs incurred to-date to estimated total costs based on the most recent information.
2. The ratio of costs incurred to date, minus unused materials at the job-site, to estimated total costs. This ratio may be further modified by engineering or architectural fees incurred, but no more than the actual percent of completion. Costs that do not signify contract performance, such as material on-site but not used or financing charges, should not be considered income.
3. The ratio of units in-place—for example, tons of gravel or sand already used—to total units to be placed. On a unit-price contract it is common to determine the basis for payments by averaging out the partially completed units. This alternative method arises out of the possibility of having a more meaningful allocation of income to measure the work performed, that can be obtained from a more appropriate ratio than costs.

For example, under the first method where the status of a contract is:

Contract price	$500,000
Estimated total cost	$450,000
Estimated profit	$50,000
Actual cost to date	$225,000
Billed to date	$280,000

the ratio of costs incurred to date to estimated costs is $225,000/$450,000, so the percentage of completion on a cost basis is 50%. With this percentage of completion the earned billings should be $250,000, resulting in an estimated profit of $25,000 after subtracting the actual costs to date of $225,000. But the billings to-date are $280,000 rather than $250,000, so the latter is debited against the former for a balance of $30,000. This overbilling of $30,000 is posted to "billings in excess of costs and estimated earnings" as a current liability (Remmen 1977).

On the other hand, if costs exceed billings it is a current asset. The difference between the two then becomes a receivable and it represents an item the contractor is entitled to as well as the profit thereon. Taking the same $500,000 contract except that $300,000 is the actual cost to date and $200,000 is the amount billed to-date, the percentage of completion is $300,000/$450,000, or 67%. Thus the contractor is entitled to the $300,000 worth of costs plus an estimated $33,333 of profit (67% of the $50,000 estimated total profit). *Thus, this cost method deals with incurred costs to the statement date, even though they are not billable until after the end of the period.*

Under the completed contract method, a contractor can defer the tax consequences associated with the profits made on a given project to the end of the project. It is not uncommon for a contractor to have a financial statement prepared using the percentage of completion method because the underwriters and bankers require it, and then prepare a corporate tax return using the completed contract method. However, underwriters typically are well-aware of this as they require contractors to provide copies of previously filed business tax returns.

Since a financial statement reflects a snapshot in time, the point at which it is prepared is significant. Two principal concerns exist: 1) contractor year end and 2) relationship of year end to the cycle of work (i.e., flow of work volume of contractor over time). With regard to year end, a contractor that has an interim financial statement prepared in November with year end in December may extract funds from the business during the month of December; hence, the November financial statement *may not* adequately represent the current financial condition of the contractor with a December year end. For surety underwriters, interim financial information is valued, but all contractors must provide a year-end statement.

The flow of a contractor's work volume also changes over the course of a calendar year. For example, in Wisconsin (where winters can be long and harsh), the peak work season occurs during the months of April to October. During the rest of the year, most contractors (e.g., highway paving contractors) significantly reduce their work load or do not perform any work at all. There exist times when contractors are periodically "cash rich" (i.e., when numerous accounts receivables come due). Thus, the financial statement *may not* adequately represent the company's financial position at some future date. For a more comprehensive description of contractor accounting issues, consult Peroutka (1991).

FINANCIAL ANALYSIS

As part of the capital analysis, the surety underwriter reviews issues related to the contractor's financial statement and performs data analysis. This analysis is a major component of the surety's investigation. The surety evaluates capital base, sufficient working capital, profitability, asset utilization, overhead expenses, notes accompanying the financial statement, schedule of completed contracts and work-in-progress, ratio and trend analysis, credit reports, the qualifications of the CPA preparing the statement, and banking credit.

Figures 16-1 and 16-2 describe the parameters a contractor's financial statement contains. A financial statement consists of a balance sheet and an income statement. The balance sheet shows that assets equal liabilities plus shareholders' equity. Figure 16-1 shows parameters that describe assets and liabilities. Shareholder's equity is the owners' claim on the assets of a firm. This consists of contributed capital (i.e., stock) and retained earnings. Contributed capital is the funds invested by the shareholders for ownership. Retained earnings are the funds the firm keeps in excess of dividends the shareholders receive. The income statement contains a firm's earnings. Appendix VIII presents a sample report of independent certified public accountants and financial statements for Example Contractor, Inc., and Chapter 18 is an extended example of the underwriting process written primarily from the surety's point of view.

The underwriter analyzes the size of the parameters on the balance sheet using dollars as units of measure. Table 16.3 presents an underwriter's view of the various balance sheet entries. This analysis evaluates balance sheet accounts as accounting principles permit, rather than treatment underwriters allow. The surety is interested in *quick loss paying power*. Thus the surety usually discounts certain items shown on the balance sheet.

Among the more significant items analyzed are inventory, notes payable to company owner or employees, company goodwill, slow account receivables, and cash value of surrendered life insurance. For example, inventory may be discounted 5% to 75% because it is overstated in value and represents items with limited worth or usefulness to the contractor. Goodwill, as another example, is removed from the asset section of the balance sheet over time because it is not a tangible, measurable item, or useful for paying bills.

Next, the surety reviews significant financial items such as a contractor's capital base and net worth. Some sureties require contractors to have a minimum of $250,000 of net worth before they consider reviewing the contractor for bonds while others may accept contractors with a net worth equal to 10% or 20% of the work program. This 10% or 20% may exclude certain assets not considered quickly realizable assets. The amount of debt the contractor has in relation to equity is also important in determining the level of risk the owners of the company have assumed.

A standard measure underwriters use involves working capital (current assets minus current liabilities). Sureties calculate a contractor's aggregate allowable work program as a multiple of their working capital. For example, a contractor with $100,000 of working capital and a multiplier of 10 can have a

ASSETS

Cash & Equivalents → **Cash & Equivalents**
All cash, marketable securities, and other near-cash items. Excludes sinking funds.

Accounts Receivable–Progress Billings → **Accounts Receivable Progress Billings**
Amounts billed on current contracts excluding r retention

Accounts Receivable–Current Retention → **Accounts Receivable–Current Retention**
Amounts held back by customers on current contracts as retention

Inventory → **Inventory**
Costs attributable to equipment, small tools, supplies and other deferred costs related to contracts in progress wherein a portion of the cost applies to work not yet performed.

Costs & Estimated Earnings in Excess of Billings → **Costs & Estimated Earnings in Excess of Billings**
The difference between the total of costs and recognized estimated earnings to date and the total billings to date.

All Other Current → **All Other Current**
Any other current assets. Does not include prepaid items.

Total Current → **Total Current**
Total of all current assets listed above.

Figure 16-1 Explanation of Contractor Balance Sheet Parameters
Source: *RMA, Annual Statement Studies 1997*

ASSETS (Continued)

Fixed Assets (net)

Joint Ventures & Investments

Intangibles (net)

All Other Non-Current

Total

Fixed Assets (net)
All property, plant, leasehold improvements and equipment, net of accumulated depreciation or depletion.

Joint Ventures & Investments
The total of investments and equity in joint ventures.

Intangibles (net)
Intangible assets, including goodwill, trademarks, patents, catalogs, brands, copyrights, formulas, franchises, and mailing lists, net of accumulated amortization

All Other Non-Current
Prepaid items and other non-current assets.

Total
Total of all items listed above

Figure 16-1 Explanation of Contractor Balance Sheet Parameters (Continued)
Source: *RMA, Annual Statement Studies 1997*

LIABILITIES

Notes Payable–Short Term	**Notes Payable–Short Term** All short term note obligations, including bank and commercial paper. Does not include trade notes payable.
Accounts Payable–Trade	**Accounts Payable–Trade** Open accounts and note obligations due to the trade.
Accounts Payable–Retention	**Accounts Payable–Retention** Amounts held back as retention in payments to subcontractors on current contracts.
Billings in Excess of Costs & Estimated Earnings	**Billings in Excess of Costs & Estimated Earnings** The difference between the total billings to date and the total of costs and recognized estimated earnings to date.
Income Taxes Payable	**Income Taxes Payable** Income taxes including current portion of deferred taxes.
Current Maturities–LTD	**Current Maturities–LTD** That portion of long term obligations that is due within the next fiscal year.
All Other Current	**All Other Current** Any other current liabilities, including bank overdrafts and accrued expenses.

Figure 16-1 Explanation of Contractor Balance Sheet Parameters (Continued)
Source: *RMA, Annual Statement Studies 1997*

LIABILITIES (Continued)

Total Current

Long–Term Debt

Deferred Taxes

All Other Non-Current

Net Worth

Total Liabilities & Net Worth

Total Current
Total of all current liabilities listed above.

Long-Term Debt
All senior debt, including bonds, debentures, bank debt, mortgages, deferred portions of long-term debt, and capital lease obligations.

Deferred Taxes
Total of all deferred taxes.

All Other Non-Current
Any other non-current liabilities, including subordinated debt, and liability reserves.

Net Worth
Difference between total assets and total liabilities. Minority interest is included here.

Total Liabilities & Net Worth
Total of all items listed above.

Figure 16-1 Explanation of Contractor Balance Sheet Parameters (Continued)
Source: *RMA, Annual Statement Studies 1997*

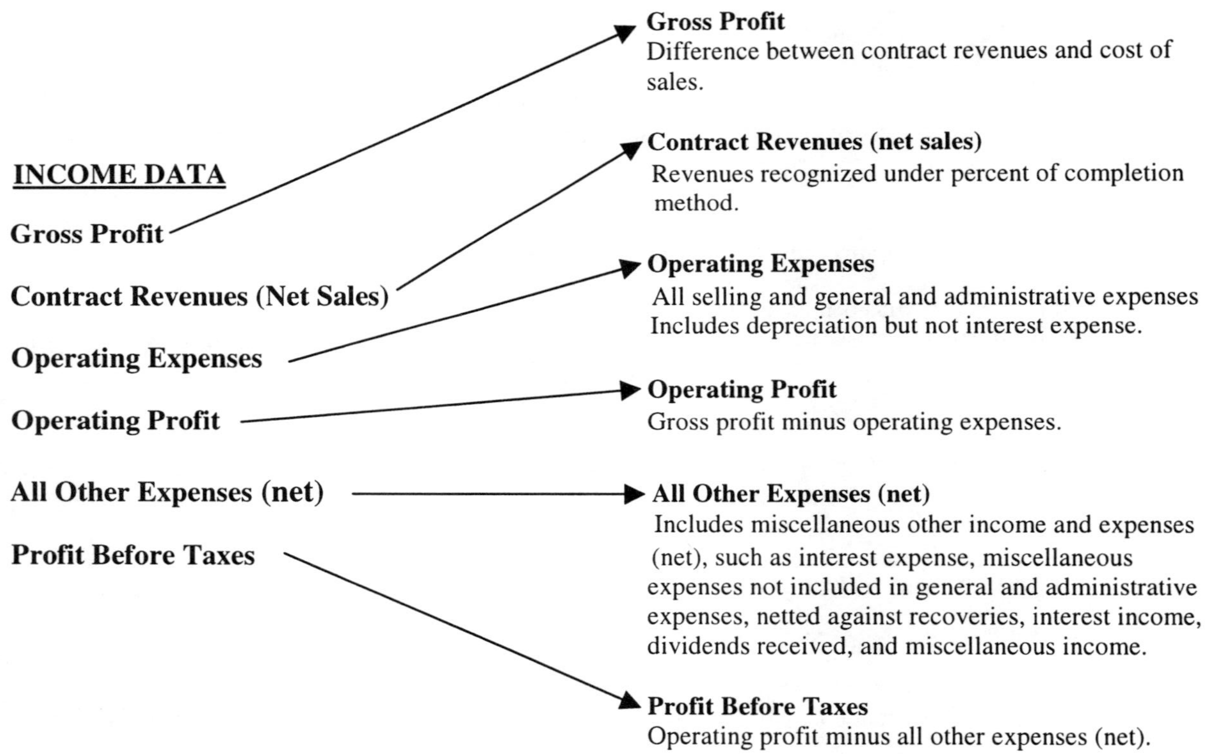

Figure 16-2 Explanation of Contractor Income Statement Parameters
Source: *RMA, Annual Statement Studies 1997*

Table 16.3 Fundamentals of Financial Analysis
Source: Unidentified Reinsurance Company

As shown on financial statement (1)	As viewed by underwriter (2)
CURRENT ASSETS	Those assets that are most liquid (i.e., cash or near cash) that can be used by Principal.
Cash	Important to get bank references – make sure cash is not pledged and amounts are accurate
Accounts receivable	Obtain aged schedule of A/R. Look for any past due items (90 days). Underwriter may discount or eliminate A/R over 90 days old.
Retainages receivable	Underwriters will generally allow as a current asset.
Notes receivable	While an accountant may view the notes as current, an underwriter will want to know who owes the money and the terms. If it appears that it will not be paid back in near future, or could not be readily sold – classify as Fixed Asset.
Inventory	Underwriter will want to know what inventory consists of and may classify a percentage as a Fixed Asset
Cash surrender value life insurance	Accountants view this as an "other" asset. Underwriters classify as a current asset since if needed the Principal can obtain the cash on short notice.
Goodwill	While accountants may include this as an asset, underwriters do not classify it as such. Goodwill is an intangible asset. Underwriters deduct it from the Net Worth.
FIXED ASSETS	Those assets not readily convertible to cash in the normal course of business.
Prepaid expense	May include the past due items and/or other miscellaneous receivables.
Real estate	Accountants view this as a current asset if held for sale. Underwriters will treat it as fixed since the money is not readily obtainable.
Equipment, furniture and fixtures	Viewed by underwriter as Fixed Asset (not liquid).

Table 16.3 Fundamentals of Financial Analysis (continued)
Source: Unidentified Reinsurance Company

As shown on financial statement (1)	As viewed by underwriter (2)
CURRENT LIABILITIES	
Notes payable—bank-other (i.e., equipment, mortgages)	If note is short term (generally due in one year or less) an underwriter will classify it as a current liability. Other Notes Payable are reviewed. Some notes are due upon demand (term unknown) so they are classified as current. Underwriters may charge 12 months of payments on long-term notes as a current liability.
Accounts payable	Try to obtain aged schedule of A/P–View terms. Trade references are helpful to determine if Principal is paying bills on time. If not is there a problem? A/P are charged in full as a current liability.
Accruals (accrued salaries, accrued expenses)	Current liability since these amounts will have to be paid shortly.
Taxes payable	Underwriter will want to understand how the Principal pays his/her taxes. Some tax obligations may be considered long-term, i.e. completed Contract–do not pay until job is complete.
LONG-TERM LIABILITIES	
Mortgage (due after one year)	Underwriter classifies that portion of the mortgage that is not due within the year as a long-term liability. Some underwriters, however, may not charge any Real Estate Mortgage payments as current.
Equipment loans (due after one year)	Equipment loans not due within the year are long-term liabilities. Sometimes if the amount is large, underwriters will question why so much equipment is being purchased. Principal may be expanding, but too much debt can lead to financial trouble.
STOCKHOLDERS EQUITY	
Capital	Investment by owners.
Surplus	Earnings accumulated from operation of business. May be a plus or minus.

total work program of $1,000,000. The multiplier depends upon type of contractor, size of work program, and amount of work subcontracted. This approach to the underwriting process has received mixed reviews. Many sureties have established a minimum amount of working capital necessary to be considered. Such values range from $100,000 to $250,000 but some sureties will accept contractors with considerably less, especially if their performance has been sound.

In addition to assets in a financial statement, physical assets are necessary to perform construction work successfully; hence, equipment, as well as other company fixed assets are of importance. Amount of general and administrative expense relative to the type and size of the firm and how well it is managed are important to the long-term welfare of the company. Thorough evaluators will further investigate how notes that accompany an audited financial statement clarify the information and impact the company. Typical notes to financial sheets include, but are not limited to, the following: 1) company's nature of business; 2) summary of company's significant accounting policies (revenue recognition method: percent complete or completed contract; cost recognition method; depreciation methods used; income tax filing status: "S" or "C" corporation; and inventory valuation: Last in First Out (LIFO) or First in First Out (FIFO); 3) accounts receivable; 4) description of long-term debts; 5) leasing arrangements; 6) major customers; 7) description of preferred stock; 8) bank financing agreement (i.e., line of credit); 9) employee benefits plans; 10) costs and estimated earnings on uncompleted contracts; and 11) contingent litigation.

Sureties closely scrutinize the profit trend of the company and the management's commitment to corporate growth. If a contractor has not made a profit in the last two out of three years, he/she will have difficulty in maintaining a relationship with a surety.

To further analyze company profitability, sureties review the *schedule of completed contracts*. Table 16.4 exemplifies this schedule, which accompanies an audited balance sheet. As shown, all projects were profitable with two exceptions: numbers 4073 and 4108. Based upon a project-by-project comparison between the *estimated* profit and *actual* profit, sureties review the schedule to assess the contractor's estimating and project management capabilities.

Another schedule that accompanies the balance sheet is the *schedule of contracts-in-progress* (also referred to as work-in-progress). See Table 16.5 for an example. This schedule contains a significant item that distinguishes construction from other industries such as manufacturing; the contractor estimated *cost-to-complete* their projects. This difference manifests itself in two parameters that a contractor's balance sheet shows: 1) costs and estimated earnings in-excess of billings (or underbillings) and 2) billings in-excess of costs and estimated earnings (or overbillings).

To compute each of these parameters, the surety takes the summation of revenues earned to-date and subtracts it from the amount billed to-date (billed to-date - revenues earned) for all contracts. For each project, revenues earned to-date is the product of the percentage completion of the project multiplied by the contract amount at award. Billed to-date is the sum of all the payment requests the owner has received.

If the amount billed to-date is *less* than the revenues earned, then the difference applies to the costs and estimated earnings in-excess of billings

Table 16.4 Schedule of Contracts Completed

	Schedule of contracts completed for year ended March 31, 1998								
	Contract totals			Before April 1, 1998			During year ended March 31, 1998		
Job # (1)	Revenues earned ($) (2)	Cost of revenues ($) (3)	Gross profit (loss) ($) (4)	Revenues earned ($) (5)	Cost of revenues ($) (6)	Gross profit (loss) ($) (7)	Revenues earned ($) (8)	Cost of revenues ($) (9)	Gross profit (loss) ($) (10)
3851	11,072	7,551	3,521	—	—	—	11,072	7,551	3,521
3852	20,096	15,891	4,205	—	—	—	20,096	15,891	4,205
3853	11,932	8,634	3,298	—	—	—	11,932	8,634	3,298
3938	6,682	4,447	2,235	—	—	—	6,682	4,447	2,235
4001	154,267	104,269	49,998	34,532	26,119	8,413	119,735	78,150	41,585
4006	10,538	6,407	4,131	7,542	5,581	1,961	2,996	826	2,170
4007	19,876	5,763	14,113	12,047	5,675	6,372	7,829	88	7,741
4073	13,934	8,120	5,814	13,494	7,408	6,086	440	712	(272)
4074	14,011	7,892	6,119	10,655	6,589	4,066	3,356	1,303	2,053
4108	64,072	61,848	2,224	18,986	14,773	4,213	45,086	47,075	(1,989)
4189	14,088	6,244	7,844	3,173	2,457	716	10,915	3,787	7,128
4190	14,198	6,274	7,924	33	20	13	14,165	6,254	7,911
4263	89,390	64,697	24,693	85,858	62,998	22,860	3,532	1,699	1,833
4283	9,247	3,467	5,780	1,253	518	735	7,994	2,949	5,045
4285	14,500	10,257	4,243	628	493	135	13,872	9,764	4,108
Totals	467,903	321,761	146,142	188,201	132,631	55,570	279,702	189,130	90,572

Table 16.5 Schedule of Contracts-in-Progress

Schedule of contracts in progress for year ended March 31, 1998							
Total contract			From inception to March 31, 1998				
Contract number (1)	Revenues ($) (2)	Est. gross profit (loss) ($) (3)	Revenues earned ($) (4)	Total costs incurred ($) (5)	Gross profit (loss) ($) (6)	Billed to date ($) (7)	Estimated cost to complete ($) (8)
3567	286,601	48,901	285,984	237,430	48,554	286,601	270
4471	6,985	1,920	3,141	2,289	852	6,985	2,776
4574	18,614	4,061	17,519	13,766	3,753	17,410	787
4584	314,781	33,577	170,199	152,090	18,109	170,301	129,114
4644	9,341	2,101	277	230	47	9,341	7,010
4665	10,670	2,400	3,990	3,093	897	10,670	5,177
4697	15,983	5,662	2,935	1,879	1,056	7,992	8,442
4819	40,673	11,979	5,811	4,496	1,315	40,673	24,198
4835	54,997	23,552	43,548	25,056	18,942	35,183	6,389
4839	13,352	5,350	3,255	2,028	1,227	7,066	5,974
4907	17,245	3,845	77	60	17	—	13,340
4971	98,637	9,689	75,778	66,718	9,060	78,802	22,230
4975	5,014	826	1,571	1,280	291	—	2,908
4983	20,694	4,403	4,662	3,591	1,071	—	12,700
5004	16,174	6,348	685	400	285	—	9,426
Totals	929,761	164,614	619,432	514,406	105,476	671,024	250,741

Table 16.5 Schedule of Contracts-in-Progress (Continued)

Schedule of contracts in progress for year ended March 31, 1998				
At March 31, 1998		For year ended March 31, 1998		
Cost & est. earnings in excess of billings ($) (9)	Billings in excess of costs and est. earnings ($) (10)	Revenues earned ($) (11)	Costs of revenues ($) (12)	Gross profit (loss) ($) (13)
—	—	—	2,573	(1,400)
—	617	1,773	2,289	852
—	3,844	3,141	13,766	3,753
109	—	17,519	152,090	18,108
—	102	170,199	230	47
—	9,064	277	3,093	897
—	6,680	3,990	1,879	1,056
—	5,057	2,935	4,496	1,315
—	34,862	5,811	25,056	18,492
8,365	—	43,548	2,028	1,227
—	3,811	3,255	60	17
77	—	77	66,718	9,060
—	3,024	75,778	1,280	291
1,571	—	1,571	3,591	1,071
4,662	—	4,662	400	285
685	—	685	279,549	55,071
15,469	67,061	335,221	—	—

account. On the other hand, if billed to-date is *greater* than the revenues earned, then the difference applies to the billings in-excess of costs and estimated earnings account.

Costs and estimated earnings in-excess of billings (or underbillings) is an asset that offsets the *understatement* of sales, net income, and ultimately equity when the actual completion of project(s) is ahead of billings. Billings in-excess of costs and earnings (or overbillings) are a liability that offsets the effects of those contracts where billings *exceed* the progress toward the completion of the project(s) (American Institute of Certified Public Accountants 1981). This is why *estimates of cost-to-complete* ongoing projects are so vital to an accurate financial profile of a company. Without such an accurate financial picture, sureties and contractors themselves cannot make sound business decisions that affect the company's future.

Another factor sureties consider carefully is the present backlog and/or a contractor's estimated cost to complete ongoing projects. These issues reveal not only the contractor's capability to perform the project under consideration, but also the contractor's ability to cover overhead expenses for the proposed project. These are salient issues of prime importance to the surety and project owner.

Ratio Analysis

Frequently, sureties use ratios they derive from selected balance sheet and income statement parameters in their financial analysis. Sureties typically focus their analysis in five areas: 1) liquidity, 2) operations (i.e., profit and sales), 3) leverage, 4) coverage, and 5) specific expense items. Liquidity ratios measure the adequacy of current assets to cover current liabilities. Operations ratios measure the effectiveness of management in conducting its business. Leverage ratios measure the level of debt pressure and how vulnerable the company is to downturns in the economic business cycle. Coverage ratios measure a firm's ability to service debt. Specific expense ratios relate expense items to net sales.

Table 16.6 contains a listing of the key ratios sureties typically calculate for a contractor's account. Sureties compare calculated ratios to the contractor's historical values and against industry averages. For example, the Robert Morris Association publishes average industry ratios that they base on two factors: 1) type of construction work and 2) annual sales (see Figure 16-3). For previous descriptions of ratio analysis and their limitations, see Halpin (1985), among others.

All sureties check a contractor's credit rating, consulting sources such as the Dun and Bradstreet Report (see Appendix IX for a sample report), Robert Morris Associates reports (see Appendix X), trade suppliers, specialty contractors, and of course the contractor's bank. From these sources, the surety culls information regarding such factors as a contractor's present/average balances, credit established (amount, rate, secured, unsecured), experience, length of business relationship, and their experience with construction. In addition, the age of outstanding accounts payable and account receivables older than 120 days are particularly telling indications of the type of difficulties a contractor is experiencing.

Table 16.6 Ratios Typically Analyzed by Sureties
Sources: "Surety Underwriting Consideration" 1989;
RMA, Annual Statement Studies 1997

Ratio name (1)	Definition (2)	Ratio range[1] (3)
	(a) Liquidity	
Quick ratio	Total quick assets (cash and Marketable securities) divided by current liabilities.	>1.0
Current ratio	Total allowable current assets divided by total current liabilities.	>1.2
Collection period (days)	Accounts receivable divided by annual sales revenue and multiplied by 365 days.	<60
	(b) Operations	
Net profit/ tangible net worth	Take net profit (before taxes) as a percentage of the owners' investment.	>15%
Net profit/ total assets	Net profits (before taxes) divided by total assets.	>1%
Net sales/ tangible net worth	Measures relative activity of invested capital.	10 to 15
Net sales/ working capital	Net sales divided by current assets minus current liabilities.	5 to 30
	(c) Leverage	
G & A/ tangible net worth	General and administrative expenses divided by owners' investment.	<60%
Fixed assets/ tangible net worth	Long-term assets (i.e., land, building, and equipment) divided by owners' investment.	10% to 40%
Total liabilities/ tangible net worth	Measures amount of debt pressure.	1:1 to 3:1
Hard debt/ tangible net worth	Measures amount of both short-term and long-term debt pressure.	1:1 to 2:1

Table 16.6 Ratios Typically Analyzed by Sureties (continued)
Sources: "Surety Underwriting Consideration" 1989;
RMA, Annual Statement Studies 1997

Ratio name (1)	Definition (2)	Ratio range[1] (3)
	(d) Liquidity	
Earnings before interest and taxes/interest	Earnings (profit) before annual interest expense and taxes divided by annual interest expense.	>1.0
Net profit + depreciation depletion, amortixation/ current maturities long-term debt	Measures the coverage of current maturities by cash flow from current operations.	>1.0
	(e) Operations	
Percent depreciation, depletion amortization/sales	Annual depreciation, amortization, and depletion expenses as a percentage of net sales.	>0%
Percent Officers' compensation/sales	Annual officers' compensation as a percentage of net sales.	>0%
Overhead/sales	Company overhead for items necessary to general sales volume.	<15%

[1] A good practice for an underwriter is to compare billings-in-excess-of-costs and costs-in-excess-of-billings from one period (or year-end) to the next, as it is a good indication of record keeping efficiency and administrative effectiveness.

	SIC 1499 MISC NOMET MNRLS (NO BREAKDOWN) 1997 (22 Establishments)		SIC 1521 SNGL-FAM HSNG CNSTR (NO BREAKDOWN) 1997 (1769 Establishments)	
	$	%	$	%
Cash	420,819	12.5	67,012	18.6
Accounts Receivable	282,790	8.4	79,982	22.2
Notes Receivable	104,363	3.1	3,963	1.1
Inventory	245,758	7.3	37,829	10.5
Other Current	171,694	5.1	58,005	16.1
Total Current	**1,225,425**	**36.4**	**246,790**	**68.5**
Fixed Assets	1,498,116	44.5	86,106	23.9
Other Non-Current	643,011	19.1	27,381	7.6
Total Assets	**3,366,552**	**100.0**	**360,278**	**100.0**
Account Payable	70,698	2.1	50,439	14.0
Bank Loans	----	----	1,081	0.3
Notes Payable	23,566	0.7	24,139	6.7
Other Current	255,858	7.6	71,695	19.9
Total Current	**350,121**	**10.4**	**147,354**	**40.9**
Other Long Term	484,783	14.4	45,395	12.6
Deferred Credits	10,100	0.3	721	0.2
Net Worth	2,521,547	74.9	166,809	46.3
Total Liab & Net Worth	**3,366,552**	**100.0**	**360,278**	**100.0**
Net Sales	5,844,173	100.0	1,517,643	100.0
Gross Profit	1,496,108	25.6	349,058	23.0
Net Profit After Tax	403,248	6.9	74,365	4.9
Working Capital	875,304	----	99,436	----

RATIOS	UQ	MED	LG	UQ	MED	LG
SOLVENCY						
Quick Ratio (times)	1.4	1.0	0.7	2.5	1.1	0.4
Current Ratio (times)	2.5	1.7	0.9	3.7	1.7	1.1
Curr Liab to Nw (%)	5.0	9.6	18.1	21.1	71.5	188.3
Curr Liab to Inv (%)	31.2	63.4	94.0	71.5	118.8	252.4
Total Liab to Nw (%)	9.8	36.2	42.4	34.3	99.8	234.6
Fixed Assets to Nw (%)	22.3	47.2	69.5	15.7	41.3	90.5
EFFICIENCY						
Coll Period (days)	31.1	36.7	51.0	9.1	25.2	49.3
Sales to Inv (times)	7.1	4.8	3.7	59.7	12.8	4.1
Assets to Sales (%)	65.3	66.9	112.3	16.8	28.7	45.7
Sales to Nc (times)	3.6	2.2	1.3	27.1	12.0	5.9
Acct Pay to Sales (%)	2.9	3.2	3.5	1.9	4.3	7.4
PROFITABILITY						
Return on Sales (%)	3.8	2.5	(0.4)	8.5	3.4	1.1
Return on Assets (%)	3.5	1.3	(1.8)	24.3	8.8	2.4
Return on Nw (%)	5.5	2.4	(2.3)	58.4	23.3	8.2

Figure 16-3 Sample Average Industry Ratios–
Robert Morris Associates

CPA Qualifications

The professional qualifications of the accounting firm preparing a contractor's financial statement is important in that construction varies substantially from other businesses. Specifically, the means of recognizing income are uniquely different in construction. Normally there is a time lag between when the contractor has performed the work and incurred cost and when he or she actually receives payment. The time lag can be anywhere from 15 to 60 days, or greater. The amount of cost incurred can also be significantly different (in a desirable and undesirable way) from what has been billed the owner. If an accountant does not understand construction, he or she may prepare the financial statement incorrectly, and thus present an inaccurate financial picture. Should this occur, the contractor and surety may make poor business decisions. When evaluating an accountant's qualifications, contractors should consider several items such as: 1) number of construction accounts, 2) knowledge of surety requirements, 3) knowledge of construction accounting methods, and 4) scope of responsibility (i.e., experience preparing financial statements and taxes, interim statements, tax planning, financial advisor, or business partner).

Banking Credit

With the exception of the large contractors, most contractors do business with one bank. The surety in general wishes to have the banker as an active member of the contractor's advising team. The surety should take note of whether a contractor does business with a banker that understands the construction industry. It is also important that the bank itself be financially healthy. Sureties should observe various aspects of the contractor-bank relationship. Unsecured credit clearly indicates the bank has a high opinion of the contractor. When securing credit the bank may often take an assignment of all assets, including contract receivables. Even if it seems self-evident that the surety has prior rights on bonded contracts, they may require that the bank acknowledge those rights in a letter.

When the bank writes a letter to the surety regarding the bank's relationship with the contractor, it should include: 1) loan specifics (i.e., interest rates, amounts, terms, and payment history); 2) average and present balances; 3) whether the contractor has established a line of credit and how it is secured, including the interest rate and amount; and 4) how long the bank has done business with the contractor and the bank's assessment of the contractor's present financial condition and character. If the surety requires a letter that goes beyond a simple outline of past experience and probable credit available based upon that experience and a review of current financial records, the bank may charge the contractor for the letter.

A strong banking relationship is a part of a solid financial history for a contractor. Surety companies want to see contractors with a strong banking relationship, but mainly with sufficient working capital to support work programs without bank borrowing. Many contractors with strong working capital positions actually generate substantial interest income from their deposits. Sureties, however, recognize that during the year there may be times when most contractors need working capital support. Thus it is critical that the surety know that the bank is willing to provide the necessary financial support to the contractor, generally short term, when needed.

Another form of bank credit that contractors may use is for the purchase of fixed assets such as equipment. It is important that long-term debt be structured to have the least possible impact on working capital. In recognition of this, many contractors use a "skip" payment plan during the "off season." Contractors must plan equipment purchases and financing and communicate those plans to the surety.

Chapter 17

BOND COST

When a surety agrees to underwrite a contractor, depending on the contractor's tangible assets and past performance, it is placed in a rate category with predetermined premium rates for each class and type of bond. In general, the better the rate the lower the premium, as there is a greater chance the contractor will perform successfully. Each year, as a contractor's fortunes change, his or her rate structure is evaluated. Sureties generally have three categories of ratings that are registered with the insurance commission of each state in which the surety underwrites bonds: 1) intermediate, 2) merit, and 3) preferred, with the preferred rate being the lowest and thus the best rate a surety can offer. Ali bondable contractors fall into one of these broad categories. A discussion of items relating to establishing and applying bond premiums follows.

CONTRACT PRICE

The basic factors that affect premiums for contract bonds are contract price, contract type, and stipulated contract length. Sample premium computations and payment terms for contract bonds follow. It is worth noting that some surety companies have implemented rating methods to allow factors such as financial strength, profit level obtained by company, and quality of management skills to enter into the premium determination process.

Contract bond premiums depend on the final contract price, i.e., the sum of money or other reimbursement that will pass from the project owner to the contractor, or the entire dollar value of work for which the contractor is responsible. At the project's close-out, the surety may adjust the premium based on actual reimbursements or the entire dollar value of the work, whichever is higher.

The surety computes premiums based on the penal sum of the performance and/or payment bonds if the penal sum is below a certain threshold percentage (usually 24% to 26%) of the total contract. The computed premium depends on the full contract price if the penal sum of performance and/or payment bond is above the threshold percentage of the total contract. For example, consider a $1 million contract at a rate of $10.00 per $1000 of contract price above the threshold percentage and $30.00 per $1000 below the threshold percentage. The sample owner has required a performance bond for 50% of the contract amount, the total of which is estimated to be $1 million. The penal value of the bond is $500,000. Using the rates listed above, the premium would be $10,000. The premium would also be $10,000 if the penal value of the bond is $1 million. Obviously, it is advantageous for an owner to require 100 percent bond coverage (Johnston 1981; "Use of Surety Bonds" 1975).

If an owner requires a bond in excess of the contract price, the premium paid remains the same. For a performance bond of 150% of the contract amount of $1 million, the penal value of the bond is $1.5 million, while the premium paid remains $10,000. On the other hand, if the owner requires a performance bond of 10% of the contract amount of $1 million (and 10% is below the specified threshold level), the penal value of the bond is $100,000 while the premium paid is $3,000.

CONTRACT TYPE

For the purpose of determining bond premium rates, The Surety Association of America (SAA) has divided construction contracts into several classifications based on the contract's inherent risk and the type of coverage. The classifications are: B, A, A-1, Supply. A brief overview of each category follows, but for a more detailed listing of each classification, see Appendix XI. For a further listing of the contracts and their corresponding classification, see the Surety Association of America's *Rate Manual for Surety Bonds* (The Surety Association of America 1987).

Table 17.1 Rate Table for Construction Contracts
Source: *Rate Manual: Surety Bonds 1987*

		------Rates[1,2]------			
Contract price ($) (1)		Class B ($) (2)	Class A ($) (3)	Class A1 ($) (4)	Supply ($) (5)
First	100,000	25.00	15.00	9.40	3.50
Next	400,000	15.00	10.00	7.20	2.50
Next	2,000,000	10.00	7.00	6.00	1.75
Next	2,500,000	7.50	5.50	5.00	1.50
Next	2,500,000	7.00	5.00	4.50	1.25
Over	7,500,000	6.50	4.50	4.00	1.00

[1] Basic rate per $1000 of contract price for first 12 months.

[2] For stipulated construction time greater than 12 months of 366 calendar days, increase the basic premium by 1% per month for each whole month over 12 months. The additional charge for time applies only if duration is stipulated in the contract documents. If no completion date is stipulated, the surcharge does not apply.

Class B contracts involve architectural building construction, related subtrades, most engineering construction, concrete and excavation work, and work performed underground or in or under water. Traditional examples of Class B contracts include construction of buildings, canals, and tunnels.

Class A contracts are typically general contracts and subcontracts of a less difficult nature than those within Class B. Examples of Class A contracts include construction forms, curbing and guttering, and roofing.

Class A-1 contracts are generally less difficult than those in Classes A or B and include furnishing and installing, installing only, or providing specific services and equipment. Examples of Class A-1 contracts include guard rails, insulation contracts, and scaffolding.

Supply contracts are generally for supplies and materials that do not become part of the realty until after the supply contract is complete. Examples of supply contracts are furnishing and delivering concrete, carpets, and pumps.

Each of these contract types has its own basic rate table. Table 17.1 provides sample rates. The minimum premium for any contract bond is $50. The maximum chargeable rate for each bond classification depends on the aggregate penalty of the performance and payment bonds. Table 17.2 shows the maximum rates for contract types per $1,000 of contract price for the first 12 months.

Table 17.2 Maximum Rates for Construction Contract Bonds
Source: *Rate Manual: Surety Bonds* 1987

Contract type (1)	Rate[1] (2)
Class B	70.00
Class A	50.00
Class A-1	50.00
Supply	12.00

[1] Maximum rates per $1000 on the aggregate penalty of performance and payment bonds.

Note that in many instances surety companies use rates that differ from those the SAA promulgates. Underwriters usually discount SAA rates for contractors who have an established record of successful operations and whose financial resources meet certain standards. Because these contractors are less likely to fail on an individual or multiple projects, they may qualify for lower premium rates. Additionally, in the current competitive market an emphasis on pricing has surfaced, and many sureties are lowering their premium rates to attract or hold onto accounts.

STIPULATED CONTRACT LENGTH

If the stipulated time for a performance of a contract exceeds one or, in some cases, two years, surety premiums may be surcharged, usually by 1% per month for each month in excess of the one or two year limit. Fractions of a month are disregarded. This premium increase offsets the risks underwriters generally associate with a longer contract duration, as more time tends to increase exposure, but it is always at the discretion of the surety whether to apply such a surcharge.

Table 17.3 Example Premium Computations

Contract price ($) (1)		Example 1 Basic rate[1] ($) (2)	Example 1 Basic premium ($) (3)	Example 2 Basic rate ($) (4)	Example 2 Basic premium ($) (5)	Example 3 Basic rate ($) (6)	Example 3 Basic premium ($) (7)	Example 4 Basic rate ($) (8)	Example 4 Basic premium ($) (9)	Example 5 Basic rate ($) (10)	Example 5 Basic premium ($) (11)
First	100,000	25.00	2,500	15.00	1,500	9.40	940	3.50	350	70.00	350
Next	400,000	15.00	6,000	10.00	4,000	7.20	2,880	2.50	1,000	—	—
Next	2,000,000	10.00	20,000	7.00	—	6.00	3,450[2]	1.75	3,500[3]	—	—
Next	2,500,000	7.50	—	5.50	—	5.00	—	1.50	3,000	—	—
Next	2,500,000	7.00	—	5.00	—	4.50	—	1.25	—	—	—
Over	7,500,000	6.50	—	4.50	—	4.00	—	1.00	—	—	—
Total basic premium		—	28,500	—	5,500	—	7,270	—	7,850	—	350
Increase for length		18%	5,130	—	—	—	—	13%	1,021	—	—
Term premium		—	33,630	—	5,500	—	7,270	—	8,871	—	350

[1] Basic rate per $1000 of contract price for fist 12 months
[2] Value is product of basic rate ($6.00) and $575,000.
[3] Value is product of basic rate ($1.50) and $2,000,000.

If the contract does not contain a stipulated time for completion, most sureties do not apply a surcharge even if the contractor believes it will take several years to complete the project. Even when the contract includes a stipulated time for completion and the surety applies a surcharge for this to the bond premium, such a surcharge is usually not incurred when the contractor exceeds the stipulated time. Whenever a contract provides a stipulated time for completion that exceeds one year, the savvy contractor will inquire as to whether a surcharge for time will be added to the premium. If so, the contractor will then be able to add this surcharge as a job cost when preparing the bid.

By comparison, the cost of a letter of credit from a banking institution, a tool contractors may use in lieu of surety bonds (especially for international projects), ranges from 1% to 2% of the contract per year, one to two times what sureties on average charge. If the beneficiary collects against the letter of credit, the letter automatically converts to a bank loan at rates above the prime rate for most contractors. On the other hand, a surety must meet its obligation to pay a claim with or without further funds from the principal. The underwriter receives no interest on funds it pays out and typically obtains only partial settlement from the defaulting contractor. Therefore, surety premiums are used to cover uncollectable losses.

EXAMPLE PREMIUM COMPUTATIONS

The following five examples illustrate the basic factors influencing contract bond premiums. Each example corresponds to a different contract type. Table 17.3 presents the computations for these examples.

Example 1 is the construction of a commercial building (Class B contract) for $2,500,000 with a stipulated duration of 30.5 months. A contract of this size would have a basic premium of $28,500. However, since the contract length exceeds 12 months, an underwriter would adjust the basic premium by 18% (30-12=18 months), bringing the total premium to $33,630.

Example 2 is a contract to furnish construction forms (Class A contract) with a total price of $500,000 for 10 months. Based on this information, the basic term premium is $5,500. The contract length is less than 12 months, warranting no adjustment of the basic premium. Therefore, the total premium is $5,500.

Example 3 is a contract furnishing and installing ornamental iron work (Class A-1 contract) with a total price of $1,075,000. The expected contract length is 14 months. Even though the expected length exceeds 12 months, the contract documents do not stipulate a particular project duration; hence this situation requires no adjustment of the basic premium. The total premium is equivalent to the basic premium, which is $7,270.

Example 4 is a contract for the supply of concrete with a total price of $4,500,000. The length of the contract is 25.5 months. The basic premium is $7,850. Since the contract length is greater than 12 months, an adjustment of 13% (25-12=13 months) makes the total premium $8,871.

Example 5 is a Class B contract with a stipulated completion time under one year. The contract price is $100,000. The owner wants to have a performance bond totaling $2,500 and a payment bond of $2,500. The aggregate coverage desired is therefore $5,000. In this case, the underwriter will charge the maximum rate (see Table 17.2). Thus, the total premium would be $350 ($5,000 at $70 per $1000). Typically underwriters also charge the maximum rate on contracts that have an indefinite quantity of work to be performed within an indefinite time period for delivery. This, however, depends on what the contract documents state and the bond amount relative to the contract price.

STATE INSURANCE DEPARTMENTS

There is no federal regulation regarding the rates and premiums surety companies charge; however, state regulations do exist. All 50 states have departments of insurance that regulate the licensed companies providing surety bonds.

States moderate bond rates through a variety of laws, rules, and regulations (Remmen 1977). The most common approaches include "file and use," where rates are legal once the company files them with the state insurance department, and "prior approval," where rates become effective only after the state insurance office approves them. All state laws have the following underlying requirements: 1) rates cannot be unfairly discriminatory, 2) rates should not be inadequate, and 3) rates should not be excessively high ("Use of Surety Bonds" 1975).

The SAA publishes a manual of advisory rates. The last time these rates were updated was 1987. Most all states have approved the rates in this manual. A surety company may use the SAA advisory rate manual by filing a document with the state's insurance department. If the surety company decides to use rates different from those in the SAA manual, which frequently occurs, it must file a separate document with the state insurance department stating the specific rates the firm will apply to that category of contractors. Companies may elect to use rates that differ from SAA rates for a variety of reasons including but not limited to: 1) market strategies, 2) competition, 3) quality of risk, 4) experience that differs from that of the industry as a whole, or 5) geographic location.

BOND PAYMENT TERMS FOR CONTRACTORS

The premiums for contract bonds are payable in advance, whether or not construction starts immediately. Independent surety agents generally deliver the fully executed performance and payment bond forms to the owner after the owner and contractor sign the contract. Typically, the agent handling the contractor's account obtains immediate payment for the bond and then forwards the premium to the underwriting company, minus his or her commission. Under special circumstances when the premium is a large amount, an underwriting company and agent may permit the contractor to pay for the bond after the contractor receives receipts from the first pay request from the owner. This exception usually applies only to large contracts. In addition, if through change orders the contract amount is increased or decreased, the premium will be adjusted at the completion of the project. As a general rule, however, the agent seeks the contractor's payment as soon as possible. Failure to receive payment may leave the agent responsible to the underwriting company for the premium owed. It is important to note as well that non-payment of premiums does not release a surety from its obligations to the owner.

PAYMENT AND PERFORMANCE BOND COST

To reiterate, rates for performance or performance plus payment bonds derive from the basic rate table. Sureties typically treat the performance and payment bonds as a single item. Sureties generally do *not* charge any additional amount beyond the premium for the performance bond for a performance plus payment bond. Factors that influence these rates include the price, length, and type of contract, as well as the filing of mechanic's liens. As a result, owners do not receive maximum benefit from buying only a performance bond. Rates for payment bonds alone derive from a different rate table and procedure, but since sureties tend to lose more money on payment bonds, the premium for only a payment bond is generally more than half of what a payment plus performance bond would otherwise cost. The bonds are packaged, since according to sureties, the courts have generally held that performance bond includes payment for specialty contractors' and suppliers' claims (Shiatte 1989).

In current practice, the rate structure that the surety industry uses for performance and payment bonds discourages obligees from requiring partial bonding (i.e., bonds that amount to less than 100% of the contract price). Performance and payment bond premiums are computed on the contract price, which they define as the total sum of money the obligee pays to the principal. In instances where the bond amount is a small percentage of the contract price (i.e., less than 25% of the contract price), the surety may still charge a premium based on the bond amount.

According to a 1975 report by the U.S. Comptroller General to Congress, the disparity between protection and cost become increasingly more pronounced as contracts grow larger. This trend is a result of the structure of the rate tables that the surety industry uses. For example, a 10% bond on a $10 million contract could cost $50,000. A 100% bond on the same size contract would cost $53,975. Premiums depend upon either 1) the contract's dollar amount or 2) the bond's penal value. If the penal value is greater then 20% of the contract amount, the contract amount determines the premium. If the penal value is less than 20% of the contract amount, the penal value determines the premium. Hence, a bond with a penal value of 20% of the contract amount costs the *same* as a bond with a penal value of 100% of the contract amount. A reduction in coverage by 80% occurs with no reduction in the premium paid (Shiatte 1989).

MAINTENANCE BOND COST

Maintenance bonds are normally issued in conjunction with performance and payment bonds. Typically, a one-year maintenance bond is included with a performance and payment bond. As such, in addition to the factors used in underwriting performance and payment bonds and determining rate structures—namely the four Cs—a surety typically uses four additional criteria when deciding to underwrite a maintenance bond: 1) rating of the contractor, 2) size and

Table 17.4 Cost of Maintenance Bonds
Source: The Surety Association of America

Class B construction contract amount[1] ($) (1)	Cost of performance and payment bonds[2] ($) (2)	Cost of 2-year maintenance bond[2] /cost plus performance and payment ($) (3)	Cost of 5-year maintenance bond[2] /cost plus performance and payment ($) (4)	Cost of 10-year maintenance bond[2] /cost plus performance and payment ($) (5)
500,000	8,500	2,300/10,500	5,750/14,250	11,500/20,000
1,000,000	14,500	4,300/18,800	10,750/25,250	21,500/36,000
2,000,000	23,500	8,300/31,800	20,750/44,250	41,500/65,000

[1] Assumes construction project is less than 12 months in duration; greater than 12 months, add 1% per month.
[2] Assumes Standard SAA Rates for Class B Construction.

complexity of the contract, 3) length of warranty period, and 4) why the owner is seeking a maintenance bond. Premium rates also depend on the market and SAA standard rates, with better rates going to the higher-rated contractors, although recently, maintenance bond premiums of all categories have been commensurate with SAA rates. In the 1990s, industry averages for performance and payment bond premiums have been much lower than the latest SAA standards set in 1987, because the market has been highly profitable and competitive and thus "soft." Maintenance bonds, in contrast, have stayed much closer to SAA rates regardless of the class of construction or rating of the contractor.

Maintenance bond premiums are charged at a lower rate than performance and payment bonds but quickly add up when extended beyond a few years. For example, based on the standard premium rates of the SAA, a 10 year maintenance bond on a $1 million Portland cement concrete spillway contract would cost $19,350, close to 2% of the total contract value. This premium would be in addition to $13,500 for the performance and payment bonds on the same contract based on SAA standard rates for Class B construction. Even based on a lower rate currently available in the marketplace, the total premium for the three bonds becomes significant, as it would probably be between $25,000 and $30,000. Table 17.4 helps illustrate the incurred costs with performance, payment, and maintenance bonds.

In addition to the rating of the contractor, the size of the contract, the type of construction, and the stipulated period of warranty determine whether a surety will issue a maintenance bond. With larger and more complicated contracts and with longer periods of warranty, the surety's exposure can become significant. However, no matter what the premium charged, it will never be enough to cover the surety's risk. With increased time, determining the cause of failure becomes extremely difficult because it may be due to any number of factors, many of which may not have been anticipated or covered under the maintenance agreement. In addition, inflation can affect the future cost of repairs, as it is unlikely that the cost will go down over time.

In instances where maintenance bonds are used in public construction, concerns exist as well in how changing administrations will interpret the intent and purpose of the bond. Elected officials and administrators change with time and many sureties are concerned with maintaining the same level of understanding as to the intent, purpose, and coverage provided by the bond, even though such continuity is often difficult to maintain. These factors weigh heavily in underwriting procedures and are only complicated without a maintenance bond form clearly separated from the performance bond.

Finally, why an owner wants the maintenance bond to begin with can be very important as well. If an owner requires a maintenance bond at the outset in the contract documents, the surety's decision will focus on the ability of the contractor to complete the construction per plans and specifications and remain in business through the period of warranty. Sometimes, however, after work is completed and before accepting a facility, an owner will require a maintenance bond. Such a decision would factor highly into the surety's decision as it would appear that the owner may have questions about the actual performance of the contractor. In such a case, the surety would weight this desire much more heavily than if the owner had requested a bond in the original contract documents, as the surety would most probably have to investigate the actual facility before deciding to underwrite the maintenance bond.

Chapter 18

UNDERWRITING CASE STUDY

Prometheus Construction Company, Inc. (PCC) has submitted through its agent, J. Hermes, a request to increase its surety credit line with a new surety, Action Bonding Company. PCC currently has an $18 million aggregate work program and seeks from Action a $20 million aggregate limit with a $10 million single job limit. In addition, PCC would like to bid on the occasional $15 million project. PCC wishes to secure this slight increase in aggregate bonding limits to pursue projects similar to but on a larger scale than those it currently constructs. PCC is not planning to enter into new fields or acquiring novel specialty trades: it is planning only to expand slightly using primarily the same labor resources, equipment, and management procedures already in place.

Action Bonding Company

Action Bonding is an approved surety listed with the United States Treasury List and as of July 1994 can write a single bond up to $50 million, meaning that it has assets of approximately $500 million. Action has been in business for over 40 years and has an outstanding profit margin due to careful and sound underwriting procedures. In the loss years of the 1980s, Action not only avoided any underwriting losses, but remained profitable.

Action is a mid-size surety associated with the larger Action Insurance Corporation, which owns a controlling interest in the bonding company. J. Hermes, PCC's agent, is considered trustworthy by Action and his judgments have proven reliable in the past: neither PCC nor any of the contractors he represents has ever defaulted. Therefore, his unqualified support speaks well of PCC. However, such a record and goodwill are far from sufficient. A thorough underwriting procedure must be completed before Action decides to accept PCC as an account.

The task before Action's underwriting department is to decide using standard industry practices and methods the likelihood of a PCC default if the proposed increases in its credit line are to take effect, and if these risks are minimal enough to justify the increased surety credit. To put it another way, what risk of default and consequent exposure would exist for Action if they underwrite the bonds for PCC to regularly bid on larger jobs?

To determine whether this risk is justifiable, Action requests the following documents from PCC to begin the underwriting process. Key items and figures are included throughout the chapter, while the complete set of materials appears in Appendix XII. They are as follows:

1. Cover letter from J. Hermes
2. Contractor's Information Record (Standard Form)
3. Accountant's Information (Standard Form)

4. Independent Auditor's report as of September 30, 1993 with additional information
5. Independent Auditor's report as of September 30, 1994 with additional information
6. Dun & Bradstreet Report
7. Personal Financial Statement (A. Prometheus)
8. A current bank Letter of Credit
9. Job Schedules.

Action, of course, can request additional information at any time in this underwriting process.

Prometheus Construction Company, Inc. (PCC)

In the late 1970s, A. Prometheus started Prometheus Construction Company in Louisville, Kentucky. The company was incorporated on July 4, 1977. Prometheus began as a builder of strip malls using a novel concrete prefabrication system. Soon, however, A. Prometheus expanded operations to include small office buildings. Under his guidance, PCC has grown quickly into a regional builder of office buildings, shopping centers, restaurants, and specialty projects such as a waste water treatment plant. It eventually opened a branch office in Lexington in 1990, but shut it down in late 1992 as it proved unprofitable.

In that same year, PCC became a Chapter S corporation, meaning that the individual owners of the firm are taxed on their earnings. PCC is fully unionized with over 25 years of experience. A. Prometheus, the president and founder, owns 100% of the company. His wife, E. Prometheus, is the treasurer, and their son, A. Prometheus, Jr., is the company's vice president. Currently, the indemnitors include two additional sons and their wives, and a daughter and her husband, but not E. Prometheus. There is currently no buy/sell agreement, but it is understood that Prometheus, Jr. is being groomed to take over the family business.

In 1994, as PCC applies for an expanded line of surety credit with Action, it was awarded close to $30 million in new contracts with $15 million of it being bonded work. PCC's previous surety is discontinuing its surety line, and therefore PCC needs a new surety. In addition, PCC would like to increase its surety base to bid on larger, bonded projects, allowing PCC to take advantage of its experience and capitalize on the possibility of increased profit margins.

PCC has good credit with Anytown Bank. The bank has issued PCC a $250,000 line of unsecured credit good for a 12-month period. PCC has also never defaulted nor failed to complete a contract for any reason, nor has any other business owned by any of Prometheus's failed. For all of these reasons, J. Hermes thinks that Action will most likely choose to bond PCC.

THE UNDERWRITING PROCESS: ACTION REVIEWS PCC

Action Bonding seeks to determine whether to accept PCC as an account, and whether the proposed $10 million single-job and $20 million aggregate limit with the understanding that an occasional $15 million project will be bid upon is appropriate. To reach its conclusions, Action uses both quantitative and qualitative methods including a thorough financial investigation. Its analysis focuses on four areas: 1) character, 2) capacity, 3) capital, and 4) continuity. The character analysis involves the evaluation of PCC's past performance as well as its fulfillment of obligations. The capacity analysis focuses on PCC's ability to provide the resources necessary to successfully complete its work program. The capital analysis evaluates PCC's financial strength and performance. Finally, the continuity analysis consists of determining the efficacy of the planned succession of PCC.

Throughout the investigation, Action uses standard industry practices discussed throughout Part 4. In the financial analysis, it relies on the accepted practices of accountants. Results are compared with industry averages and levels to normalize findings. Whenever possible, a quantitative analysis is employed by the surety. However, as discussed above, due to the subjective nature of the construction industry in general and the difficulty of weighing such factors as reputation and relative strength of a particular management hierarchy, in the end, parts of the surety's decision are always subjective. Still, as the reader will see, at every step of the review Action determines what information it needs to complete its evaluation of character, capacity, capital, and continuity, never relying on incomplete or contradictory information.

Character Analysis of PCC

Character is the sum total of a contractor's performance to-date, commitment to meet obligations, and the reputation of both the company and personnel in relation to owners, designers, architects, engineers, subcontractors, suppliers, creditors, other sureties, and unions. Areas of concern for a surety are general performance, alleged or actual bid rigging, fraud, prior failed projects and/or businesses, and potential for litigation.

In the case of PCC, there was little information initially provided with which to assess the character of the firm, but even with this limited information, there are more than a few causes for concern. J. Hermes and Anytown Bank both wrote favorably of the character of A. Prometheus and PCC. The underwriting review team notes this as significant: good relationships with creditors and agents can only help in the case of a default, and, more importantly, can often prevent one from occurring.

In the information PCC provided directly to Action in the Contractor's Information Record (CIR), found in Appendix XII, A. Prometheus contends that neither he nor any of his businesses has ever defaulted or failed to complete a project for any reason. While most probably true, discrepancies arose between information provided in the CIR and information found in both the Dun & Bradstreet Report (D&B) for PCC and in the independent auditor's statements from Hercules & Smith, PCC's accountant. The discrepancies stem from two areas: 1) the nature of any litigation involving PCC as the defendant and 2) net income for 1993. Since all information in the D&B reports was provided voluntarily by PCC, such discrepancies were not expected and were therefore immediately causes for concern.

The CIR reports no liens outstanding against work, no serious controversies over payment for labor or materials on any project, and no judgments, suits, or claims listed as pending against PCC. However, the D&B report and information from the auditor, Hercules & Smith, tell a different story.

It seems that at the time of the request for bonding, PCC has 10 liens/suits currently open or pending, with a total value estimated by Hercules & Smith to be in the range of $100,000 to $499,999. Since the total net income for the two years of credit reports submitted, listed at $153,131 (1994 income minus 1993 deficit) falls within this range, the nature and details of these suits is vital if Action is to determine the financial condition and character of PCC. (For detailed quantitative information, see "Independent Auditor's Report," especially "Financial Statements," in Appendix XII).

Another example of conflicting information deals with these same income figures. The D&B report lists net income for fiscal year 1993 as $181, 535, while the audited financial statements declare a net loss of $68,467. In addition, there was a substantial increase in professional fees ($61,735 to $106,302) from 1993 to 1994. This increase may reflect an increase in legal fees to deal with the liens/suits pending, but from the information provided, it is impossible to know for certain. Without more information, any determination as to the nature of the discrepancies and thus the general character of PCC, would be premature. To more accurately assess PCC's character, the underwriting review team needs to immediately begin a more detailed investigation into these financial and legal discrepancies. Legitimate liens would not necessarily reflect negatively on the contractor—perhaps PCC has come into conflict with a particularly pesky subcontractor on several jobs—but for Action to feel comfortable accepting PCC as an account, the exact nature of each and every lien (if there are any) must be known.

Questions raised and information requested by the team in order to complete the review of character include:

1. A thorough investigation into why a discrepancy exists between the information in the CIR and in both the D&B report and the information provided by Hercules and Smith. Specifically, the nature of any and all liens and litigation pending, and how much money was earned or lost in 1993 needs to be clarified.
2. Why was there an increase in professional fees of 72% from 1993 to 1994?

To accomplish this, the team needs to begin interviewing owners, engineers, architects, bankers, suppliers and subcontractors who have worked with PCC, and if at all possible, with any litigants who have claims against PCC.

Capacity Analysis of PCC

The measure of a contractor's resources necessary to successfully complete projects and desired work programs is its capacity. Within this measurement are many resources including management expertise, equipment, labor, material, and financial reserves. An assessment of PCC's capacity requires in-depth investigations into each of these five resources. Possible and usual sources of information are architects, engineers, owners, subcontractors, suppliers and bankers who have worked with or for the contractor—in this case, PCC.

PCC employs 30 people, using four work crews. Management of the company is primarily controlled by the family members A. Prometheus

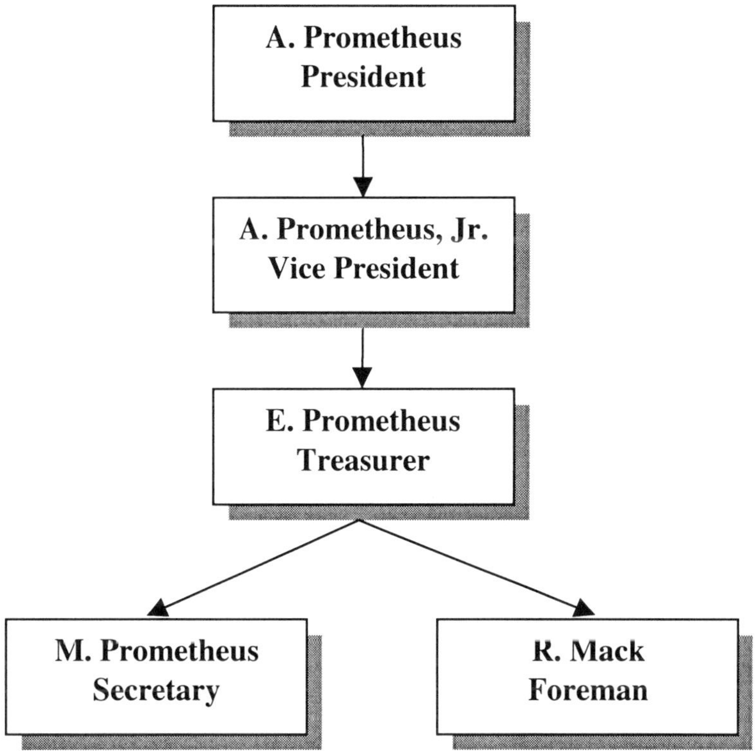

Figure 18-1 PCC Organization Chart

(president), A. Prometheus, Jr. (Vice President), and Mrs. E. Prometheus (Treasurer). In addition, M. Prometheus is the only secretary for the firm, and R. Mack is the foreman (Figure 18-1).

Experience levels of these key personnel vary significantly due to respective ages. The vice president is only 25 years old. The Action review team is not immediately comfortable with this level of experience and requires further information regarding his qualifications.

No information aside from the managerial hierarchy provided in the CIR was given regarding the corporate structure of the company (Figure 18-1 is for the reader and was not originally provided by PCC to Action). For instance, how does PCC handle project management and estimating? An organizational chart showing the structure from the president down to superintendents, and/or a detailed explication of these and other key decision-making processes, are necessary to further analyze the capacity of PCC.

Questions raised and information requested by the team in order to complete the review of capacity included:

1. What are the experience levels and/or training of key personnel?
2. A detailed explication of individual roles in the corporation and on individual projects is needed including how bids are estimated.

Table 18.1 Ratios for PCC

Ratio (1)	1994 (2)	1993 (3)	Ratio range (4)	Comments (5)
		(a) Liquity		
Quick ratio (quick assets/ current liabilities)	1.09	1.20	>1.00	PCC has adequate funds in liquid assets to cover debt
Current ratio (current assets/ current liabilities)	1.24	1.26	>1.20	The margin of safety available to cover any possible shrinkage in the value of current assets is barely sufficient
Collection period (contract receivables/ net sales) *365	76.40	86.70	<60 days	Improving, but still too long; may point to inefficient billings, or worse, an overbilling or underbilling problem
		(b) Operations		
Net profit/ tangible net worth	14.09%	-4.66%	>15%	PCC did not get any return on capital in 1993 but did get a reasonable return in 1994
Net profit/ total assets	3.76%	-1.27%	>1%	PCC had a net profit loss relative to total assets in 1993; appears to be improving in 1994
Net sales/ tangible net worth	12	11	10 to 15	The relative activity of invested capital if reasonable
Net sales/ working capital	19	16	5 to 30	Well within industry averages

Table 18.1 Ratios for PCC (Continued)

Ratio (1)	1994 (2)	1993 (3)	Ratio range (4)	Comments (5)
		(c) Leverage		
General and administrative/ tangible net worth	37.40%	40.74%	<60%	Well-within industry average
Fixed assets/ tangible net worth	21.45%	19.51%	10% to 40%	The contractor is not investing significant monies in the fixed assets
Total liability/ net worth	2.75	2.67	1:1 to 3:1	Total liabilities relative to net worth are on the high side; may mean financial trouble
Hard debt/ tangible net worth	0.07	0.09	1:1 to 2:1	Long-term debt pressure is low
		(d) Coverage		
Earnings before interest & taxes/interest	9.95	(0.80)	>1.0	1994 is much improved
Net profit + depreciation/ current maturities	2.63	(2.10)	>1.0	1994 is, again, much improved
		(e) Specific Expense Items		
Depreciation/ net sales	0.07%	0.10%	>0%	May mean an old equipment fleet
Officers compensation/ sales	0.62%	0.73%	>0%	Appears reasonable
Overhead/sales	3.11%	3.77%	<15%	Appears to be a lean organization without excessive overhead

Capital Analysis of PCC

The thrust of this evaluation is based on the financial strength and performance of PCC. Financial statements, namely the balance sheet and income statement, were closely scrutinized and financial ratios are presented below. The review team compared these ratios with industry averages published in Robert Morris Association. Attention was also given the schedules of general and administrative expenses and to schedules of completed and ongoing projects.

Ratio and Trend Analysis of Financial Variables

When compared to industry averages, key financial variables can indicate a great deal about the financial condition of a contractor. Examining financial ratios and doing a trend analysis of key variables can expose potential areas of concern and quantitatively reveal a contractor's current financial position.

Ratios can be divided into five categories: 1) liquidity, 2) operations, 3) leverage, 4) coverage, and 5) specific expense items. Each of these categories is comprised of several ratios that reveal a portion of the financial picture of PCC. What follows are several detailed charts revealing the ratios from 1993 and 1994 and their definitions, whether they are below or above industry averages, and brief commentary by the underwriting team.

An initial concern is raised with the liquidity ratios presented in Table 18.1. Although above the minimum acceptable range, the margin of safety on the quick and current ratios is questionable and may indicate a problem with capital management. Additionally, the collection periods of accounts receivable improved from 85 days in 1993 to 76 days in 1994, but this is still 27% more than the desired 60-day collection period. Is there a problem with overbilling/underbilling? To answer this, the team will have to wait until it performs an analysis of the schedule of work-in-progress.

Further review of the financial ratios yields several important "red flags." The loss incurred in 1993 means several things. First, PCC got no return on its capital in that year and had a profit loss relative to its total assets (see Table 18.1). While both of these improved in 1994, the total liabilities relative to PCC's net worth (2.67 for 1993 and 2.75 for 1994) is on the high side, as anything over a 3-to-1 ratio could mean that a contractor is having any number of troubles, be they managerial, financial, labor, or legal.

PCC is also not investing significantly enough in fixed assets, which again, could signify a specific problem with a phase of company management or could be a systemic failure of the organization. However, the relative activity of invested capital seems reasonable (net sales divided by tangible net worth). The long-term debt pressure is low (hard debt divided by tangible net worth), and the company shows many signs of being healthy. Administrative costs relative to tangible net worth is well-within industry averages (< 60%), and officer compensation and overhead relative to sales appear in good to very good ranges. The general and administrative expenses are running at approximately 3% when viewed as overhead on an $18 million work program, which is very reasonable. Furthermore, expenses were even slightly reduced due to the closing of the Lexington branch office in 1993, which appears to have been a wise decision. However, the review team requires more information to determine this for certain.

Trend analysis of financial variables is conducted to assist in evaluating the firm's financial stability. Specific ratios are calculated and normalized to gross annual sales according to Severson et al. (1994). The variables are then compared to the statistical outcomes for claim and non-claim contractors on surety bonds, the results of which are shown in Table 18.2. To perform a reliable trend analysis, three years of contractor data is needed, but in this case, PCC has only provided two. Action will most likely request another year of financial data.

Results of the trend analysis for PCC point to concerns, as many categories are in the claim range. For instance, in the asset section of Table 18.2, underbillings in 1994 were too high. Accounts payable, a liability, adversely affected the liabilities to sales ratios. Shareholder's equity to sales is also in the claim region; however, this is not a concern due to the large amount of retained earnings.

Table 18.2 Trend Analysis of PCC

Financial variables (1)	Year 1 (1993) (2)	Year 2 (1994) (3)
(a) Assets		
Cash/sales	0.032 *	0.046
Accounts receivable/sales	0.214 *	0.241 *
Inventory/sales	0.000	0.000
Underbillings/sales	0.023 *	0.007
Total current assets/sales	0.275	0.299
(b)Liabilities		
Accounts payable/sales	0.202 *	0.206 *
Notes payable/sales	0.006	0.009
Total current liabilities/sales	0.222 *	0.238 *
Total long-term debt/sales	0.006	0.009
Total liabilities/sales	0.228 *	0.247 *
(c) Stockholders equity		
Stock/sales	0.001	0.001
Retained earnings/sales	0.082	0.092
Total stockholder's equity/sales	0.083 *	0.093 *
(d) Income statement		
Cost of sales/sales	0.960 *	0.978 *
Gross profit/sales	0.040 *	0.022 *
Net income before taxes/sales	0.012	(0.004) *

* Denotes a value close to that of a claim contractor as defined by Severson, et. al. (1994)

The income statement is the most troublesome (see Appendix XII). Cost of sales are too high, and thus profit margins are too low, yielding poor performance on net income. This finding, coupled with the findings in the capacity analysis, make it vital for PCC to furnish a detailed explanation of its bid preparation mechanisms.

PCC states in the CIR that it has been operating as a Chapter S Corporation. This arrangement is common in the construction industry. In an S corporation, the stockholders of the business in question are taxed personally on their individual tax returns for any returns based on their share of the company's earnings. This can be a concern to a surety because the owners may take money from the company at any time. The Independent Auditor's Report sent by Hercules & Smith showed that since the fiscal year ending on September 30, 1994, $337,821 has been distributed to shareholders, while the net income for this period was $502,274. This constitutes a 67% distribution of net income to shareholders for the three-year period from 1992 to 1994.

Such a high distribution is of particular concern for a surety when a contracting company seeks to grow, as expansion requires a fair amount of capital investment. Since PCC desires an increase in bonding, Action would like to see as much of the earnings as possible left in the company to reduce the amount needed to finance the inevitable costs of expansion. During the 1993-1994 period, 126% of net income was distributed. Furthermore, according to the information provided by A. Prometheus, he is the sole shareholder. If this is so, what plans are in place for his untimely departure from the head of company operations? Will the transition be a successful one? Does A. Prometheus, Jr., have the experience to replace his father? Will Action be severely "hit" with a default? These concerns must be dealt with in order for Action to make a positive recommendation for the increased bonding limits.

Schedule Of Contracts

An underwriter can tell a great deal from looking at a contractor's schedules of contracts completed and contracts in-progress. The underwriter can judge how accurately the contractor estimates the cost of projects, and how well they manage costs in the field. In addition, from the contracts in-progress, the whether the contractor is underbilling, overbilling, or has a significant backlog is readily noticeable.

A perusal of the PCC's work schedules (Tables 18.3-18.6) reveal that profit margins (mark-ups) are adequate and that jobs are being completed for the dollar value expected. This leads to the conclusion that accurate estimates are being prepared but also to a discrepancy. How are mark-ups established, since they range from 0.35% to 17.74% on completed projects over the two years of data? The review team is uncomfortable with the mark-ups below the general administrative overhead of 3%—three out of six jobs in 1994—and with the yearly average of only 3.64%, just 0.64% against administrative overhead costs. In 1993, the mark-up average was a much more profitable 8.76%, almost six percent over overhead. Why the change? Does it represent a trend? Is this something Action should be concerned with?

Accounting

The overall internal accounting system is deemed excellent by the auditing accountant. Controls are in place to monitor contract change orders and prepare monthly "in-house" financial statements. The recently implemented (June of 1994) preparation of interim financial statements on a semi-annual basis to evaluate "estimated costs-to-complete" is a plus. Increasing this procedure to quarterly or monthly would be encouraged.

Table 18.3 Schedule of Contracts Completed for PCC, Year Ended September 30, 1993

Contract (1)	Revenues earned ($) (2)	Cost of revenues ($) (3)	Gross profit ($) (4)	Mark-up (%) (5)
RC-133	1,885,982	1,773,267	112,715	6.36
RC-1395	486,570	445,910	40,660	9.12
RC-141	515,197	484,841	30,356	6.26
RC-142	550,624	522,362	28,262	5.41
RC-143	724,458	657,969	66,489	10.11
RC-145	664,055	564,019	100,036	17.74
RC-147	946,751	860,890	85,861	9.97
Small contracts	49,959	45,312	4,647	10.26
Totals	5,823,596	5,354,570	469,026	8.76

Table 18.4 Schedule of Contracts Completed for PCC, Year Ended September 30, 1994

Contract (1)	Revenues earned ($) (2)	Cost of revenues ($) (3)	Gross profit ($) (4)	Mark-up (%) (5)
RC-137	9,508,048	9,283,885	224,163	2.41
RC-138	6,589,936	6,566,791	23,145	0.35
RC-146	1,086,513	996,931	89,582	8.99
RC-148	4,182,233	3,733,882	448,351	12.01
RC-149	4,387,907	4,277,262	110,645	2.59
RC-155	87,630	82,887	4,743	5.72
Small contracts	111,245	100,910	10,335	10.24
Totals	25,953,512	25,042,548	910,964	3.64

On the down side, in spite of such in-house procedures, the team is again unsure how other financial decisions such as assigning profit margins to projects is handled. The discrepancy noted above in mark-ups may reflect market pressures or may be due to some other factor. Upon further investigation, as was first indicated in the financial ratio analysis, overbilling/underbilling values have been identified leading to a question of efficient billing to owners. How efficiently and promptly does PCC bill owners for completed work? What is the billing process? Again, more information is required to render a decision.

The $250,000 line of unsecured credit from Anytown Bank is a strong testament to its confidence in PCC. The review team was curious, however, to learn the exact terms and conditions upon which this authorization is based. At

Table 18.5 Schedule of Contracts-in-Progress For PCC; Year Ended September 30, 1993

Contract	RC-137	RC-138	RC-146	RC-148	RC-149	Totals
Estimated Total Contract						
Revenues ($)	9,508,048	6,616,122	1,083,771	4,117,869	4,320,136	25,645,946
Gross profit ($)	215,741	23,431	94,857	446,816	195,599	976,444
Expected mark-up (%)	2.32	0.36	9.59	12.17	4.74	3.96
Year ended September 30, 1993						
Revenues earned ($)	4,451,909	3,878,419	988,920	1,723,068	2,232,124	13,274,440
Gross profit ($)	713	(114,408)	86,555	186,964	101,062	260,886
1993 mark-up (%)	0.02	(2.87)	9.59	12.17	4.74	2.00
From inception to September 30, 1993						
Revenue ($)	9,495,611	6,505,011	988,920	1,723,068	2,232,124	20,944,734
Gross profit ($)	215,456	23,039	86,555	186,964	101,062	613,076
Mark-up (%)	2.32	0.36	9.59	12.17	4.74	3.02

Table 18.6 Schedule of Contracts-in-Progress For PCC; Year Ended September 30, 1994

Contract (1)	RC-150 (2)	RC-151 (3)	RC-152 (4)	RC-153 (5)	RC-154 (6)	Totals (7)
Estimated total contract						
Revenues ($)	403,709	4,392,665	5,367,520	5,170,647	4,306,461	19,641,002
Gross profit ($)	61,013	183,650	244,218	222,053	158,993	869,927
Expected mark-up (%)	17.80	4.36	4.77	4.49	3.83	4.63
Year ended September 30, 1994						
Revenues earned ($)	24,868	3,282,037	4,128,321	3,331,841	3,124,058	13,891,125
Gross profit ($)	3,758	137,541	188,229	143,574	115,647	588,749
1994 mark-up (%)	17.80	4.37	4.78	4.50	3.84	4.43
From inception to September 30, 1994						
Revenue ($)	24,868	3,282,037	4,128,321	3,331,841	3,124,058	13,891,125
Gross profit ($)	3,758	137,541	188,229	143,574	115,647	588,749
Mark-up (%)	17.80	4.37	4.78	4.50	3.84	4.43

what rate of interest is this loan, and what happens if at the end of the payment period, presumably 12 months, PCC is unable to meet the principal or the interest?

Questions raised and information requested in order to complete the review of capital include:

1. Underbillings in 1994, liabilities to sales ratios, and cost of sales were high or too high.
2. Although it seems to have been a wise decision to close the Lexington branch office, more information is needed regarding why it was opened in the first place and why business there is unsuccessful.
3. Mark-ups ranged from 0.35% to 17.74% on completed projects. The team needs more information as to how estimates are prepared in general. For jobs earning below the administrative mark-up of 3% (one-half of the bonded jobs in 1994), how were they specifically prepared?
4. There is some concern as to the distribution of income, especially considering recent payments and the desired expansion. The review team requires information about the succession of the company and the exact arrangement whereby the officers have been compensated.
5. Collection periods improved slightly over the two years of financial information provided, but information from other years is necessary to determine if any serious problems exist.
6. Concerns with overbilling and underbilling have been raised, and coupled with the varying profit margins and mechanisms for determining such mark-ups, the team requests more detailed information on billing procedures and at least three more years of financial information. In addition, as determined in the capacity analysis, a detailed explanation of how bids are prepared and monitored is also necessary.
7. How did Anytown Bank arrive at the $250,000 unsecured credit limit granted PCC and what are the specific terms of this loan?

Continuity Analysis of PCC

Continuity is the measure of a company's ability to continue operations in the event of the loss of key personnel. A planned succession is vital in an industry such as construction where a select few individuals control the workings of the operation and any personnel loss will inevitably affect the contractor's ability to complete on-going projects as well as secure future projects.

PCC did not provide the Action review team with any information as to the transition of ownership nor leadership, as indicated above in the capacity and capital reviews. This multiple determination of this problem is important as it shows that the four analyses overlap in many ways, and troublesome information revealed in one category will most likely appear elsewhere. Transition of ownership is more than just a plan for the future. A company's particular proposal, or lack thereof, speaks to its ability to anticipate, cope, and prepare for eventualities in the construction industry.

PCC's seeming lack of a formalized plan is a concern to Action as there is only a $100,000 life insurance policy on A. Prometheus with PCC as beneficiary. No information regarding the beneficiary(ies) of personal assets was disclosed either, leading to the assumption that Mrs. E. Prometheus will receive most or all of any personal funds, provided A. Prometheus is not intestate. This presents a problem: Mrs. Prometheus will not indemnify herself, meaning that in the case of A. Prometheus's incapacitation and the eventual default of PCC, Action might be left without indemnity and the possibility for increased exposure. The possibility also exists for the transfer of assets into her name in the event of insolvency. The lack of clear information on these necessary items raises some possible concerns.

Another concern relates to the vice president's age and education or training, as mentioned above in the character and capacity sections. Since he is only 25 years old, does he have the experience and/or management training to carry on the business? The team needs to know more information about his training and education before deciding any further action.

Questions raised and information requested by the team in order to complete the review of continuity include:

1. Detailed succession chart with explanation of key personnel roles
2. Detailed information regarding the beneficiaries of personal assets
3. Why won't Mrs. E. Prometheus indemnify herself?
4. What training and other qualifications does A. Prometheus, Jr., posses?

CONCLUSIONS AND DECISIONS

Based on its review of the information provided initially by PCC, J. Hermes, and Hercules and Smith, Action needs to thoroughly investigate several important matters. Concerns regarding the four Cs—character, capacity, capital and continuity—have also been identified as have the means to acquire the necessary information to make an informed decision. The two years of financial information are deemed insufficient to establish with any comfort the general state of PCC's finances, especially considering that a large loss seems to have been incurred in one of the two fiscal years presented (1993). Action has had no trouble with PCC in the past and this is significant, but they would like to review financial information from 1990, 1991, and 1992.

The following are the salient concerns facing the Action underwriting review team:

1. Will PCC be able to continue its operations in the absence of the owner, A. Prometheus?
2. Why were profit margins so erratic and for the most part low?
3. What is the organizational structure of the company and how has it been used?
4. Can newly implemented accounting practices improve collection periods and reduce the probability of claim to the surety?

5. Will A. Prometheus continue to make withdrawals from retained earning?
6. Is there a low-cost solution to the legal claims currently open and pending?
7. Is the risk associated with a 54% increase in the single job bonding limit in the best interest of both Action and PCC?

To make a positive recommendation to increase PCC's line of credit to a normal single-job limit of $10 million and an aggregate work program of $20 million with the occasional $15 million project, answers to the questions raised in the preceding analysis are necessary. The list is divided into two sections: 1) the information needed from PCC and 2) the information Action needs to acquire elsewhere before arranging a meeting with A. Prometheus and the officers of PCC.

1) From PCC:

1. Information on all liens, suits, and claims from the contractor's point of view including an explanation of why Action was not informed of their existence on the CIR.
2. Provide an explanation of the increase in professional fees from 1993 to 1994.
3. Provide an organizational chart showing the structure and experience levels/training of employees from president through superintendents.
4. Provide both a short and long-term business plan.
5. Respond to the over/underbilling problems and long collection periods.
6. Supply a basis for job mark-ups and explain the existing wide range.
7. Explain why the Lexington branch office was initially opened and recently closed.
8. Present the planned succession of ownership and leadership in the case of incapacitation or death of key personnel.
9. Provide additional financial statements for the fiscal years 1990, 1991, 1992.
10. Supply a copy of all current union trade agreements.

2) From Other Sources:

1. Collect information on open and pending liens, suits, and claims from the perspectives of the involved accountants, lawyers, and if possible, plaintiffs.
2. Contact architects, engineers, subcontractors, suppliers, owners, and bankers to obtain an assessment of character and capacity.
3. Contact Anytown Bank to discuss further the $250,000 unsecured line of credit.

With a strong performance record, an unsecured credit line from Anytown Bank, and outstanding recommendation from J. Hermes, the underwriting review team was quick to pursue its questions. At the end of the in-house review, the Action underwriting team requested and received the necessary information from PCC, J. Hermes, Hercules & Smith, Anytown Bank, and from other key personnel including several of the litigants who had indeed filed liens on PCC projects.

Action had a successful site visit to PCC's Louisville office arranged by J. Hermes. There they found out more about A. Prometheus, Jr.'s experience and plans to take over the family business and how bids are prepared. After learning that PCC had settled or was in the process of settling all of its liens (it turns out that on one of the underbilled projects, several subcontractors were unpaid and were slow in hearing about the owner's difficulties with money), Action performed a new trend analysis with five years of data and found that Action could indeed handle an increased line of credit. They promptly rendered a favorable decision and within a few weeks, PCC successfully bid on its first $10 million job bonded by Action.

LESSONS

This review demonstrates several things:

1. ***Problems are not often isolated; they are manifest in several guises.*** In the above example, the need for an organized succession plan, a detailed hierarchy, and a thorough explanation of bidding procedures came up time and again in the analysis. These areas of concern were located indirectly in some cases through comparisons and ratios, and directly through the character and capacity analyses.
2. ***Incomplete or incorrect information will be eventually corrected and/or discovered.*** The conflicting information as to the nature of liens and litigation against PCC was found rather easily in the D&B report and the CIR. Since PCC offered the information to D&B, it seemed likely that the omission on the CIR was unintentional, as turned out to be the case. Still, it was a discrepancy and had to be dealt with.
3. ***The capital analysis is by far the most detailed and rigorous.*** In today's market, financial solvency is extremely important. The surety thus spends a great deal of its time reviewing the contractor's finances, meaning that up-to-date, complete, and thorough records are vital to securing bonding. Solid accounting methods are not only beneficial, for the bonded contractor they are a must.

Part V

SURETY BOND CLAIMS

In construction, defaults are inevitable. Even with careful planning, a solid managerial structure and experienced workers, weather conditions, the economy, and accidents can cause even a good contractor to fail. When default occurs on a bonded contract, a claim ensues whereby the surety fulfills its obligation to the obligee. A claim may be brought against the surety by the obligee or other parties that are protected under the bid, performance, and payment bonds that the surety has issued. When the surety receives a claim notification it has an obligation to investigate the matter and seek to determine its liability and that of the contractor as soon as possible.

Indeed, the first thing surety does upon notification of a default is to read and understand the bond, contract, and any underlying agreements. Since after establishing bonding limits and a work program it is usually up to the independent agent to authorize particular bids for particular projects, the surety is often unaware of what projects a contractor is working on. Immediately upon being notified of default, the surety wants to familiarize itself with the following:

1. What was bonded?
2. Who are the parties of the agreement, bond, and contract?
3. What are the terms and conditions of the bond and agreements?

The surety will immediately want to determine the facts, issues, and options available to all interested parties. This can only be accomplished through a thorough investigation that usually entails an in-house review of all documents and a visit to the contractor's office(s) and job-site(s). What follows is a comprehensive examination of how the claim process works with numerous examples and two extended case-studies; one involving the fictional firm of PCC, Inc., and another involving the fictional multi-million dollar contractor Zimbrick International.

Chapter 19

CONSIDERATIONS IN HANDLING SURETY BOND CLAIMS

Since the contractor is responsible for having read the contract documents carefully and having visited the job site as appropriate, bid bond claims are uncommon. Thus, it is unlikely that the contractor will not want to enter into the contract if the owner accepts the bid. However, in the event of a bid default in which the contractor refuses to enter into the contract or fails to provide the required performance and payment bonds, the obligee may bring a claim against the surety. Usually, the terms of the bid bond provide that the surety will pay the obligee for the difference to the next higher bid or the penal amount of the bid bond, whichever is less. There have been instances where the principal is excused and the surety released from its liability under the bid bond. Usually arithmetic mistake or gross error in estimating (e.g., forgetting to include the cost of an HVAC system) in the contractor's bid, if supported, has occasioned such release.

When claims are made against a payment bond, the surety must investigate to determine whether the claimant is in fact protected by the bond, verify the amount, and confirm that the claim is for services or materials that were supplied to the bonded project. If everything is in order, the surety discharges its liability by paying the claimant the proper amount.

By comparison, a performance bond claim can be much more complex in nature. Before determining which course of action to follow, the surety must investigate the overall situation as quickly as possible. Below is a description of the process and a review of the sureties' rights, remedies, and defenses.

SURETY'S RIGHTS, REMEDIES, AND DEFENSES

When the surety is notified of a declared default or the formal termination of a contractor's rights to complete a contract, a surety must first determine whether or not the situation provides any defenses that justify a complete or pro tanto discharge. Such defenses include: 1) invalid bond, 2) technical defenses such as late notice, 3) material alteration of contract including gross overpayment or change in scope of the contract without notice to surety (changes resulting in a contract different from what the surety originally bonded, i.e., materially changed the nature or *quantity of work* so that original contract ceases to exist), and 4) any other acts committed that may have benefited the principal but compromised the surety, and thus derogated the contract.

If the surety has no defenses with which to discharge its obligations or if it has some defenses that may aid in obtaining concessions from the obligee in the negotiation process, the surety proceeds to identify the course of action to terminate as much liability as possible. Generally, overpayment of the contract results only in a pro tanto discharge to the extent of the overpayment. If an owner or its representative has allowed front-end loading, the result may be gross overpayment by the owner. It is important that the owner carefully analyze work-in-place and progress payments to avoid overpayment or risk losing whatever monies are in excess of a contractor's progress in the case that the contractor fails.

For example, an obligee terminated the contractor and demanded the surety complete the final 60% of the contract. The contractor had already been paid 80% of contract funds, and thus only 20% of the project proceeds were available to complete 60% of the project. Because owners or their representatives (architects and engineers) must abide by the terms of their contract and monitor percent of project completion, the court held the owner accountable for the amount of overpayment to the contractor (i.e., 40%) in addition to discharging the surety's performance bond compensatory to the overpayment amount.

The surety's rights to undispersed contract funds arise from two sources: 1) the contractor's indemnity agreement and 2) common law that states that as guarantor of the contractor's performance and of payment to labor and material suppliers, the surety is entitled to these undispersed funds. Thus, the courts have upheld the surety's equitable rights to those funds in such cases as Pearlman v. Reliance Insurance Co., 371 US 132 (1962). These funds are for the payment of labor, equipment, and material costs incurred as a result of the construction process. This means that these funds *are not available* for compensating other creditors such as banks.

Depending on whether the owner is a public body and the state in which work is completed, there are federal, state and local regulations for the time period an obligee has to notify the surety of its obligation. The federal government has no limitations, but many private owners must file within a certain amount of time after the completion of construction. An obligee should be familiar with all applicable statutes, which vary from state to state. If an obligee fails to make a claim within the period prescribed by local law, the action is called a "late notice" and is usually sufficient grounds to terminate the surety's obligations. This is primarily because of one of the fundamental principles of suretyship: the owner has an obligation to mitigate in any loss. The owner must monitor the progress of a contractor and cannot expect a blank check from the surety if, not having done so, the contractor fails.

A surety's protection against losses due to paying the principal's (contractor's) obligations takes the form of the general indemnity agreement (GIA) signed by the owner(s) of the contractor firm and any other signatories, usually including their spouse(s). For a sample agreement see Appendix I. The GIA is the foundation of the relationship between the surety and contractor and it specifically outlines each party's respective rights. In addition to indemnification, the surety typically has rights to demand collateral such as the assignment of indemnitor's property, to become involved in completion of contracts, to be furnished information from the contractor, and to demand that they be given power of attorney for the contractor. As well, the GIA provides several tools to the surety when a contractor defaults.

In theory, the purpose of the GIA is to protect the surety from financial loss stemming from contractor default. In effect, the construction company owners are requested to pledge their personal assets as well as the corporate assets as sources of funds to pay loss or exposure incurred by the surety. In some instances, the principal may fail to have the necessary financial resources to prevent the surety from paying a loss. The agreement does, however, provide the right to obtain the cooperation of the principal in mitigating any loss by providing the surety with the necessary information such as job cost and the cooperation of the principal in retaining key personnel to continue work on uncompleted projects should a default occur.

FIVE STEPS OF SURETY PARTICIPATION IN CONTRACTOR DEFAULT

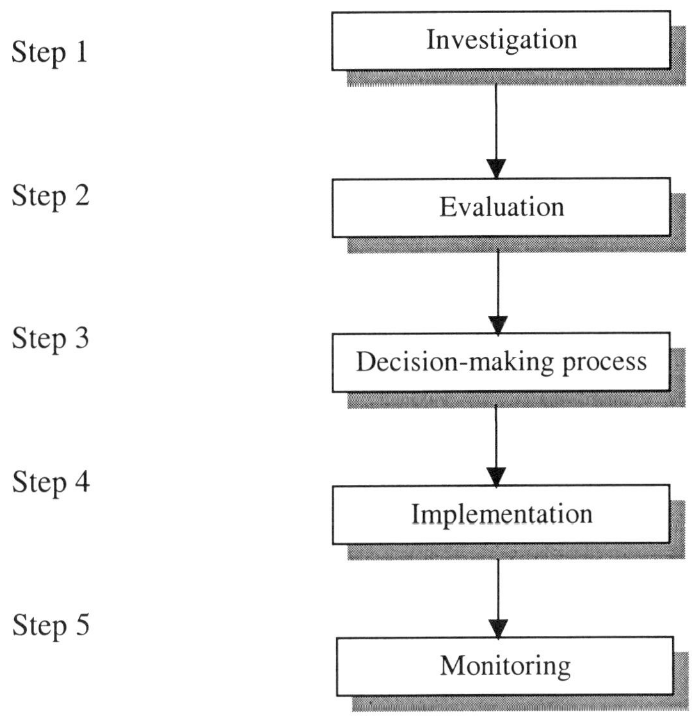

Figure 19-1 Five Steps Used by Sureties to Discharge Liability in a Contractor Default Situation

In the event of a performance bond claim by an owner alleging contractor default, the surety generally expends considerable effort gathering relevant facts and assessing its liability. The engagement of this effort is to identify the course of action to determine and minimize the surety's exposure. Figure 19-1 shows the effort consists of five steps:

1. *Investigation*—gather all necessary information to determine exposure,
2. *Evaluation*—analyze and weigh optional methods of conveying any performance bond obligations,
3. *Decision-making*—select option or options to complete all bond obligations and limit surety exposure,

4. *Implementation*—implement and set in place a course of action to follow, with appropriate out clauses should a better option for the surety present itself,
5. *Monitoring*—observe progress with option selected.

An in-depth discussion of each step follows.

Step 1. Investigation

The surety investigation consists of two parts: 1) in-house underwriting review and 2) on-site investigation. The necessary information needed to begin the investigation should be readily available within the surety's office and is where the investigation should commence. The in-house underwriting investigation involves thoroughly reviewing the contractor's underwriting file. It is a custom and practice for this file to include detailed credit reports and specific items such as prior bonds written, uncompleted contracts, subcontracts, and other contract documents that may include plans, specifications, and general conditions, and periodic job status reports received by the surety from the owner/obligee. The status inquiry report usually notes if the performance is satisfactory, whether any bills remain unpaid, and the adjusted contract amount. In addition, it includes the general indemnity agreement, bond application, corporate and personal financial statements, as well as the financial status of completed and uncompleted work, along with any and all correspondence.

The GIA should be with a Uniform Commercial (Code-1) Financing Statement. This all-encompassing business code is by nature complicated and detailed. In the case of a default, a UCC financing statement is usually filed by the contractor's creditors such as banks holding the title of contractor equipment, but as the exact specifications vary from state to state, it is best to consult and review the particular requirements where the indemnitor and project reside.

After the in-house review, if not simultaneously, the surety usually obtains more complete, timely, accurate, and useful information through its field investigation. The field investigation is conducted by an experienced representative of the surety and consists of visiting the contractor's office and job sites. During the visit to the contractor's office, an assessment as to the financial condition of the contractor's total work program (i.e., bonded and non-bonded jobs) is obtained. Data relevant to such an assessment includes percent of construction complete for each project, detailed accounting, job correspondence, pending and approved change orders and time extensions, purchase orders, job site survey, claims for extras, and time extensions for items such as weather and changed field conditions, as well as a review of all subcontractor files, whether bonded or not.

The surety representatives interview key personnel with the contractor's organization in addition to checking general and administrative overhead expenses and office systems such as accounting, scheduling, estimating, cost control, payroll, and payroll taxes. Depending upon the circumstances of the default, the surety may retain competent surety consultants, including outside legal counsel, engineers/contractors, construction consultants, and accountants to determine the surety's exposure by evaluating the potential loss in all its ramifications. Photographing property, equipment, and job-site(s) can be very helpful in the investigation.

If during any time in the investigation the contractor or any records become unavailable to the surety investigators for any reason, such as the contractor's disappearance or missing or incomplete records, it becomes especially important for the obligee to cooperate with the surety. In such a situation, the owner cannot expect the surety to make payment without having first thoroughly investigated the claim. Furthermore, it is in the best interest of the owner to do so because any information they provide the surety will help expedite the claim process and ultimately the settlement. Goodwill between the two parties will prove valuable both in completion of the structure and settlement.

The most significant parts of the field investigation are 1) accounting investigation and 2) job-site survey. The accounting investigation invariably includes a spreadsheet containing all accounts payable, accounts receivables, and corporate assets. *This spreadsheet should include both bonded and non-bonded jobs.* For each separate job, the claim personnel should determine the outstanding debt for each supplier or subcontractor and any related correspondence. Specific items would include outstanding material suppliers and specialty contractors both for determination of debts due them but also original requirement on contract amount completed and amount still to be done, both dollar-wise and physically. In addition, equipment leases, other rental equipment, miscellaneous tools, suppliers, equipment, fuel, equipment maintenance, payroll, union benefits, worker's compensation, insurance premiums, other fringes, and payroll taxes need to be determined and taken into account. The surety's investigators must also research outstanding loans due banks and/or financial institutions, or from officers in the organization. In addition, investigators must determine each individual and the total of all real estate mortgages, purchase orders, equipment, mortgages, and lease purchases. Noted also would be secured and unsecured working capital loans.

The tax liability of the contractor is another major and important fact. It is particularly important if the Internal Revenue Service (IRS) files a tax lien on the contractor's property, called alternatively a "set-off" or an "off-set." The timing of liens is extremely important and may be a factor in priority of lien. The IRS may assert its claims against property by taking such assignment of federal, state, and local tax withholdings, as well as any Federal Insurance Contributions Act (FICA) taxes that may be owed. If such taxes are due and if the Federal Government is obligee, it may consider initiating a set-off lien against the bonded jobs of the delinquent contractor. If a set-off is initiated, the surety must asses the contractor's inventory, assets, and materials on the job, determining total value and who has rights to reclaim what. In addition, the surety must determine the extent and period covered by insurance, especially worker's compensation, builder's risk, general liability/umbrella, auto, equipment floaters, life, health, and pension plans. However, set-offs are in general uncommon.

On each project, the surety's investigator(s) must review the receivables and their respective ages. Specific items of interest are earned but unpaid progress estimates (when they are due and why they have not been paid); materials stored on job site; earned retention; unearned contract balances; change orders (approved and unapproved, and completed and uncompleted); outstanding loans to and from construction company's officers; miscellaneous accounts receivables; and insurance claims. Regarding loans the officers may have made to their company, the surety should require the officers to subordinate their rights to the surety. As described above, the Federal Government as obligee may claim the right to set-off against contract funds on the projects for which taxes are owed, depending on the timing of the liens, if any. However, the surety has a prior right to these funds in the event they sustain a performance bond loss, but if the government is the owner/obligee, this may require considerable litigation.

Of interest to the surety are the available corporate and personal assets available for offsetting losses the surety incurs as a result of the contractor's default. Assets may take the following forms: cash-in-banks, cash-on-hand, certified bid checks, accounts receivable including completed contracts, earned estimates, retained percentages, labor and materials (not yet billed), materials and supplies on-hand (inventory), cash surrender value of life insurance, marketable securities, other accounts receivable, notes receivable, other securities (non-marketable), prepaid expenses, land and buildings, other real estate, equipment, furniture, and other assets.

Maintaining Status Quo

Normally the surety seeks to maintain job momentum and continuity during their investigation. A timely infusion of indemnitor or surety money may enable the principal to complete the work, avoiding a shutdown of the job and a declared default. What frequently occurs during the surety's investigation, however, is that the contractor cannot meet its payroll. When this happens, subcontractors may abandon the job if they have not received payment, if forthcoming payment seems unlikely, or if work to be done exceeds payments to be made. Major suppliers will demand cash on delivery (COD) for critical supplies and materials. In such emergencies the surety may choose to advance the contractor funds, and this is called "look-see" money. Subsequently, the surety may choose to assume control of the contract funds either formally or informally. *The contractor should clearly understand that such courses of action are stop gap measures and in no way commit the surety to a course of action to which the surety has not committed itself.* Sureties are often reluctant to forward any more money than is absolutely necessary, as any dispersed funds may impinge upon their relationships with reinsurers. In such situations, the surety usually requests a letter from the contractor stating that the contractor cannot meet its obligations and requests that the surety provide financial assistance. This helps to avoid characterization of advances by the surety as "voluntary" contributions not within the bond penalty balance.

Step 2. Evaluation

The surety can choose from several options in determining and handling their exposure, including: 1) financing the contractor, 2) reletting the contract, 3) completing surety—take-over and relet, 4) obtaining bids and arranging a contract between the owner/obligee and the completing contractor, 5) buying the bond back, and 6) doing nothing. The surety will make every attempt to select the option that best mitigates its loss, but once it has selected an option, the surety may find itself committed to this course of action even if the result is a loss greater than the penal amount of the bond. Thus, a surety will complete a thorough investigation before committing itself to any particular course of action and review the general advantages and disadvantages of each specific option. Table 19.1 presents the six choices a surety has on notification of default. The

Table 19.1 Advantages and Disadvantages of Alternatives to Complete a Defaulted Project
Source: Adapted from Cushman and Stamm 1984

Alternative to complete project (1)	Advantages to surety (2)	Disadvantages to surety (3)
Finance defaulted contractor	No profit factor for remaining work No mobilization costs No loss of job momentum No learning curve Claims are preserved Reduced consideration necessary for salvage Need collateral from third parties	No exoneration of bond penalty Increased monitoring costs Inability to fix losses Increased overhead The "10 percent" completion problem Increased taxes
Secure another contractor/relet contract	Provides a clean break with problem Losses/Exposure become fixed early Eliminates danger of exceeding penal limit of bond	May be more costly Procedure demands fixed price approach Can be time consuming Not desirable when project is near completion No preservation of claims
Completion by surety/take-over and relet	Greater control over the completion process	Irrevocable commitment required by surety
Completion by owner	Surety avoid financing project to completion Surety has more time for investigation Surety avoids in-house administration costs	Opens up surety to suit for breach of its bond obligations Surety can be blacklisted by owners and architects/engineers as unacceptable for future Loss of control over the completion of the work
Buy bond back	Releases surety from any potential warranty problems Provides a clean break with the problem Exposure immediately fixed, at or below the penal limit of bond	May prove unnecessarily costly
Do nothing	Problem may resolve itself without unnecessary action	Opens surety up for breach of its bond obligation May alienate the obligee and/or owner

surety must evaluate all options prior to arriving at a decision, and the following sets out each option.

It should be noted that the surety may negotiate with the obligee at any time, serving to make the claim process a highly complex endeavor. The surety will do everything within its power to mitigate its loss, while the obligee will do everything within its power to recoup whatever loss it has or will incur. If a deal can be struck, chances are that it will, as all parties want the same result: a satisfactory resolution to the problem. This subjectivity thus prevents a systematic outlining of how such a process plays out "in the real world." As one long-time surety veteran says, "There is no template." Certain protocol and options available to each party are covered under law and thus lend themselves readily to explanation and interpretation, and wherever possible, common deviations from the normative are explored. However, the reader should remember at all times that an actual default will play out like a good jazz number—as a variation on a theme.

Financing the contractor

The surety exercises caution when considering whether to finance a defaulted contractor. Reinsurers typically require sureties to consult them before deciding whether to finance a defaulted contractor. The surety carefully selects which defaulting contractors to finance because the success ratio must be high. Table 19.2 lists the primary advantages and disadvantages a surety might consider when determining whether a defaulted contractor may be a good candidate for financing. One useful rule of thumb is that when 80% or more of the contractor's work program is unbonded, the surety should *not* provide financing.

If a surety decides to finance a defaulted contractor, the surety must consider numerous issues in order to first ensure a successful completion. The issues include: 1) types of financing, 2) tools for financing, and 3) special requirements. Below is a discussion of each issue.

Types of Financing

To fulfill a financing arrangement, the surety usually either provides funds directly or guarantees a bank loan to the contractor. The loan guaranty may be a fixed sum or a revolving line of credit. The following factors may be considered by the surety when deciding whether to use the loan-guarantee method in providing funding:

1. The surety's involvement may not be readily apparent, allowing them to present a facade of "business as usual" with the bank and presumably with any reinsurers as the third parties aware of surety involvement. Without this facade, businesses may try to take advantage of a financially weak principal. On the other hand, it may be more advantageous to have owners and creditors aware of the surety's involvement, as this may make them more flexible in any compensation they are to receive. Additionally, the surety's representing its contemplated course of action to the owner may prevent a default, assuming one hasn't already occurred.

Table 19.2 Advantages and Disadvantages in Financing Defaulted Contractor

Advantages (1)	Disadvantages (2)
1. Sufficient equity in assets with which the surety may be collateralized or "salvage made easier." Most contractors would prefer to complete their work and avoid their collateral in return for financing and generally will agree to liquidate their assets as required to pay off their loan.	1. Monies expended in financing do not generally reduce surety's liability under performance, exposing the surety to the risk of exceeding the bond penalty.
2. A timely infusion of money enabling the contractor to complete the work, avoiding a shut down of the jobs and the resultant disruption of all subcontracts and purchase orders that are incidents of default. This means that job momentum and continuity can be preserved. Thus, there can be a substantial savings in time, and time translates into money.	2. Payment of non-bonded obligations, which includes overhead and capital expenditures. *As a practical matter, it is not possible to exclude non-bonded obligations.*
3. Provides the surety the potential opportunity to underwrite additional bonds in the future. Add work to be completed within existing work program. Improves cash flow and helps meet company and project overhead. Supports idea that the contractor can survive re-establishing both creditability and financial stability and eventually reverting to a normal underwriting relationship. Risky and diverts surety's attention from number one priority of terminating the performance and payment bond liabilities by concerning themselves with determining the profitability of this new bond.	3. Uncertainty as to amount of loss. This uncertainty can be reduced by a thorough job analysis and cash flow projection made during investigation. There is, however, the possibility of exceeding these projections.
4. Completion by a new contractor may be considerably more costly because of remedial work and mobilization hence, avoiding increased cost of completion by another contractor.	4. Where bonded work is substantially uncompleted or not yet commenced. No defective work minimizing cost to complete. Easier to get subcontractors to complete for their original price. If completing contractors can be found among original bidders of recent origin and their knowledge of contract requirements can save time and money. Tail-end let downs and contractor prolonging surety financing considerably increasing loss. May be especially true with contractor's employees. Time strain on claim department staff. Too many un-bonded projects making this alternative uneconomical.
5. If bonded work is substantially completed, delays involved in reletting plus liquidated damages and increased costs may be so substantial that it may be advantageous to finance.	
6. Too many bonded jobs making it impossible to relet projects due to logistical and economic considerations. If the total work program has 80% or greater of the projects bonded, financing the contractor may be a cost effective alternative.	

2. The surety may be reluctant to advance funds when assets are available from indemnitors that can be liquidated in an orderly fashion. Through GIAs signed by contractors and their families, all assets are guaranteed the surety and may be used to continue contractor operations.
3. Instead of immediately asking an indemnitor to liquefy holdings, a bank loan may be preferable if it appears the contractor can survive and can pay the going interest rates the bank loan may require. If the contractor does not seem likely to survive the interest charges and the surety is likely to have a loss, it may be to the best interest of the surety and of all parties involved to finance directly, thus saving the interest cost. This would also be an incentive to eschew financing at any odds.4.

   If the Federal Government is obligee, an assignment to a bank may be necessary as the Assignment of Claims Acts prohibits set-offs against assignee banks. In effect, the assignment to a bank forestalls a set-off lien by the Federal Government. Current law states that on projects where the federal government is the obligee, it may set-off against a surety sustaining loss under its payment bond but not against a surety sustaining a loss under its performance bond. The exception may apply to a performance bond to the extent of earned but unpaid progress estimates.
5. An assignment directly to a surety is not legally enforceable against the federal government, and therefore, the assignee must be a financial institution.

Tools of Financing

The tools available to help make the financing arrangement succeed are: 1) loan and security agreement, 2) acknowledgment of default, 3) voluntary letters of default, 4) letters of direction, 5) joint-control agreement (also referred to as a trust agreement), 6) agreement with depository bank if not a party to joint control agreement, and 7) cash flow projection.

In an agreement under which the surety has guaranteed a loan or advanced a specific sum of money without electing to implement joint control, the surety must reserve the right at any time to direct payment of the loan proceeds. Figure 19-2 shows the essential elements that this agreement should include.

Each obligee would receive a particular form combined with specific instructions as to the assignment of contract funds in the event of a breach of the financing agreement by the principal, or if the surety elects to terminate financing. The surety may need to secure such letters for the future and would thus maintain these letters on file.

The essential provisions of a joint control agreement between the surety and the contractor are as follows:

1. A schedule of all bonds and contracts for which payments are to be made out of the joint control account including the qualifying language such that if an unlisted bond or contract comes to light later, the agreement applies to that job as well and will be added.

Acknowledgment of default.

Request for financial assistance.

Agreement to furnish collateral security described in an attached schedule and unincorporated into the agreement.

Specify the use of the loan proceeds for payment of labor and materials furnished in the prosecution on contract described in an attached schedule and other expenses and overhead necessary in the contractors operations.

Surety may direct payment or contractor shall notify surety on a regular basis of the application of the loan proceeds.

Specifically refer to type of payments, i.e. labor, materials, withholding taxes union fringes.

Surety's access to books and records and demands for accounting.

No waiver of surety's rights under bonds, contracts, and indemnity agreement.

Surety's right to undisbursed funds in the event of a breach.

Signature of all indemnitors.

Surety should have the right to make last withdrawal from account and close account.

Figure 19-2 Essential Elements of a Loan and Security Agreement

2. The agreement specifies that the contractor has directed all owners to remit to the surety all contract funds now due or to become due and such monies to be deposited into the noted account.
3. The agreement grants Power of Attorney to surety to endorse checks on behalf of the contractor.
4. The agreement provides for dual signatures: one representative of the contractor and one of the surety. The agreement should also contain language allowing surety to make withdrawal or payment upon the signature of a representative of the surety alone should no representative of the contractor sign the

withdrawal. Checks that draw out all funds in the account coming under joint control can be signed by the surety alone.

5. All deposits should have trust fund restrictions to be exclusively for labor and material bills or contracts listed within a schedule and on only those for which the surety may have liability under bonds and other expenses necessary to complete the contracts. Furthermore, all monies deposited shall be held to indemnify surety from all loss, cost, and expense, in connection with having executed bonds.
6. Trust fund theories may protect against attacks by assignee banks and bankruptcy trustees upon funds on deposit. The courts will usually designate a properly established and operated joint-control account to be a trust account and, as such, not subject to the claims of third parties. The surety's unqualified right to withdraw and the payment of non-bonded obligations or non-job obligations may erode the trust fund concept.
7. The agreement should expressly reserve right of surety to cease financing at any time at its discretion. *In short, the agreement should clearly communicate to the contractor that the surety is not making an irrevocable commitment to finance.*

The courts have treated a properly established and operated joint account as a trust fund, not subject to attack by third parties. Usually three accounts are set up: 1) joint control, 2) payroll, and 3) petty cash. In each of these accounts, the surety should maintain the minimum amount practically feasible.

An agreement with the depository bank if not a party to joint control agreement is necessary. This is to ensure the following: 1) the account is properly credited, 2) there is no right of set-off, 3) no other accounts in the name of the contractor should be allowed except a payroll and petty cash account with approval by the surety, and 4) the bank acknowledges the surety's unqualified right to withdraw all funds or deposit without the consent of the contractor. The contractor should never be permitted to deposit contract funds into his or her personal account.

Carefully drawn cash flow projections are the surety's best tool for determining the cash requirements on a monthly basis. To ascertain the cash requirements, the surety analyzes the following for each job: 1) contract amount, 2) contract balance based on projected monthly project completions, 3) account payables, 4) subcontractors, 5) cost-to-complete the contract, 6) overhead, and 7) liquidated damages. All of these items are considered in the development of a monthly cash flow.

Consider the following sample analysis of the information collected and analyzed to project the cash flow of a fictional defaulted contractor. Table 19.3 presents the results of the first step—compiling the background information related to the contractor's bonded jobs. This information assists in the preparation of a contract analysis sheet, a sample of which is presented in Table 19.4. This analysis results in a net surplus or deficit for each bonded project. Table 19.4 contains several items necessary to calculate the surplus or deficit. Table 19.5 shows the contract balance summary, and Table 19.6 shows aged accounts payable. Table 19.7 provides the summary of the subcontracts for each project, while Tables 19.8 through 19.12 provide specific subcontractor details for each contract. Table 19.13 presents a summary of the liquidated damages per project.

The surety must project a contractor's expenses and income for each project over time (i.e., cash flow projection). Table 19.14 shows the summary of a projected cash flow. To prepare a statement of cash flow, the surety must total on a monthly basis the following items: general company overhead, notes payable (Table 19.15), accounts payable, liquidated damages (Table 19.13), and direct construction costs (Table 19.16). The surety can then compare the total monthly cost expenditures to the projected income for each project (Table 19.17). This allows for early detection of cost overruns in any area and for swift, appropriate action.

Special Requirements

It is possible that the contractor may have assigned the contract funds to one or more banks prior to the surety's involvement. The surety has prior rights to use of these funds for completion of the project. The surety should meet with the bank(s) immediately, clarifying the surety's jurisdiction over contract funds. This may require negotiation with the bank for written consent allowing, in accordance with the surety and contractor agreement, the surety's use of any surpluses in the surety's bank. The bank may agree to a moratorium on monthly payments for previous loans. If the assignee bank is unreasonable, financing may not be possible. It is inadvisable for a contractor to give a bank authority over or input into the direction of payments.

Of concern to the surety is the potential tax liability that a defaulted contractor represents. Five situations exist in which the surety is responsible for the contractor's failure to withhold taxes or to pay taxes (Mrozek 1978):

1. When surety has executed Miller Act payment and performance bonds.
2. When executing non-Miller Act bonds, when bonded contract expressly obligates principal to pay withholding taxes and bond is conditioned upon performance of all contract provisions.
3. When surety has exercised "exclusive" control over principal's (contractor's) employees and payment of wages to those employees.
4. When surety directly pays wages of its contractor's laborers or, alternatively, indirectly funds the payment of those wages knowing that the contractor will not withhold and pay the necessary payroll taxes. In the latter situation, the surety's responsibility is limited to 25% of the monies advanced.
5. When the surety willfully failed to pay the taxes, even if the surety did not necessarily have exclusive control of payment of the principals bills.

Regarding previous taxes incurred, the timing of liens filed by the IRS is important. Also, timing is a factor on other liens for taxes due IRS (income, etc.). Regarding a federal tax lien, federal statute grants a lien in favor of the U.S. Government upon "all property and rights to property" for delinquent taxes. If a progress payment is a right to property held by the contractor, then the tax lien has

Table 19.3 Background Information on Bonded Projects as of January 31, 1998

Job # (1)	Project name (2)	Location of project (3)	Obligee/ owner (4)	Bond number (5)	Effective date of bond (6)	Bond penalty ($) (7)	Claim number (8)	Estimated completion date (9)	Percent complete (%) (10)
127	Oklahoma County	Oklahoma County, Oklahoma	Oklahoma DOT	57966	1/10/95	7,286,793	451107	Complete	100
132	White River	Marion & Baxter Const., Oklahoma	Arkansas State Highway Commission	61275	2/25/96	7,566,183	451102	6/30/98	75
133	Hall County	Hall County, Texas	State of Texas	62340	6/16/96	3,074,808	451108	Complete	100
134	Hot Springs	Hot Springs, Arkansas	Arkansas State Highway Commission	62682	7/01/96	2,279,438	451103	4/30/98	77
135	Ellis County	Ellis, Texas	State of Texas	62848	8/08/96	3,882,638	451104	9/30/98	54
136	Hempstead County	Hempstead Arkansas	Arkansas State Highway Commission	62871	7/24/96	6,862,729	451105	9/15/98	68
137	Pottawatomie	Shawnee, Oklahoma	Oklahoma DOT	65964	9/14/97	3,064,965	451106	12/31/98	18

Table 19.4 Contract Analysis as of January 31, 1998

Job # (1)	Project name (2)	Contract amount ($) (3)	Contract balance ($) (4)	Aged accounts payables ($) (5)	Amount of sub-contracts ($) (6)	Estimated cost-to-complete ($) (7)	Total direct costs (5)+(6)+(7) ($) (8)	Overhead ($) (9)	Liquidated damages ($) (10)	Total costs (8)+(9)+(10) ($) (11)	Net surplus (deficit) before advances (4)-(11) ($) (12)
127	Oklahoma County	7,278,393	63,878	213,644	—	1,000	214,644	—	—	214,644	(150,765)
132	White River	7,592,561	1,924,588	47,304	303,516	1,891,076	2,241,896	133,272	29,400	2,404,568	(479,980)
133	Hall County	3,050,997	—	107,352	—	5,000	112,352	—	—	112,352	(112,352)
134	Hot Springs	2,279,438	527,080	67,396	18,042	487,694	573,132	34,370	—	607,502	(80,422)
135	Ellis County	3,882,638	1,981,254	104,541	60,230	2,027,735	2,192,506	142,902	—	2,335,408	(354,154)
136	Hempstead County	6,567,865	1,906,949	222,469	184,070	1,665,415	2,071,954	117,368	76,200	2,265,522	(358,573)
137	Pottawatomie	3,064,865	2,549,225	50,477	91,114	2,372,015	2,513,606	167,588	4,620	2,685,814	(136,589)
	Other Nonbonded Payables	—	—	333,739	—	—	333,739	—	—	333,739	(333,739)
Total		33,716,757	8,952,974	1,146,922	656,972	8,449,935	10,253,829	595,500	110,220	10,959,549	(2,006,575)

Table 19.5 Contract Balance as of January 31, 1998

						Balance of contract funds			
Job # (1)	Project name (2)	Original contract amount ($) (3)	Change order amount ($) (4)	Revised contract amount (3)+(4) ($) (5)	Payments received to date ($) (6)	Total (5)-(6) ($) (7)	Earned not paid ($) (8)	Earned retention ($) (9)	Unearned ($) (10)
127	Oklahoma County	7,286,793	(8,400)	7,278,393	7,214,514	63,879	63,879	—	—
132	White River	7,566,183	26,378	7,592,561	5,667,972[1]	1,924,589	—	—	1,924,589
133	Hall County	3,074,808	(23,811)	3,050,997	3,050,998	—	—	—	—
134	Hot Springs	2,279,438	—	2,279,438	1,752,357[1]	527,081	—	—	527,081
135	Ellis County	3,882,639	—	3,882,639	1,901,385	1,981,254	—	100,078	1,881,176
136	Hempstead County	6,862,728	(294,862)	6,567,866	4,660,915	1,906,951	—	—	1,906,951
137	Pottawatomie	3,064,865	—	3,064,865	515,640	2,549,225	—	27,139	2,522,086
Total		34,017,454	(300,695)	33,716,759	24,763,781	8,952,979	63,879	127,217	8,761,883

[1]Payments received subsequent to January 31, 1998: Job #132-White River, $21,526 on 2/2/99; and Job #134-Hot Springs, $41,455 on 2/2/98.

Table 19.6 Aged Accounts Payable as of January 31, 1998

Job number (1)	Project name (2)	Current ($) (3)	31-60 days ($) (4)	61-90 days ($) (5)	Over 90 days ($) (6)	Total (3)+(4)+(5)+(6) ($) (7)
(a) Bonded projects						
127	Oklahoma County	(29,842)	—	190,503	52,984	213,645
132	White River	4,419	146,908	89,184	110,310	350,821
133	Hall County	—	53	23,272	84,028	107,353
134	Hot Springs	3,223	13,799	11,203	57,214	85,439
135	Ellis County	31,697	28,876	69,817	34,382	164,772
136	Hempstead	18,310	114,993	182,936	90,302	406,541
137	Pottawatomie	12,646	53,222	75,725	—	141,593
Subtotal		40,453	357,851	642,640	429,220	1,470,164
Percentage (%)		3	24	44	29	100
(b) Non-Bonded projects						
	Non-Job	25,794	8,161	(13,921)	46,509	66,543
107	Choctaw, Oklahoma	—	—	—	2,266	2,266
109	Ellis County, Oklahoma	—	—	—	25,346	25,346
123	Caddo County, Oklahoma	—	—	—	—	—
Subtotal		25,794	8,161	(13,921)	74,121	94,155
Percentage (%)		27	9	(-15)	79	100
Total						1,564,319

Table 19.7 Subcontractor Summary as of January 31, 1998

				Balance of subcontract			
Job number (1)	Project name (2)	Reference (3)	Revised subcontract amount ($) (4)	Total ($) (5)	Earned not paid ($) (6)	Earned retention ($) (7)	Unearned ($) (8)
132	White River	Table 19.8	4,608,460	2,160,158	303,517	250,620	1,606,021
134	Hot Springs	Table 19.9	417,478	329,945	18,042	—	311,903
135	Ellis County	Table 19.10	3,522,615	2,181,942	60,230	68,987	2,052,725
136	Hempstead County	Table 19.11	2,835,527	1,076,302	184,071	188,081	704,150
137	Pottawatomie	Table 19.12	971,907	966,282	91,115	—	875,167
Total			12,355,987	6,714,629	656,975	507,688	5,549,966

Table 19.8 Subcontractor Detail as of January 31, 1998
Job Number 132–White River

					----------Balance of subcontract funds---------			
Subcontractor name (1)	Original subcontract amount ($) (2)	Change order amount ($) (3)	Revised subcontract amount (2)+(3) ($) (4)	Payments received to date ($) (5)	Total ($) (6)	Earned not paid ($) (7)	Earned retention ($) (8)	Unearned ($) (9)
Bridges & Company[1]	1,735,035	—	1,735,035	1,426,957	308,078	136,684	—	171,394
Patrick Fence Company	28,488	—	28,488	26,985	1,503	—	1,420	83
Maybee Corporation	75,858	—	75,858	50,857	25,001	—	1,295	23,706
Southern Guardrail[1]	15,180	—	15,180	—	15,180	—	—	15,180
Twin Lakes Quarries	1,473,397	—	1,473,397	216,513	1,256,884	166,833	21,883	1,068,168
Wilson Bridge Company	1,280,502	—	1,280,502	726,990	553,512	—	226,022	327,490
Total	4,608,460	—	4,608,460	2,448,302	2,160,158	303,517	250,620	1,606,021

[1] Bridges & Company and Southern Guardrail are bonded.

Table 19.9 Subcontractor Detail as of January 31, 1998
Job Number 134–Hot Springs

					---------Balance of subcontract funds---------			
Subcontractor name (1)	Original subcontract amount ($) (2)	Change order amount ($) (3)	Revised subcontract amount (2)+(3) ($) (4)	Payments received to date ($) (5)	Total ($) (6)	Earned not paid ($) (7)	Earned retention ($) (8)	Unearned ($) (9)
Jack Wood Construction	44,359	—	44,359	28,268	16,091	12,727	—	3,364
M U, Inc.	23,709	—	23,709	12,984	10,725	—	—	10,725
Pavement Specialists, Inc.	15,019	—	15,019	—	15,019	—	—	15,019
Smith Erosion Control	8,088	—	8,088	—	8,088	—	—	8,088
Stafford Enterprises	23,763	—	23,763	3,052	20,711	5,315	—	15,396
Stud Welding, Inc.	31,476	11,753	43,229	43,229	—	—	—	—
Thurl Haber	57,593	—	57,593	—	57,593	—	—	57,593
Thompson Fence Co.	10,700	—	10,700	—	10,700	—	—	10,700
Wilson Bridge	191,019	—	191,019	—	191,019	—	—	191,019
Total	405,726	11,753	417,479	87,533	329,946	18,042	—	311,904

Table 19.10 Subcontractor Detail as of January 31, 1998
Job Number 135–Ellis County

Subcontractor name (1)	Original subcontract amount ($) (2)	Change order amount ($) (3)	Revised subcontract amount (2)+(3) ($) (4)	Payments received to date ($) (5)	---------Balance of subcontract funds--------- Total ($) (6)	Earned not paid ($) (7)	Earned retention ($) (8)	Unearned ($) (9)
Black Diamond Fence	33,470	—	33,470	—	33,470	434	23	33,013
C. Ray Dement Concrete[1]	989,191	(471,832)	517,359	491,491	25,868	—	25,868	—
Dan Buckner, Inc.	471,832	3,750	475,582	116,316	359,265	18,386	6,122	334,757
Erosion Control, Inc.	23,429	—	23,429	—	23,429	—	—	23,429
Frontier Foundation, Inc.	27,514	—	27,514	—	27,514	—	—	27,514
H&H Foundation[2]	27,198	—	27,198	25,838	1,360	—	1,360	—
Hwytex	1,858,081	—	1,858,081	688,102	1,189,979	—	34,437	1,155,542
J.W. Fields, Inc.	198,809	—	198,809	16,562	182,248	36,205	—	146,043
Municipal Services	23,540	—	23,540	22,363	1,177	—	1,177	—
Southwestern Contractor	288,088	—	288,088	—	288,088	—	—	288,088
Traffic Systems	49,545	—	49,545	—	49,545	5,205	—	44,340
Total	3,990,697	(468,082)	3,522,615	1,340,672	2,181,943	60,230	68,987	2,052,726

[1]C.Ray Dement Concrete was replaced by Dan Buckner, Inc.
[2]H&H Foundation was replaced by Frontier Foundation, Inc.

Table 19.11 Subcontractor Detail as of January 31, 1998
Job Number 136–Hempstead County

					Balance of subcontract funds			
Subcontractor name (1)	Original subcontract amount ($) (2)	Change order amount ($) (3)	Revised subcontract amount (2)+(3) ($) (4)	Payments received to date .($) (5)	Total ($) (6)	Earned not paid ($) (7)	Earned retention ($) (8)	Unearned ($) (9)
B.J. Crown Construction	58,122	(53,122)	5,000	5,000	—	—	—	—
Biddle Sandblasting	30,268	—	30,268	16,600	13,668	—	—	13,668
C.W. Smith	42,499	—	42,499	—	42,499	—	—	42,499
Herzog[1]	2,288,598	—	2,288,598	1,610,938	677,660	34,502	181,015	462,143
Hills, Inc.	45,445	—	45,445	2,590	42,854	—	136	42,718
Southern Guardrail	75,050	—	75,050	7,885	67,165	5,930	885	60,350
Pavement Specialist	238,553	—	238,553	110,936	127,617	113,742	4,760	9,115
Traffic Systems, Inc.	110,113	—	110,113	5,275	104,839	29,898	1,284	73,657
Total	2,888,648	(53,122)	2,835,526	1,759,224	1,076,302	184,072	188,080	704,150

[1]This subcontractor was bonded.

Table 19.12 Subcontractor Detail as of January 31, 1998
Job Number 137–Pottawatomie County

Subcontractor name (1)	Original subcontract amount ($) (2)	Change order amount ($) (3)	Revised subcontract amount (2)+(3) ($) (4)	Payments received to date ($) (5)	---------Balance of subcontract funds--------- Total ($) (6)	Earned not paid ($) (7)	Earned retention ($) (8)	Unearned ($) (9)
Action Safety	33,470	—	33,470	—	33,470	434	23	33,013
American Earth Mover	989,191	(471,832)	517,359	491,491	25,868	—	25,868	—
Don Willis Landscaping	471,832	3,750	475,582	116,316	359,265	18,386	6,122	334,757
Frontier Foundation[1]	23,429	—	23,429	—	23,429	—	—	23,429
J.C. Contractor	27,514	—	27,514	—	27,514	—	—	27,514
Stout Surveying	27,198	—	27,198	25,838	1,360	—	1,360	—
T.J. Campbell[1]	1,858,081	—	1,858,081	688,102	1,189,979	—	34,437	1,155,542
Western Guardrail	198,809	—	198,809	16,562	182,248	36,205	—	146,043
Wilson Bridge	23,540	—	23,540	22,363	1,177	—	1,177	—
Total	3,990,697	(468,082)	3,522,615	1,340,672	2,181,943	60,230	68,987	2,052,726

[1] These subcontractors are bonded.

Table 19.13 Liquidated Damages as of January 31, 1998

Job # (1)	Contract completion date (2)	Anticipated completion date (3)	Contractual liquidated damages ($/day) (4)	Excess contract days (3)-(2) ($) (5)	Liquidated damages to date (4)*(5) ($) (6)	Anticipated future damages ($) (7)	Total liquidated damages ($) (8)	Notes (9)
132	5/03/98	6/30/98	840	35	—	29,400	29,400	Potential dangers: backcharged to sub-contractor
134	5/06/98	4/30/98	630	None	—	—	—	
135	5/27/98	9/30/98	600	127	—	76,200	76,200	Backcharges to Sullivan
136	11/30/98	10/30/98	630	None	—	—	—	
137	11/30/98	12/31/98	210	22	—	4,620	4,620	Backcharges to Sullivan
Total					—	110,220	110,220	

Table 19.14 Summary of Project Cash Flow

Cost item (1)	1998 February ($) (2)	March ($) (3)	April ($) (4)	May ($) (5)	June ($) (6)	July ($) (7)
General Overhead	44,000	35,000	24,000	24,000	24,000	24,000
Notes payable	64,000	60,000	60,000	17,000	13,000	13,000
Accounts payable	750,000	1,053,900	—	—	—	—
Liquidated damages	—	—	—	—	—	—
Direct job costs	779,844	891,558	1,079,063	1,049,342	998,957	949,810
Total costs	1,637,844	2,040,458	1,163,063	1,090,342	1,035,957	986,810
Income projection	753,530	875,631	1,174,428	1,247,200	1,183,726	1,024,227
Net surplus (deficit)	(884314)	(1164827)	11,365	156.858	147,769	37,417

Cost item (1)	August ($) (8)	September ($) (9)	October ($) (10)	November ($) (11)	December ($) (12)	1999 January ($) (13)	Total ($) (14)
General Overhead	24,000	18,300	18,300	18,300	18,300	18,300	290,500
Notes payable	13,000	13,000	13,000	13,000	13,000	13,000	305,000
Accounts payable	—	—	—	—	—	—	1,803,900
Liquidated damages	—	—	—	—	—	110,220	110,220
Direct job costs	832,596	700,898	474,990	237,202	237,202	218,475	8,449,937
Total costs	869,596	732,198	506,290	268,502	268,502	359,995	10,959,557
Income projection	837,236	623,877	488,479	254,923	254,923	234,797	8,952,977
Net surplus (deficit)	(32,360)	(108,321)	(17,811)	(13,579)	(13,579)	(125,198)	(2,006,580)

Table 19.15 Summary of Notes Payable as of January 31, 1998

Loan Company Name (1)	Loan Number (2)	Loan Amount ($) (3)	Rate (%) (4)	Payment ($/month) (5)	Starting Date (6)	Balances as of 1/20/98 ($) (7)	Security (8)
CIT	32211	33,500	10.00	1,081	12/05/96	21,645	Equipment
CIT	32376	51,615	10.29	1,672	04/25/97	38,829	Equipment
CIT	32183	1,582,117	10.00	43,099	11/15/96	1,106,248	Equipment
CIT	32585	31,341	10.52	1,019	09/30/97	27,553	Equipment
CIT	32542	30,750	10.53	1,000	08/23/97	26,272	Equipment
Subtotal	—	—	—	47,871	—	1,220,547	
Payments						(739,547)	
Total						481,000	
GMAC	22963	17,300	2.90	502	10/04/96	9,796	1984 Cadillac
GMAC	66485	13,006	6.90	311	07/12/97	11,325	1988 Chevy Pick up
GMAC	25232	11,445	14.50	316	01/17/95	3,233	1985 Chevy Pick up
GMAC	25643	11,445	14.50	316	01/31/95	3,233	1985 Chevy Pick up
GMAC	55352	12,559	2.90	365	10/19/96	7,104	1985 Chevy Pick up
GMAC	14325	11,705	1.35	317	02/23/94	314	1984 Chevy Pick up
GMAC	17142	11,700	1.35	317	05/08/94	1,232	1984 Chevy Pick up
GMAC	23363	11,047	1.44	305	11/05/94	2,552	1985 Chevy Pick up
GMAC	25109	11,410	1.44	315	11/04/95	3,192	1985 Chevy Pick up
GMAC	44953	8,760	9.99	221	05/01/96	5,348	1986 Chevy Pick up
GMAC	44099	16,620	1.30	445	04/11/96	10,056	1983 Suburban
GMAC	48541	11,222	6.90	345	07/02/96	5,588	1986 Chevy Corsica
GMAC	62086	12,492	5.90	379	04/04/97	9,240	1987 Chevy Pick up
Total	—	—	—	4,454	—	72,213	
First American National Bank	5140	100,000	10.50	2,034	05/07/97	21,051	Equipment
First American National Bank	7324	265,771	10.50	4,466	05/07/97	143,762	Equipment
First National Bank	5515	—	12.75	2,500	01/31/94	27,199	
United Federated Savings & Loan	—	225,000	12.00	2,729	06/03/93	190,125	Office Building
Total	—	—	—	64,054	—	935,350	

Table 19.16 Cash Flow Projection-Construction Costs

Job #	Project Name 1998	January ($) (1)	February ($) (2)	March ($) (3)	April ($) (4)	May ($) (5)	June ($) (6)
127	Oklahoma City	—	—	1,000	—	—	—
132	White River	123.538	180,555	266,083	323,098	323,098	257,584
133	Hall County	—	—	5,000	—	—	—
134	Hot Springs	138.835	150,284	102,409	78,264	17,902	—
135	Ellis County	225,304	225,304	225,304	225,304	225,304	225,304
136	Hempstead	179,808	198,088	2,329,457	247,896	251,631	236,689
137	Pottawatomie	112,359	137,327	149,810	174,780	181,022	212,233
Total		779,844	891,558	1,079,063	1,049,342	998,957	949,810

Job #	July ($) (7)	August ($) (8)	September ($) (9)	October ($) (10)	November ($) (11)	December ($) (12)	Total ($) (14)
127	—	—	—	—	—	—	1,000
132	266,080	133,040	—	—	—	—	1,891,076
133	—	—	—	—	—	—	5,000
134	—	—	—	—	—	—	487,694
135	225,304	225,304	225,304	—	—	—	2,027,736
136	128,978	92,868	—	—	—	—	1,665,415
137	212,234	249,686	249,686	237,202	237,202	218,475	2,372,016
Total	832,596	700,898	474,990	237,202	237,202	218,475	8,449,937

Table 19.17 Cash Flow Projection-Income

Job #	Project Name 1998	February ($) (1)	March ($) (2)	April ($) (3)	May ($) (4)	June ($) (5)	July ($) (6)
127	Oklahoma City	—	—	63,879	—	—	—
132	White River	125,727	183,755	270,796	328,824	328,824	280,468
134	Hot Springs	100,260	137,499	147525	105,989	35,807	—
135	Ellis County	220,139	220,139	220,139	220,139	220,139	220,139
136	Hempstead	186,651	186,651	311,085	404,410	404,410	295,531
137	Pottawatomie	120,753	147,587	161,004	187,838	194,546	228,089
Total		753,530	875,631	1,174,428	1,247,200	1,183,726	1,024,227

Job #	August ($) (7)	September ($) (8)	October ($) (9)	November ($) (10)	December ($) (11)	January 1999 ($) (12)	Total ($) (13)
127	—	—	—	—	—	—	63,879
132	270,796	—	—	—	—	—	1,924,588
134	—	—	—	—	—	—	527,080
135	220,139	220,139	220,139	—	—	—	1,981,251
136	118,212	—	—	—	—	—	1,906,950
137	228,089	268,340	268,340	254,923	254,923	234,797	2,549,229
Total	837,236	623,877	488,479	254,923	254,923	234,797	8,952,977

been attached and the dispute over priority between the Federal Government and the surety is largely determined by the principle of *first in time, first in right* (i.e., "first come, first served") as that principle is set forth in federal statutes. However, if progress payment is an expectation—a contingent right of anything else not amounting to "property"—then there is nothing to which an IRS lien can attach and the surety prevails. It should be noted that state law determines what is property or a property interest, and thus varies quite widely.

Some courts have held the contractor in breach of contract when the contractor has failed to pay labor and materialmen and the owner had the right to withhold and apply. This bars the contractor's right to recover under the contract, and since he or she has no contract right, he or she has no property interest. Thus, an IRS lien cannot be attached. Again, the contractor's property interest to which an IRS lien can attach is determined by state law.

At times the surety may obtain excessive control over the contractor's entire operation, but according to Marilyn Klinger, Esq., no court in recent history has found a surety "liable for interfering with the contract between the principal and the obligee." Writing in *Suretyscope*, a trade journal for the surety industry, Klinger (1997) cites several cases from the 1950s to the present, all of which end up vindicating the surety from any wrongdoing in its dealings with a defaulted or near-defaulted contractor. In the three-party relationship of bonding, if the surety is justified in perceiving that the contractor is in dire straits, even without the principal's consent, sureties have managed to gain control of contracts.

However, effectively becoming the contractor's "alter ego" is not without trouble for the surety. In such a situation, a surety may assume parts of the principal's liabilities. This certainly applies to the surety's direct tax liability to the IRS as previously discussed. As a general rule, however, the surety can and does avoid alter ego liability to third party creditors (such as unbonded creditors) for insurance premiums, equipment payments, and specialty contractors. Still, it is an option available to a surety and might, in special circumstances, be employed.

In a case in New York, a surety placed its own representatives in charge of the daily business of the principal and acted as directors of the company. Even so, the surety managed to escape liability to subcontractors. Generally the surety *is not* liable for obligations of the debtor other than that in which they agreed to assume. Historically, the surety has received fair consideration for its security interests.

A surety *may not* use surplus funds from one job and apply them to a deficit on another. The only exception occurs when the obligee on two projects is the same and the obligee has the right to set-off an excess on one job against the deficit on the other. The surety is subrogated to that right of the obligee and can do likewise.

The surety who provides financing acquires the same rights by subrogation as the surety who completes performance or pays bills after a declared default. Whatever the surety outlays to complete performance is a loss under the performance bond even to the extent that the surety pays subcontractors and material suppliers in the course of completing performance.

Although the financing surety is generally able, by subrogation, to defeat conflicting claims to the contract funds to the extent of its losses, important exceptions exist in addition to those previously discussed. The purpose of acquiring and perfecting contractual security interest in contract funds is to protect against these exceptions. The contract funds alone may not suffice to compensate the surety for its losses. Although the indemnity agreement gives a security interest in equipment and materials stored on the job site on bonded projects, in order to prevail against third parties, the surety must perfect that security interest by filing under the Uniform Commercial Code (UCC). Sureties may prevent voidable preference for up to a three-month period. A financing surety may be able to prevail even when it perfects its security rights within three months of bankruptcy. Surety rights of subrogation can put them at the head of the line before general creditors to receive surpluses on its bonded projects as well as other projects.

The financing surety's involvement in the affairs of its principal creates two problems. First, if that involvement is great enough, it may render the surety the alter ego of the contractor as previously stated, or at least liable for situations that it would not otherwise be liable for under its bonds. Second, the principal itself may claim the surety "dominated" its affairs, although as was noted above, such a claim would be difficult to prove in court.

Other Options Available to Surety

The surety's decision not to finance may be made easier when there has been a declared default by the obligee and when the contractor has: 1) disappeared, 2) become otherwise unavailable, 3) filed bankruptcy, 4) has no desire to complete the project, 5) experiences death or disability of key employees, or 6) may be clearly incapable of doing the work. Also, as a general rule, if a job is less than 50% physically complete or, again, if a contractor has anywhere near 80% of unbonded work in his program, than the surety will most likely *not* finance, but instead give top priority to reletting the contract to another construction contractor. Under any of these circumstances, the surety has several options: 1) reletting the contract, 2) completing surety, or taking-over and reletting, 3) buying bond back, 4) obtaining bids and arranging contract between owner/obligee and completing contractor, and 5) doing nothing. These options are weighted by the surety prior to making a decision, and are often made in consultation with the obligee. The following section elaborates on each of these options.

Reletting Contract

In addition to bringing to the surface the defenses available to the surety with regard to its performance bond obligations, the investigation (step 1) should also reveal whether there are any other serious disputes existing concerning the owner's conduct or interpretation of contract documents. If the contractor has been overpaid or if serious change orders were approved without the surety's consent, for example, this will generally help a surety decrease its exposure. In addition, contracts with public authorities may not always relet in whole or in part without authority's consent, and then only by offering it for re-bid.

Before exercising any of the remaining options, the surety must obtain bids for project completion. The surety should most likely express to all interested parties that it may reject any and all bids at its discretion. A variety of sources will yield the names of perspective contractors, including:

1. Other contractors who originally bid the job. The obvious advantage is that they are already familiar with the project and the contract documents.
2. Owner, architect, or engineer.
3. Word of mouth. Almost invariably, contractors contact the surety offering to complete the project. The surety should have no difficulty obtaining competent contractors, especially when new construction projects are scarce.

The surety then selects a contracting strategy, depending upon the percent of construction complete and the quality of construction in-place, among other factors. Several types of contracts exist, including: 1) lump sum or fixed price, 2) cost plus overhead and percentage of profit, 3) cost plus a fixed fee, 4) cost plus percentage of profit with price ceiling, 5) negotiated contract, and 6) force account.

When a job has not yet commenced or is in early stages of completion, the surety will generally sign a lump sum or fixed price contract with a completing contractor. This affords the surety certainty in identifying and subsequently transferring its liability. Other contract types may be necessary where the job involves a substantial amount of remedial work, underground construction, or inadequate or substandard quality of work. The surety overseeing the completion of a project on which the original contractor defaulted must be sure that the contractor completing the project obtains performance and payment bonds.

Sureties may use a force account (i.e., time and materials) on a subcontract bond where the prime contractor agrees to finish the defaulted specialty contractor's work. From the surety's perspective, this is only advisable when little work remains and it appears likely that the prime contractor can complete the project for the balance remaining from the original contract amount due under the original subcontract agreement. In addition, parties entering into a contract that uses a force account should understand that the surety can terminate the agreement at any time if after further analysis of the situation the surety determines it can make more favorable arrangements. A force account arrangement may serve as a useful interim device that permits construction to continue while the surety completes it evaluation.

Whenever reletting a job, failure to use the existing subcontractors may substantially increase the cost of completion. Using the prior subcontractors eliminates the possibility of one or more of them making a quantum meruit claim (i.e., as much as he deserves or the amount invested). Also, it eliminates the problem of 1) someone else having to assume responsibility for work already performed, 2) the mobilization costs for a new subcontractor, and 3) the time spent finding appropriate substitutes. This is one of the major reasons sureties require many contractors, especially under-capitalized ones, to use bonded subcontractors.

Completing Surety

Completing surety is another strategy sureties commonly employ when the original contractor defaults. Generally, this option is the only one acceptable on Federal Government contracts as well as by other public obligees. However, on such publicly funded jobs the surety may not be able to act without the consent of the obligee and must put the completion contract out for public bid. In most cases, absent an agreement with the owner, a surety who takes over a defaulted job is liable for the cost of completion *whatever it amounts to* even if it exceeds the penal sum of the bond. Thus, the key to success when using this option is a formal takeover agreement and a carefully negotiated and drafted relet contract with the completing contractor. After all, if the new contract is not skillfully worded, this takeover gives the obligee a blank check from surety if the amount expended is not deductible against the penal sum of the bond.

One important incentive for takeover is the opportunity to obtain concessions from the owner that may not be obtainable otherwise. From the owner's standpoint, takeover by the surety is far and away the most desirable way to complete a defaulted contract. Some possible concessions that may be negotiated by the surety include: 1) liquidated damages, 2) delay claims and extras by original contractor, 3) other disputes regarding interpretations of plans and specifications, 4) increased unit prices on certain items such as oil prices on asphalt paving jobs, especially where contract provides for such, and 5) reduction of retainage.

Surety Tenders Completing Contractor to Obligee

At times the surety cannot negotiate a satisfactory takeover agreement or is unable to buy the bond back. In such cases, the surety may tender a completing contractor to the owner/obligee with the surety agreeing to pay the obligee the increased price to complete the remaining work *upon completion of the job.*

State and local governmental agencies typically argue that they cannot accept this option or buy back because of competitive bidding requirements under statutes applicable to public jobs. Sureties have successfully argued that the applicable statutes refer to public bids with expenditure of public funds. Since no public bids are necessary in a relet situation because *no additional public funds are necessary as the excess monies come from the surety*, courts have ruled in favor of sureties using these options.

Buy Back Bond

In many cases, buying the bond back may be the most desirable option because it fixes the surety's exposure and releases the surety from any potential warranty problems. Simply put, the surety pays the obligee *up front* the amount of the calculated increased cost to complete beyond the contract balance plus the contract balance. This option is often more difficult to implement with owner obligees and is more readily done on subcontract bonds where the prime contractor is the obligee. Generally, a prime contractor prefers to have the money up front. This option, however, is also used when the surety has defenses to their performance bond liability as discussed earlier in this section, or when a project has just broken ground.

Do Nothing

This option is exercised when the surety does not believe the contractor is in default. This frequently happens when an over-zealous owner panics or when it is not clear cut as to whether default has occurred. Also, a surety may choose to do nothing when it is obvious from contractor documents that an owner severely overpaid or paid for substandard construction. In addition, when a contractor files for bankruptcy, a surety may be unable to proceed under the bond until the court decides a course of action. Doing nothing usually forces the owner to take legal action to enforce the surety's responsibility under the bond. However, such action would only be deemed appropriate by a surety after an investigation had yielded such evidence warranting no action at all, at least for the time being.

Step 3. Decision-Making Process

In a claim situation, after identifying the options suitable for further study and gathering and evaluating data, the surety can initiate the decision-making process. That process, step 3, involves weighing the optional methods of terminating performance bond obligations. The selection of the option to terminate the liability is the outcome of this step.

Step 4. Implementation and Step 5. Monitoring

Implementation involves deciding on an option for terminating performance bond liability as soon as possible. The following section explores how a surety finances a contractor. In addition, Chapter 20 explores how a surety deals with a contractor defaulting on multiple jobs in which each of the options available to the surety are exercised. Monitoring involves making sure that the discharge of responsibility, whether through financing the contractor or buying back the bond, is handled efficiently and effectively. This process can be as simple as documenting payment or as complex as monitoring multiple project completions, depending on which option the surety selects for each bond.

Financing Arrangement

If the surety has decided to finance the contractor, its representatives must determine the nature of expenditures acceptable from the joint control account. The surety must then specify with precise detail what is to be paid out of the account. Specific items to consider include:

1. Trade payables—invoices should be allocated to specific jobs.
2. Net wages to employees, both those who work on and do not work on bonded projects.
3. Federal, state, and local withholding taxes and FICA.
4. Union-related matters such as dues, health and welfare, pensions, and vacation benefits.
5. Equipment rental and/or purchase payments (conditional sales) as to construction equipment, whether or not the particular equipment is being used on bonded projects.
6. Fringe benefits such as hospitalization and pension plans for open shop employees.
7. Insurance of various types: insurance specifically related to the bonded projects (and generally required by the contract documents), insurance on other projects, and general insurance, including general liability, workman's compensation, key man life, job risk, and health insurance.
8. General office overhead, such as rent, mortgage payments, utilities, office equipment, miscellaneous office supplies, maintenance, postage, and audits.
9. All loans, particularly to an assignee bank of the contractor, and any tax liens in existence.

The surety should establish and keep at its home office a special bank account report file. Specifically, the surety should maintain the following three files: 1) "special bank account reports" without documentation; 2) "special bank account reports" with documentation, including copies of checks, invoices or any other information documenting payment; and 3) file containing monthly bank statements and canceled checks. The surety should reconcile all statements and keep running accounts on each job.

The surety should also establish petty cash and payroll accounts, the funds for which come only in the form of a check drawn upon the joint control. Other specifics include:

1. Copies of payroll accounting sheets should support payroll account. It may be advisable to engage a payroll company to handle payroll obligations.
2. The surety and the contractor should agree on a ceiling for the petty cash account. Before funding the account, the surety should ensure that receipts and invoices document all expenditures by setting administrative procedures for payment of items.

Because issuing payroll checks drawn on a joint control account is cumbersome and impractical, these accounts should have as few authorized signatures as possible. In addition, the surety should obtain special types of checks with corresponding stubs for itemizing withholding items. A petty cash account will expedite emergency situations that arise when the signatory for the surety is not available to sign checks. Offers of bonuses to key personnel of the contractor, payable upon completion of the project on time and within budget, serve as effective incentives to key employees.

Financing Arrangement Operations for Completing Surety

Together, the surety and the contractor must establish day-to-day procedures regarding the surety's financial and job site supervision. The degree of the surety's involvement depends upon the size and complexity of the case and any other factual circumstances. It is extremely important that the contractor do everything within its power to ensure diligent execution of the work. If that is not the case, then the surety must consider terminating the financing arrangement.

Experience has taught surety industry professionals that in any financing situation, communication between the surety and the contractor should be constant, thorough, and forthright. If the surety is financing the contractor's entire operation, then the surety may need to place personnel in the contractor's office on a full-time basis.

As the surety's involvement becomes more limited, so does the monitoring supervision. The surety should summarily investigate any and every glitch in communication that develops. Such problems usually develop when an event occurs that the contractor does not want the surety to know about. The surety must insist upon prompt replies to all inquiries and should calmly respond to minor snags and major problems alike, focusing on amicable problem resolution, not combative confrontation and accusations. The surety should consider ceasing to finance any uncooperative contractor. The worst mistake a surety can make is to commit to financing a project to completion, but then fail to pay sufficient attention to what occurs during the course of that financing. A summary of the performance on each job should be required usually weekly, and certainly not less frequently than monthly, so that delays and overcosts can be promptly discerned.

Depending upon the circumstances surrounding a default, the surety should take stock of its own available internal resources and possibly consider employing outside consultants. Some suggested guidelines and instructions regarding the use of outside consultants follow:

1. Outside consultants should not be employed without obtaining the consent of the proper authority within the surety's organization.
2. Consultants should not be hired to perform functions that are within the surety's own capacity to perform within allowable time and reasonable costs.
3. The surety must first perform a factual investigation and analysis of the situation.
4. Consultants should receive instructions specifying the scope of the assignment, the optimal number of personnel the task requires, the rates of compensation, and the minimum qualifications of each person in accordance with the nature of the assignment. In addition, consultants need to understand the task at hand and to whom they report.
5. If the surety requires written reports, it should analyze the reports carefully. The form and content of the reports should be laid out by the surety ahead of time. Since such reports may become subject to the discovery process in litigation, the surety should be aware of potential damage if the information is insufficient, improperly or inadequately analyzed, or inaccurately stated.

SPECIAL CONCERNS AND EFFECTS OF BANKRUPTCY UPON SURETY

Some special problems that may confront a surety upon contractor default include the following questions:

1. What effect does the contractor's indebtedness to a bank have on the surety's course of action?
2. What is the surety's liability for federal withholding and other state taxes?
3. How should a surety proceed when a solvent principal's contract is terminated?
4. What effect will contractor bankruptcy have on the surety's course of action?

Regarding the contractor's indebtedness to the bank, the question involves the security interest that the banks have, including the pledge of accounts receivables. This affects the surety's decision whether or not to finance the defaulted contractor. Regarding the surety's tax liability as discussed above, there are five instances in which the surety is liable for withholding taxes. The surety must also perfect its security interest by filing a lien against the contractor's property. Additionally, the surety must watch federal tax liens attaching to contractors' property.

Depending on the circumstances, when a solvent principal is terminated by the obligee, the surety will always conduct an independent investigation, even though it may decide in the end to do nothing. In some instances the surety may act contrary to the principal's position. When the principal and obligee both perceive the other has breached its contract, this situation becomes even more complicated.

When a contractor files for bankruptcy, it complicates the claim process significantly. Specifically, three issues generally surface: 1) legal and equitable rights of surety to contract funds, 2) surety's decision on how to terminate its bond liabilities, and 3) obligee's rights. While filing for bankruptcy may be somewhat beneficial to the contractor, it is almost always detrimental to the owner and the surety. A contractor should be advised to consider that such an action might strain its relationship with its surety, and thus the surety's counsel should be sought at every step of the procedure. For an extended example of what can happen when this occurs, see Chapter 20, "Claim Case Study: The Defaulting of Zimbrick International."

By paying the bankruptcy filing fee and filing a petition in chapter 7 or 11 under which it seeks relief, a contractor commences a voluntary case and satisfies an order for relief. The order for relief initiates the bankruptcy provisions of the respective chapter and effectuates an automatic stay from collection efforts and foreclosure actions against the debtor. In other words, by filing a chapter petition and paying the requisite fees, the debtor obtains at least *temporary* protection from its debts. Then, depending on the chapter, the debtor may proceed to liquidate, to attempt repayment, and/or to reorganize. It is important to remember that filing for bankruptcy introduces additional parties into decision-making positions: the trustee in bankruptcy and possibly the credit community.

The immediate effects of a contractor filing a bankruptcy petition include the following:

1. Automatic stay that enforces a pre-petition obligation of the debtor, thwarts any litigation against the debtor, impacts the right of the surety to issue a stop payment order to owner, and impacts the owner's right to terminate the contract with the contractor.
2. Negates the enforceability of "ipso facto" or bankruptcy clauses in construction contracts.
3. Rejects the assumption and assignment of executory contracts.
4. Voidable preferences come into affect and the surety does not receive preferential treatment.
5. Impact on the surety's rights under its indemnity agreement with the principals of the construction company.

6. Effect upon surety's common law rights to the contract funds under non-bankruptcy law including issues such as contract funds as property of the debtor's estate, unearned contract balances versus retainage and earned but unpaid progress payments, and when debtor-in-possession if allowed assumes a bonded contract.

The automatic stay mechanism is pro-debtor because it gives the debtor temporary relief from a creditor's collection efforts, but readers should note that this relief is not permanent. The purpose of the automatic stay is not only to grant the debtor protection and time so that it may consider the future of its operations, but also to prevent the dismantling of the debtor's property in such a fashion as to prioritize some creditors over others (Jewler 1989). Specifically, the automatic stay prevents:

1. Implementation of all pre-petition processes or proceedings of an administrative or judicial nature, or the recovery of a claim that was or could have been brought before commencement of the case
2. Enforcement of pre-petition judgments against the debtor or his property
3. Acts to obtain possession of estate property
4. Acts to create, perfect or enforce property liens which would secure a creditor's pre-petition claims
5. Acts to create, perfect or enforce any lien against estate property
6. Acts to collect, assess or recover pre-petition claims
7. Set-off of monies owed by a creditor to the debtor prior to the debtor's filing, against any claim by the creditor against the debtor
8. United States Tax court proceedings (11 U.S.C. §362(a), 1988).

Business transactions that occur 90 days prior to the filing of a bankruptcy may become void. "Voidable preference" is the term for such occurrences. This prevents the bankrupt company from favoring selected creditors. Creditors receiving payment during this 90-day period may be required to return any payments received from the contractor. This may affect the surety's payment bond status.

Under the revised bankruptcy law of 1986, the obligee cannot terminate a contract with a contractor that has filed for bankruptcy. Prior to this law, owners invoked the ipso facto clause in contracts, which permitted the owner to terminate a contract with a contractor that had declared bankruptcy.

The assumption of executory contracts by a debtor-in-possession or trustee in bankruptcy involve: 1) the problem of the right by the surety to object to assumption of its bonded jobs, 2) cure of defaults if any in existence at time of assumption, 3) the right by estate to assume only contracts that will benefit estate, 4) compensation to parties who have suffered consequential damages prior to assumption, 5) provision of adequate assurance of future performance, and 6) surety's liability under its bond when contract assumed.

A surety's practical options for resolving problems associated with contractor bankruptcy include: 1) invoking the surety's right to object to or accelerate assumption and 2) financing the contractor during bankruptcy proceedings. When the surety refuses to do anything until the bankruptcy court decides whether to assume the contract, the obligee must make every effort to have influence in decisions. The negotiation hats should always be on, and in such a case, the owner must attempt to remain cordial with the surety. If the two can agree on what needs to be done, the chances of a successful outcome greatly increase.

Chapter 20

CONTRACT SURETY: EFFECTIVE CLAIM MANAGEMENT

When faced with a default, especially one precipitated by a hostile owner or lender, a contractor may have scant room to maneuver. When payrolls cannot be met or when project completion requires a timely infusion of money, the surety is one of the first sources the bonded contractor might consider. As one long-time claims specialist puts it, "the surety is the only game in town when the contractor is in trouble." Unfortunately, due to the complex nature of the bonded construction cycle where the surety is the guarantor of present and outstanding projects and the contractor wants to continue bidding, the surety is usually the last to know of any financial troubles.

As has been discussed throughout Part 5, when the surety does hear of problems, it attempts to obtain as much information on the project and financial state of the contractor as quickly as possible. While the investigation is being pursued, the surety will often provide funding contingent upon the surety's final decision to keep project momentum. Such "look-see" money is understood to be available only until the surety has acquired the information it needs to pinpoint its exposure. In essence, such money allows the surety to buy time.

Contractor problems may be external—a surge in the cost of materials, an oversight in an estimate, or problems with the contractor's other businesses—and thus, with a reorganization of finances, the contractor may continue to be a viable and responsive on-going builder. But the problems can just as easily be internal. For example, when a contractor starts "buying" jobs in order to keep money flowing in, this is a major sign of more serious concerns. To determine the reality of the situation, a surety has to thoroughly investigate, and investigations take time. Investigations also increase a surety's exposure as they increase the money invested in a project, but even so, it is common for the surety to provide "look-see" funding while gathering and assessing information.

To illustrate how a the claim process works, what follows are examples of bond claims. The first two deal specifically with maintenance bond claims, while the latter two are extended examples of contractor default. Since maintenance bonds are increasingly being required by owners, it behooves contractors and owners alike to familiarize themselves with how maintenance bonds are structured and how claims under these bonds play out. In addition to the claim examples here, there is further analysis of the use of warranties and maintenance bonds in Chapter 26.

Using the criteria set forth in this book, the first extended example of contractor default is a fictional case study involving Prometheus Construction Company, Inc. (PCC)—the same company we saw qualify for surety credit with Action Bonding in Part 4. In this scenario, PCC defaults with 10 jobs in varying stages of physical completion. Much like in the underwriting example, this scenario unfolds from the perspective of the surety, but this time, it focuses on the other side of the surety enterprise—the claim department.

The second extended example is also a fictional case study of contractor default. This one deals with Zimbrick International. What the author hopes to demonstrate with here is the manifold interplay of forces that can result when a large contractor defaults, especially when that contractor files for bankruptcy. This case is based on extensive interviews with surety industry experts and shows how competing interests can make defaulting a costly, time-consuming process.

MAINTENANCE BOND CLAIMS

In general, maintenance bond investigations take longer and are more complicated than performance or payment bond claims. For starters, since they occur after a project has been completed, many of the original participants may no longer be available to assist with the investigation. These include not only the personnel of the contractor, but of subcontractors, suppliers, and owners. Furthermore, as mentioned in Chapter 11, it becomes extremely difficult to determine who or what is at fault even a few years after a project has been completed (defective materials, workmanship, improper use of the facility by owner, etc.), and replacement costs generally increase over time. Especially with public entities where political forces are more at work, maintenance bonds may also be interpreted inconsistently (i.e., as a right to free maintenance).

Due to the temporal distance from the original contract, a maintenance bond claim is also much more likely to involve litigation, increasing both the owner's and surety's cost. This should be taken into account by obligees when considering whether to require a maintenance bond and for what duration. However, much of the legal mire of maintenance bond claims arise from the obligee's general misunderstanding of what the bond covers. It is paramount for the contract documents to clearly state what is covered and under what operating conditions and external environments. It behooves owners to require separately executed maintenance bonds, meaning that the maintenance bond form is physically separated from that of the performance and payment bond form. Even with clearly authored bonds, however, settling can become a significant problem. Consider the following actual cases involving the standard one-year warranty included in a performance bond.

Example 1

A surety issued a performance bond for a subcontractor installing a rubber roof membrane on a building in Kentucky. Under the performance bond, the workmanship of the roof was guaranteed for one year. In addition, the roof itself carried a 10-year manufacturer's warranty. A few years after the installation, the roof began to leak and continued to do so even after repairing, eventually necessitating replacement by the owner. At such time, the subcontractor was out of business. The owner notified the surety covering the subcontractor, claiming it was responsible for the original faulty installation. The claim occurred seven years after the work had been performed, well past the one-year maintenance guarantee period.

It appeared to the surety that the owner had a legitimate case against the roof manufacturer, but not against the surety. However, a court upheld that the surety was responsible for the damage and was required to pay the full penal sum of the bond for the roof, nearly $1 million. The reason cited was that the owner never officially approved the workmanship of the subcontractor as "fit," and thus it became the surety's responsibility to compensate the owner for losses incurred due to supposed poor performance.

Example 2

Another maintenance bond claim case involved a sewage charcoal filter on a project worth roughly $100,000. When a claim of poor performance arose from the overseeing division of the Wisconsin municipality contracting the work, the surety believed it was not at fault. To prove this, however, it had to invest over $6,000 in photocopying fees alone, plus over $90,000 in engineering consulting fees. Ultimately, the surety spent more than the penal sum of the bond in investigating the claim. Even though justified in their original appraisal, the surety lost a significant amount of money.

The nature of maintenance bond claims requires the surety to carefully consider whether to underwrite them in the first place. The surety will look carefully into the duration, the amount, and the reasons for the bond. In addition, it will take into account the contractor's rating and performance and the length of time the surety has had a business relationship with the contractor, as well as the other insurance products provided to the contractor. It is very important for an owner to understand what is covered in a maintenance bond, and as such, it is highly advisable for the owner to require separate documents for any maintenance bonds over the normal one year of coverage provided by the performance bond. In addition, because of the increased cost of such guarantees, an owner should weigh its options carefully.

CLAIM CASE STUDY: THE DEFAULTING OF PCC

In 1994, after receiving the necessary information from PCC, litigants, banks, previous owners, subcontractors, engineers, architects, and unions, Action increased PCC's bonding limit with the stipulation that Mrs. Prometheus indemnify herself, which she did. After several years of significant growth, PCC has found that growth might not always be the best thing for a contractor. Now under the control of A. Prometheus, Jr., PCC has gone through several significant changes; unfortunately, not all of them have been good. What follows is the story of what happens when a technically capable contractor grows too large too fast, and how its surety deals with the contractor if and when it defaults.

Introduction

PCC has 10 jobs in its work program worth $20 million, the maximum bonding limit allowed by Action Bonding since it was accepted as an account four years ago. All of the bonded jobs are in varying stages of physical completion, and all have different obligees and architects. PCC has further increased its line of credit with Anytown bank to $2 million, a loan to be indemnified by the corporation. A. Prometheus and A. Prometheus, Jr. signed the bank loan, but are not personally indemnified. While PCC owns 50% of its equipment and leases the balance, several of its jobs require special equipment that must be procured, most likely through leasing. The corporation had built up close to $5 million in reserves a few years ago, but now, after using the money to acquire large swaths of the countryside outside of Anytown, USA, which Prometheus, Jr. says, "is ripe for developing," cash reserves are minimal.

Two years ago, a significant change occurred at PCC. After years of working with their father, the two older sons and the single daughter left the company. The three Prometheuses agreed to have their share in the company bought out for $250,000 apiece—$50,000 up front and $20,000 each year for the next 10 years. As a matter of course, they also wanted off the general indemnity agreement (GIA). The GIA with all the stock holders, including A. Prometheus, his wife, their four children and spouses, was thereafter amended. About this time, A. Prometheus, at the age of 65, decided to retire.

As part of an arrangement reached within the family, A. Prometheus handed the reigns of the company over to the youngest Prometheus, A. Prometheus Jr., who had been the Vice President for three years since he graduated from college. The Action underwriter overseeing the file agreed to take the brothers and sister and their spouses off the GIA, but insisted that A. Prometheus and his wife continue to be indemnitors along with A. Prometheus, Jr. However, since half of the jobs were bid when the whole family was indemnified, technically, the entire family is still indemnitors for those jobs. These projects happen to be the five oldest in PCC's 10-job work program. For any jobs bid after the brothers and sister officially left PCC—that is to say, the five newest jobs—only A. Prometheus, his wife, and A. Prometheus, Jr., are listed as indemnitors.

The Root of All Trouble

PCC subcontracts out 50% to 70% of its work, and it is a stipulation of its agreement with Action that PCC use only bonded subcontractors on major work. When A. Prometheus was still President of the company, only bonded subs were used. Thus, PCC's five oldest jobs have, and continue to use, only bonded subcontractors. On the five later jobs, however, unbonded subcontractors have been increasingly employed. This is because once in control of PCC, Prometheus, Jr. decided to pursue development. It is there, he thought, that large profits existed. Developing has been part of his master plan to diversify PCC's construction program in the bullish economy, and in good part necessitated the decision to increase its bonding limits four years ago, when as a college student Prometheus, Jr. oversaw the company accounts.

To finance his expansion plans, Prometheus, Jr. slowly but surely began drawing on PCC's cash reserves to purchase new land to improve. To help keep PCC in the black, he increasingly used non-bonded subcontractors and leased more and more equipment, as he perceived both as inexpensive solutions to PCC's cash problems. To avoid emptying the coffers, he has purposely underbid jobs, passing on some of the perceived savings to the owners in an effort to keep cash flowing in. This way, Prometheus, Jr. believes, PCC can eat its cake and have it too. What Prometheus, Jr. failed to take into account is the unguaranteed nature of unbonded labor. He also failed to realize that cutting corners in construction will, in the end, increase costs for everyone involved.

Unfortunately, his desire to grow has and will cost PCC dearly. It remains to be seen how much it will cost Action, but if the claim personnel have anything to do with it, the amount will be as small as possible.

The Trouble Becomes Apparent

On a routine inspection of one of PCC's jobs, architect Sarah Smith notes that the work is severely behind schedule. Even more troubling, the subcontractors do not appear to be returning to the site. Upon talking with some of them, she finds this is because they are slow in getting paid. When she notifies the owner, H&H Management, they decide to notify Action Bonding and call a meeting with PCC to find out why there is so little performance.

When Ms. Smith meets with PCC, Prometheus, Jr. informs her that PCC has cash flow problems because they can't collect their retainage on their older jobs, which is really a falsehood because he used the money to purchase more land to develop. Prometheus, Jr. assures her that he will rectify the situation, which he in part does by getting several of the subcontractors to go back to work by promising them more work on a bid that was accepted just last week.

On another PCC project, this one an 85% complete office building, a plumbing subcontractor, Gonzalez Plumbing, has filed a mechanic's lien against the owners for $35,000 because he has not been paid either. The notice was received by the dual-obligees on the bond, First Bank and Development, Inc. Work on the project has halted because other subcontractors cannot perform until the piping has been laid. This has caused a backlog and may cause a serious delay in the completion of the building, which is slated to be turned over in less than four months and already has 90% occupancy. As is common for many owners to do, whatever the problem, First Bank and Development, Inc. requests that as writer of both the performance and payment bonds Action promptly attend to the situation.

Around the same time, rumors about slow pay and PCC's financial problems start spreading around the construction community of Anytown, especially among subcontractors. Several of the subcontractors on other PCC jobs have the same problem: they cannot perform because either they have not been paid or they are waiting for other work to be completed first. Several of the subcontractors' sureties get nervous as well when they hear of developments around PCC, especially when learning that PCC has gone to unbonded subcontractors on its newest jobs. These sureties pick up the phone and notify Action as well.

Immediately, then, Action is forced to enter the fray.

Action's Response

When the Action underwriter is notified of PCC's problems with its subcontractors, he is supposed to turn over the information to the claim department. Often, however, the underwriter will investigate to see if something can be done to quickly rectify the situation. The first thing one will do is to familiarize oneself with the bond, the contract, and any information pertaining to that specific problem.

When J. Johnson, the Action underwriter "holding" the PCC file, is notified of the lien filed by Gonzalez Plumbing, the stalled performance for H&H and First Bank and Development, Inc., and the concerns of the bonded subcontractors' sureties, he immediately checks the PCC documents he has on file. Gonzalez has worked for PCC for several years but has not been retained on the last five jobs PCC has bid. When the underwriter contacts the plumber, he finds out that he was replaced by an unbonded subcontractor who underbid him by a measly 10%. Since it is a stipulation of PCC's agreement with Action that it use 100% bonded subcontractors for major work, this piece of information does not sit well with Johnson. Since he begins to suspect that PCC has been trying to save money by using unbonded subcontractors, he next checks the bid sheets (Table 20.1).

Table 20.1 Bid Spread on Five PCC Projects

PCC project number (1)	PCC's accepted bid price ($) (2)	Next lowest bidder ($) (3)	Percent under bid (%) (4)
5	1,500,000	1,800,000	20.00
7	3,000,000	3,500,000	16.67
8	650,000	710,000	18.33
9	2,000,000	2,225,000	11.25
10	4,000,000	4,550,000	13.75
Total	—	—	16.00

Johnson's hunch proves correct, as he finds that PCC's last five bids, including one awarded last week, have been seriously low—by as much as 20%—averaging about 16% lower than the second-lowest bidder. Furthermore, upon checking into the subcontractors on the H&H job, Johnson finds that they are all unbonded as well. This is in stark contrast to the previous pattern where PCC was never off by more than a few percent from any other bidder on any job they were awarded. Worse still, the under-bidding has progressively worsened, leading Johnson to the inescapable conclusion that PCC has been "buying" jobs. At this time, realizing that some serious problems are looming, Johnson contacts his friend in Action's claims department, A. Adams, and says, "Let's have a chat."

When A. Adams gets the file in his hand, he immediately looks at the financial report and the GIA. In addition to a recent transfer from liquid capital to property rights, Adams notes that two GIAs exist and wonders how such an arrangement was approved. He quickly notices that the acquisition of land coincided with the departure of the former indemnitors and the promotion of the Vice President to President. Thus, once of Adams's first priorities is to find out whether those who got out before PCC started buying jobs saw the problems approaching. To answer these and other questions, Adams promptly calls up the principal, PCC, and says, "We need to talk."

Adams arranges a meeting with Prometheus, Jr. and PCC's indemnitors, including his father, mother, older siblings, and their spouses, since they are still indemnitors on the five oldest jobs. Adams reviews the complete underwriting file, including bid sheets, prior mechanic's liens, specifications sheets, equipment titles, and leases. Adams confirms Johnson's suppositions that PCC has been buying jobs and he wants to find out what has happened to a this once-healthy company. If there is any hope of keeping it viable, action must be taken swiftly.

When Adams meets with the indemnitors in a few days, there is much confusion. The older siblings thought they were no longer associated with PCC, and A. Prometheus had no idea his son had acquired so much land or that was buying jobs to foot his speculation. Adams explains to them that they are all still indemnitors on the five oldest jobs and that the father, wife, and youngest son are also indemnitors on the five newest jobs. When Adams informs the older siblings that the money they were to receive in the next years as part of their buyout arrangement may not be available—$20,000 each year—tempers boil.

Adams asks PCC if it will voluntarily put itself in default, but as CEO, Prometheus, Jr. insists that PCC is still solvent. He manages to calm his family down, which is no small feat, and convince them to allow him to carry on, albeit with less land acquisition and a more brass-tacks mentality. For the time being, this is the extent of the surety's ability to influence PCC. However, as it is becoming clearer and clearer that default is inevitable, the surety does have other options. What follows are three common scenarios that may play out in such a claim situation.

Scenario One: No Official Default (Not Yet Anyway)

Lack of performance costs the surety money. Every day that a job sits idle, as do the jobs for H&H Management and Development, Inc., the surety's exposure inevitably increases. Until an official default, however, the surety technically cannot do anything other than try to influence the contractor and obligee. It takes an official default by either party for the surety to step in. Since there has been no official defaulting and since PCC refuses to place itself in default, Adams decides to contact G. Jackson, the architect who notified Action of the Gonzalez Plumbing lien and First Bank and Development, Inc.'s representative on that job.

Jackson is at first combative, demanding that Action keep the work progressing. In addition to the lien and subsequent backlog, specialty equipment required for the project has not been procured and PCC has made no arrangements to order it. Furthermore, it is unlikely that PCC will even be able to obtain it. After talking with Big Bob's Tools, PCC's main equipment supplier, Jackson learned PCC is in arrears and probably lacks the money for a even a down payment. Jackson wants to know what Action can do to help speed things up.

Adams informs the angry architect that at this point he can only listen. Until there is a defaulting by an owner, there can be no action taken by the surety. If Action attempts to interfere in any way, PCC could sue the surety for contract interference, a favorite strategy for broke contractors, Adams informs Jackson. Jackson becomes insistent that Action do something, to which Adams replies again that he needs a declared default to step in. It suddenly dawns on Jackson that it is up to him to act, and after conferring with First Bank and Development, Inc., the obligee he represents, he immediately drafts a letter defaulting PCC with demand upon the surety.

The more help obligees and their agents give sureties, the better they are able to help expedite any claim processes. In addition to the empowering letter, Adams asks for and receives copies of contract documents, change orders (COs), and payments from Jackson's files. With these documents in hand, he then contacts Prometheus, Jr. and informs him of PCC's default on the First Bank and Development, Inc. project. However, Prometheus, Jr. still insists PCC is solvent on this and the rest of its jobs. To determine the accuracy of this statement, Adams requests the records for all jobs even though only one is in default. Furthermore, he asks Prometheus, Jr. how he expects to meet his payrolls to continue work on the First Bank and Development, Inc. job and several others. He also requests that the indemnitors, including Prometheus Jr.'s older brothers and sisters, his father, and his mother, provide certified letters stating that they will pay any obligations, or that they post collateral.

Although it might be readily apparent to many within the construction industry in Anytown, USA, technically, at this stage, PCC is only in default on the First Bank and Development, Inc. job. When PCC refuses to go into voluntary default, Action immediately apprises J. Hermes of the situation. As his insurance company has failed to receive any premiums from PCC for worker's compensation insurance in several months—often a major sign that a contractor is in trouble as it is one of the first things a contractor fails to pay on time—Hermes is forced to cancel the policies and notify the affected architects. All the other architects are therefore forced to halt work, as there is no insurance in place, and all put PCC in default. Action is now effectively in control.

Scenario Two: See Scenario One and Be Reasonable

A surety can only take over a contractor's jobs when the contractor is in default or when the contractor asks the surety for assistance. As demonstrated above, a surety can take actions that may lead to the defaulting of a contractor for just such a purpose. Often, however, they won't have to. When it becomes obvious to the surety that the contractor is in dire straits, it is usually the case that the contractor is fully aware of the trouble. In such a situation, the surety may prevail in convincing the contractor that it needs help managing its workload and day-to-day operations. The contractor would then turn over to the surety any information it has on its jobs—all subcontractors, change orders, and invoice numbers—while fully cooperating with the claim team. This way, the contractor may be able to avoid a public defaulting, which might seriously hamper the contractor's personal reputation, especially in a smaller community such as Anytown, USA.

In addition, a contractor in trouble may voluntarily ask the surety for assistance. If this occurs, the surety should require a letter from the contractor stating that it is initiating this process. Assistance may include help with long-term financing or a temporary infusion of cash. Such financing can be official, meaning that the management of surety and any interested parties such a reinsurers are informed of the arrangements, or it can be unofficial. Such less-overt financing is often called "look-see" money, and sureties do not finance large amounts or very often with "look-see" money as it can become very costly for them down the road, especially in their dealings with reinsurers. It is often a stipulation in treaties signed between sureties and reinsurers that any financing, or even any claim decision, be first cleared by the reinsurer (see Chapter 7).

In this scenario, Action does not offer PCC any "look-see" money. After learning of the First Bank and Development, Inc. default, Adams calls in the Prometheus family and informs them of the looming collapse. After they all go home to "sleep on it," the family makes a collective decision to voluntarily put PCC in default, allowing Action to come in the next day to begin investigating the files. This decision has an advantage for both PCC and Action: cooperation. Both entities will be well served by the other's expertise and will be able to maximize whatever resources are available. This has the further advantage for J. Prometheus and his wife, who have an unblemished reputation in Anytown as straight-shooters and patrons of local little league, women's softball, and community arts.

Scenario 3: Bankruptcy

Whether Scenario One, Two, or neither is chosen, there is always the possibility that PCC may file bankruptcy *at any time*, allowing the court-appointed debtor-in-possession to decide what jobs PCC will complete, in what order, and how they will either restructure or dismantle themselves. Bankruptcy complicates the claim process immensely as all parties involved must retain counsel throughout a complex legal process. In some situations, however, if a contractor perceives its surety or lenders as aggressors, then bankruptcy may be perceived as a good strategy (see Zimbrick International case study below). Needless to say, bankruptcy would change everyone's roles in the claim process. For our purposes, PCC does not declare bankruptcy.

Job-By-Job Resolution, or Let's Make a Deal

If either Scenario One or Two is chosen, eventually, Action will be in control of PCC's job schedule. At this stage, it is always preferable to the surety to have the defaulted contractor as an ally, but the process works basically the same whether the contractor is friendly or hostile. For each of the defaulted jobs, the surety must analyze and decide the best individual course of action. The surety will have to hire engineers, architects, accountants, and other contractors to determine costs-to-complete. The better the records are and the more help they receive from the contractor and owners, the quicker this process will be resolved, making a claim situation better for all parties involved.

Within PCC's 10-job workload, Adams and his team find nine jobs in varying levels of physical completeness and a single job whose bid has just been accepted. Especially in claim situations, sureties prefer bonded subcontractors. With bonded subcontractors, a surety immediately knows large portions of its exposure since the subcontractor has agreed to a specific, guaranteed price. On PCC's earlier jobs, it is relatively easy for the claims team to determine the exposure amount, but hardly a straightforward endeavor. On the later, less-physically complete jobs with mostly unbonded subcontractors, the situation is much more complex. Assignment of work to new subcontractors may involve a great amount of negotiating and "deals" with old subcontractors, thereby increasing time, estimations, and finally, costs.

As a rule of thumb, the surety can expect a 40% to 50% additional cost on the remaining cost-to-complete. For example, if a contractor claims its work program will cost $10 million to complete, the surety can expect an additional $4 million to $5 million loss in costs. The Action claims team determines that it will cost $13 million to complete the 10 jobs, meaning that they stand to lose anywhere from $1.3 million to $5.2 million in costs. Technically, they can recover all of these funds from the Prometheus clan and PCC through the GIAs, but in reality, a surety will be unable to collect the full amount in situations like these. Making the sum as small as possible is thus the goal of Adams and all claim departments.

What follows is a job-by-job analysis. It just so happens that five of the six options available to a surety upon a default are used to meet their obligations on the PCC bonds. These are: 1) finance the contractor, 2) do nothing, 3) takeover and relet, 4) buy back the bond, 5) relet and walk away. The six option, obtaining bids and arranging a contract between the obligee and completing contractor, was avoided in this scenario by Action.

Finance Contractor

Action decides to finance PCC to complete its two oldest jobs, both projects that are 80% to 90% complete. One of the projects is a $1.5 million office building and the other is the $2 million building for Development, Inc. Action promptly pays Gonzalez Plumbing, who in turn withdrawals his mechanic's lien and completes the work. At this stage, J. Prometheus has returned as supervisor of day-to-day operations of PCC and wants to help put this whole episode behind him. He ensures the full cooperation of all staff including his son, who has overnight become much more humble. Action gathers the field crew and subcontractors and reviews all current work and plans for completion. Together, they formulate a schedule with weekly goals.

It is determined that the balance of the contract funds are insufficient for both projects. Action secures a bank loan from Anytown Bank for $500,000, guaranteed by Action, which is placed in a trust account. For the time being, Action will pay the interest and make sure that work on the two jobs is on schedule by having weekly meetings with the on-site managers, J. Prometheus, and J. Prometheus, Jr. These are the only jobs of the 10 that PCC will complete.

Table 20.2 PCC Project Resolutions: Finance PCC

Project (1)	Surety option selected (2)	Percent complete (%) (3)	Total contract price ($) (4)	Estimated cost-to-complete ($) (5)	Contract balance unpaid ($) (6)	Potential loss to Action ($) (7)
1	Finance PCC	80	2,000,000	650,000	400,000	250,000
2	Finance PCC	90	1,500,000	400,000	150,000	250,000
Subtotal	—	—	3,500,000	1,050,000	550,000	500,000

Do Nothing

On a $2.35 million job for a local industrialist that is 75% complete—a small fireworks factory—it turns out that the architectural firm (as an agent to the owner) grossly overpaid PCC for the work done in an effort to keep the work going. The company wanted to be able to ship out for the Fourth of July selling season, and they perceived any delays as costly. All told, over 100% of the contract value has been paid out to keep PCC building, as there were several significant COs.

One of Adams's team finds several letters in PCC's files stating that the architect knew he was overpaying, but as he was offered a bonus for an on-time completion, the architect wanted to keep job momentum. Adams informs the architect and owner that he will bring in three contractors and ask them to provide estimates for the cost-to-complete. He then informs them that Action is not going to complete the project; because of the architect's negligence in monitoring payment, Adams asserts that the architect's errors and omissions policy will have to be used. This makes the architect very nervous, as he was in error, and if a claim is filed from this $2.35 million job under his E&O policy, his premiums will increase dramatically—that is, if he can even manage to pay them. To avoid such a disaster, the architect has to choose between paying for the remaining construction by contracting the lowest bidding contractor or allowing a claim on his E&O policy. Either way, it is out of Action's hands (Table 20.3).

Table 20.3 Project Resolutions: Do Nothing

Project (1)	Surety option selected (2)	Percent complete (%) (3)	Total contract price ($) (4)	Estimated cost-to-complete ($) (5)	Contract balance unpaid ($) (6)	Potential loss to Action ($) (7)
3	Do nothing	85	2,350,000	500,000	—	—
Subtotal	—	—	5,850,000	1,550,000	550,000	500,000

Take-Over and Relet, or Finishing Surety

On a $1 million contract that is 60% percent complete for a the City of Anytown, Action is effectively forced to keep close ties with the job. Anytown wants its new women's shelter completed on time, and it wants Action to have a heavy hand in overseeing the work as further guarantee. Since Anytown is a good customer of numerous other product lines offered by Action's parent company, Action Insurance, management decides that it will take the role of finishing surety.

There are $400,000 in funds remaining to complete the contract in the account. Action determines through its consulting engineers that it will cost approximately $600,000 to complete. Since Action, like all sureties, is not in the construction business, it agrees to hire a subcontractor who will effectively become the general contractor. It will be the subcontractor's responsibility to provide workmen's compensation insurance, which Action underwrites. Under an agreement reached with Anytown, Action agrees to spend up to $300,000 to complete the work, making the total money available for the project $700,000. In addition, all the current work is to be accepted as fit, and any expenses above the $300,000 ceiling will be paid for by Anytown. The deal is a successful one, as within six months—only a month behind the original schedule—400 of Anytown's battered, abused, and homeless women are in a safe and sturdily constructed environment, and it cost Action only $250,000 (Table 20.4).

Table 20.4 PCC Project Resolutions: Take-Over and Relet/Finishing Surety

Project (1)	Surety option selected (2)	Percent complete (%) (3)	Total contract price ($) (4)	Estimated cost-to-complete ($) (5)	Contract balance unpaid ($) (6)	Potential loss to Action ($) (7)
4	Take-over and relet	60	1,000,000	600,000	400,000	250,000
Sub total	—	—	6,850,000	2,150,000	950,000	750,000

Buy Bond Back

Since sureties generally have connections with architects, owners, and contractors, they can sometimes find suitable situations that pull together several various strands. A surety tries to find marriages when a claim arises; if a shot-gun marriage presents itself, a surety will most often jump at it. Mystery costs will undoubtedly rise with time, so a surety tries to determine its exposure and get out as quickly as possible. On the $1.5 million contract that PCC was just awarded, Action pays the owner $75,000, the penal value of the 5% bid bond. Since Prometheus, Jr. was in charge of estimating the bid, PCC was over 20% lower than the nearest bidder, who came in at close to $2 million. This made the penal value of the bond a much more attractive alternative than trying to perform the contract with insufficient funds.

On another job, a $2 million commercial building 80% complete, Action finds a solution immediately acceptable as well, this time on its performance and payment obligations. The Jefferson Group has commissioned duplicate buildings adjacent to each other on Anytown's capitol square, one to PCC and one to Saturn Contractors. Action consults Saturn and learns that the contractor will perform the work for between $700,000 to $750,000. Action arranges to meet with The Jefferson Group and agrees to pay them $400,000. In addition to the $400,000 left in account, this would give The Jefferson Group $800,000 to finish the project, a seemingly generous settlement. Over time, however, this turns out to be a good deal for Action: with significant COs, The Jefferson Group has to pay over $1 million to complete the second building.

On a third project, this one a $3 million, 50% complete structure with 50% of the funds still remaining, the architectural firm of BFD claims that the work performed by PCC is shoddy and unacceptable. Action again brings in three contractors to estimate the cost-to-complete, asking BFD to describe the poor quality work. When it turns out that the estimated cost to remove the substandard construction is $1 million *over* the contract balance, Action informs the architect that they are responsible for the poor quality work because they failed to monitor the progress.

BFD, in consultation with its law firm of Fine & Dandy, decides to take Action to court, as it appears that the surety is unwilling to perform its obligations per the performance and payment bonds. Action, upon contacting BFD's insurer, arranges a meeting. Since no one wants to have this error on the books, Action agrees to pay 30% of the estimated cost-to-complete, while BFD and its insurer agree to pay 35% each. It is now the responsibility of the architect and obligee to find a new contractor and put up a new bond (Table 20.5).

Table 20.5 PCC Project Resolutions: Buy Bond Back

Project (1)	Surety option selected (2)	Percent complete (%) (3)	Total contract price ($) (4)	Estimated cost-to-complete ($) (5)	Contract balance unpaid ($) (6)	Potential loss to Action ($) (7)
5	Buy bond	0	1,500,000	1,500,000	1,500,000	75,000
6	Buy bond	80	2,000,000	750,000	400,000	400,000
7	Buy bond	50	3,000,000	2,500,000	1,500,000	750,000
Sub total	—	—	13,350,000	6,900,000	4,350,000	1,975,000

Relet and Walk Away

On a relatively new $650,000 job 30% complete, this one for H&H Management, the engineering firm Action hired as a consultant estimates that it will cost roughly $600,000 to complete. All current work is deemed satisfactory by the engineer and obligee. There is approximately $450,000 left in the account.

Action contacts one of the contractors it underwrites, Theseus General Contracting (TGC), to bid the job. Finding all the work acceptable as well, the contractor says it can complete the work for $700,000 using all the current subcontractors. H&H Management wants the payments scheduled, as a typical contract at this stage would be, but Action has another idea. The surety offers to pay TGC $650,000 up front and write a new bond for the remaining construction, premium-free. TGC and the owner agree, and Action has managed to relet this contract for $200,000 over original contract funds without having to monitor any progress.

Action has to deal with another newer job, this one only 20% complete. The contract is for $2 million, and there is $1.6 million left in the account. Again, action contacts TGC. This time, having some idea what Action has in mind, TGC offers to complete the contract for $250,000 over the remaining contract balance for a total of $1.85 million. Action could employ an engineering firm to come up with another estimate and then bring in other contractors, but it first tries to negotiate with TGC.

Realizing the $250,000 figure is probably inflated, Action settles with TGC on $1.8 million paid in installments, which is exactly $200,000 over the remaining contract balance. Again, Action agrees to write the bonds at no charge to TGC. This option may prove more costly down the road for TGC, but again,

Action only has to be concerned that TGC does not default. Since TGC has a work program of over $30 million, assets of over $6 million, and has a preferred rating with Action, it is a pretty good bet that it will not. TGC does not, and Action ends up paying $200,000 for the completion of this job.

PCC has just broken ground on a $4 million strip mall with an atrium playground for children. Action immediately contacts the next lowest bidder. Roughly 15% of the work has been completed by PCC and it appears to be in good shape. In a rare instance of underbilling (for PCC), only 10% of contract funds have been paid out. In addition, there are minor mobilization costs as much of the work is being performed by subcontractors who have agreed to stay on-board. The new contractor therefore agrees to take the job for the remaining balance of the contract. As part of the arrangement, it is their responsibility to procure bonding. Action pays the owner $400,000, or 10% of the bond, and has no more contact with the project (Table 20.6).

Table 20.6 PCC Project Resolutions: Relet and Walk Away

Project (1)	Surety option selected (2)	Percent complete (%) (3)	Total contract price ($) (4)	Estimated cost-to-complete ($) (5)	Contract balance unpaid ($) (6)	Potential loss to Action ($) (7)
8	Relet	30	650,000	600,000	450,000	200,000
9	Relet	20	2,000,000	1,850,000	1,600,000	200,000
10	Relet	15	4,000,000	3,600,000	3,600,000	400,000
Total	—	—	20,000,000	12,950,000	10,000,000	2,775,000

Resolutions

In the end, Action ended up paying roughly $2.8 million to finish the work, which was roughly 20% of the original cost-to-complete. This did not include engineering fees, accountant fees, and other miscellaneous costs. All of these added up to nearly $500,000, which means Action's total costs ran to the tune of $3.3 million, which was about 25% of the original estimated cost-to-complete. As far as losses go, this one was not so bad, as a rough estimate for surety costs is from 30% to 40% of cost-to-complete.

Action reached an agreement whereby they received $1 million in cash from A. Prometheus and his wife and recovered much of their costs in the sale of the equipment and property of PCC. Finished with the construction business, A. Prometheus, Jr. is currently enrolled in law school, as he became fascinated with the earning potential and lifestyle of an attorney-at-law through the drawn-out and complicated defaulting of PCC. In addition, after carefully consulting with J. Jackson, the underwriter, Action's Adams decided that there was little that could have been done to predict PCC's downfall. J. Hermes, for his part, could have notified Action of PCC's failure to pay its premiums, but he too is not to be blamed.

Table 20.7 clearly summarizes the costs for each project, as well as Action's job-by-job and total exposure. Note that the administrative costs for monitoring and managing the claims, including consultants fees, which

Table 20.7 Resolutions of All 10 PCC Projects

Project (1)	Surety option selected (2)	Percent complete (%) (3)	Total contract price ($) (4)	Estimated cost-to-complete ($) (5)	Contract balance unpaid ($) (6)	Potential loss to Action ($) (7)
1	Finance PCC	80	2,000,000	550,000	400,000	250,000
2	Finance PCC	90	1,500,000	400,000	150,000	250,000
3	Do nothing	85	2,350,000	500,000	—	—
4	Take-over/relet	60	1,000,000	600,000	400,000	250,000
5	Buy bond	—	1,500,000	1,500,000	1,500,000	75,000
6	Buy bond	80	2,000,000	750,000	400,000	400,000
7	Buy bond	50	3,000,000	2,500,000	1,500,000	750,000
8	Relet	30	650,000	600,000	450,000	200,000
9	Relet	20	2,000,000	1,850,000	1,500,000	200,000
10	Relet	15	4,000,000	3,600,000	3,500,000	400,000
Total	—	—	20,000,000	12,950,000	10,000,000	2,775,000[1]

[1] Loss figure does not include $500,000 in professional fees. Total loss would be equal to almost $3.3 million.

significantly raised the total exposure, are not included in the total exposure figure. Another $500,000 was depleted coordinating the resolution of the 10 jobs.

In a claim situation, a surety has to wear many hats. In this example, the surety managed to recoup most of their expenditures with some informed claim decisions and through the GIAs, but this in not always the case. In the end, perhaps Action should have followed a long-standing bit of surety lore and maybe none of this would have happened: never underwrite a contractor with an initial in place of a first name, and never underwrite a contractor with a deity in the company title. According to legend, PCC already had two strikes against it.

CLAIM CASE STUDY: THE DEFAULTING OF ZIMBRICK INTERNATIONAL

In 1998, Zimbrick International was considered robust. With over 70 years of building experience, the contractor seemed poised to continue prospering into the foreseeable future. After a period of unprofitability in the 1980s, Zimbrick had turned itself around. Concentrating energy on its in-house engineering and construction services, Zimbrick sold off all its non-construction subsidiaries in 1993, netting profits close to $150 million. With this money, Zimbrick paid off all its debt, wrote off uncollectables, and paid a sizable dividend to shareholders—the first since 1987. That year also marked the promotion of long-time Zimbrick employee Wayne Patrick to chief executive officer.

Under Patrick's guidance, Zimbrick netted $536 million in 1997. He steered Zimbrick away from the low-profit public works projects that had been a staple of the firm since the 1930s into what looked to be the more profitable—and more risky—private sector. Patrick's plan was to build large specialty projects such as pulp mills and generator plants. While Zimbrick proved profitable with many of its jobs, problems developed after 1996. Several huge, multi-million dollar contracts were uncollected or deemed uncollectable in the space of a few months, and all at once, the problems started mounting.

First, when the private owner seemed unable to pay for construction services, Zimbrick agreed to accept deferred payments for a Plano, Texas, assisted-living complex. In exchange for this flexible payment schedule, Zimbrick was to receive an extra $6 million upon completion. After all work was in place and the facility was ready to be turned over, the developers filed for bankruptcy before Zimbrick was properly paid. This left Zimbrick with a $60 million account receivable. The contractor expected to see all the money when the facility was eventually purchased, as they were confident it would be, but in combination with two other huge defaults, Zimbrick's creditors saw things differently.

Next, a hydroelectric plant near Lima, Peru, completed late in 1997, went over budget because of intense inclement weather including severe flooding lasting much of the spring and summer. Zimbrick's acceptance of a fixed-price contract raised the final cost of the $99 million contract to $155 million, but even still, it expected to recoup much of this cost. Finally, Zimbrick was expected to lose over $25 million on a $72 million contract to build a multi-purpose factory in New Jersey. This third troublesome project was eight months behind schedule due to the discovery of hazardous materials on the site that had to be excavated. However, once again, Zimbrick expected to recoup a good portion of incurred costs—if not through the owners, through the depositor of the materials and the courts.

As late as April 1998, Zimbrick's prospects looked good enough for Patrick to declare to company shareholders, "The financial strength of Zimbrick International is solid." Until the last pieces of a buy-out were finalized in early 1999, Patrick continued to speak confidently about Zimbrick's prospects even as pressure continued to build from its creditors and surety. Then, on July 8, after Zimbrick was unable to meet the terms on a $67 million loan from a five-bank consortium led by First Trust Bank due on May 31, 1998, all Zimbrick assets, which totaled close to $20 million, were subsequently frozen. With no access to its bank funds, Zimbrick International filed Chapter 11 in U.S. Bankruptcy Court in Dallas, Texas, on July 10, 1998.

Patrick and those close to him claimed the filing strategic: it was Zimbrick's answer to what would be described in the popular press as a "fight to the finish" and by Zimbrick in court proceedings as a "preposterous" decision by the banks. Patrick claimed throughout that the bankruptcy filing was primarily a tool to recover the frozen funds it needed to continue operating, which included wrongly seized joint-venture monies. Zimbrick argued that with sufficient time, it could meet all debt and accrued interest. In addition to the $20 million seized by the banks, the contractor expected revenue from its three galling projects totaling over $130 million. Unfortunately for all involved, the timing of these expected payments forced the banks to call the outstanding loan and Zimbrick to rigorously avoid default, setting the stage for a lengthy and complicated ordeal.

The bank consortium was unwilling either to extend the $67 million loan period or lend any other money to Zimbrick, even with what appeared to be backing by its surety. Furthermore, the consortium wanted Zimbrick to cease work permanently so the firm could be dismantled to repay its debt. One of the banks involved in the original loan, Righteous Savings and Trust, eventually sold its debt for 75 cents on the dollar to cut its losses as litigation entered its fifth month—over six months since the original loan was due.

During the litigation process, Zimbrick was eventually granted access to the $20 million in frozen funds by U.S. Bankruptcy Court Judge P.F. Sloane. However, without adequate funding, Zimbrick would be unable to continue working on its almost $1 billion aggregate work schedule. Since 31 of the company's 47 total projects were bonded, Zimbrick's surety had an obvious vested interest in resolving any problems—be they financial, legal, or both—as soon as possible.

From the beginning, Zimbrick's surety, The Surety Group, stood behind the contractor. The surety seemed willing to provide funds under an agreement to finance Zimbrick and continue writing bonds subject to review and approval. With court-approved debtor-in-possession financing, the surety advanced Zimbrick nearly $58 million in funds from roughly July to December 1998 to continue work on a portion of its work program. Without this money, Zimbrick would have been unable to remain even marginally viable. However, even such a sum was insufficient for Zimbrick to pull itself out of its legal and financial mire.

As court proceedings piled up, it soon became apparent that Zimbrick would be unable to continue operating as an autonomous entity. Zimbrick needed more than just a piece-meal loan from its surety, and without creditors, the necessary funds seemed impossible to secure. At this time, Zimbrick International was unofficially put on the market by The Surety Group.

After Thanksgiving, rumors of a sale or merger began to circulate and were reported in the major construction and financial weeklies. The consortium of banks still wanted Zimbrick to cease work entirely so the process of selling off its assets could begin. However, with more money involved than the banks and with multiple obligations to multiple owners, The Surety Group wanted as much of the work completed as possible.

In January, 1999, a possible sale of Zimbrick to Caltex Contracting was widely reported in the press. However, few with close connections to either Zimbrick, its surety, or the banks would reveal any particulars. The deal eventually fell through because of an inability to agree on what to do with the several contracts Zimbrick stood to lose significant money on. Nonetheless, another arrangement quickly materialized.

In a deal that speaks volumes to the realities of the construction and surety industry, a contractor with close ties to The Surety Group absorbed Zimbrick into its operations. The Trident Construction Company of Wheeling, West Virginia bought Zimbrick's name for $2.25 million with another $9 million in future profits. Zimbrick officially became a subsidiary division of Trident with a stipulation that any Zimbrick employees who chose to could join the Trident payroll. Trident would also be able to bid for new work with the Zimbrick name.

The Surety Group undertook to handle Zimbrick's 31 bonded projects with roughly $300 million of work left to complete, and the banks retained control of Zimbrick's 16 unbonded jobs and 21 joint ventures. This arrangement allowed the ailing contractor to remain somewhat viable, if less independent, but even as the deal transpired, Zimbrick continued to point to profits expected in the near future. From its joint ventures alone, Zimbrick expected to earn close to $75 million over the next several years. As to its financial viability, right through to the end, Patrick blamed the banks for not allowing Zimbrick to continue working.

Through several large projects, including the Plano apartments and Peruvian hydroelectric plant, Zimbrick International became severely taxed. Although profitable on many of its projects, the contractor had proved that its estimating systems were suspect. After the banks refused to extend the period of the loan and then seized Zimbrick's remaining assets, the contractor was forced to file bankruptcy. At this time, with a running deficit and no access to funds, the surety became Zimbrick's lifeline. In today's multi-billion dollar construction market, when banks are hostile and when jobs need to be completed, a bonded contractor in trouble has nowhere else to look. Zimbrick accepted—or perhaps was forced to accept—its merging into another company, releasing its autonomy and a 70-year history. Behind the scenes, Patrick revealed that being sold was clearly not Zimbrick's first choice. However, as litigation piled up, he admitted that it was only a matter of time before something had to give.

On a defaulted project or with a defaulted contractor, the surety must step in and quickly determine its exposure while simultaneously fulfilling its obligation under the bond. When it became apparent that Zimbrick's creditors were not going to allow it to continue operating and after the surety determined that Zimbrick could not overcome the situation, an arrangement was worked out to allow the resolution of multiple bonded and unbonded projects. It should come as no surprise that Trident and Zimbrick share the same surety. Unofficially, The Surety Group brokered the deal between the two companies. Sources close to The

Surety Group say that losses have reached $150 million and that it is unclear how much of this loss might be recouped. However, the surety is in business to handle defaults when and if they arise. It speaks highly of Zimbrick executives and The Surety Group that in the end, no Zimbrick employee lost out on pay or a job because of the larger company's failure to overcome: every Zimbrick employee was welcome at the new subsidiary at Trident. Currently, it remains to be seen what the future holds for Zimbrick.

Part VI

PRACTICAL CONSIDERATIONS REGARDING THE SURETY INDUSTRY

In essence, a surety is in business to prequalify potential construction clients, selecting those that it finds to have minimal or acceptable levels of risk. An assessment of potential clients generally entails examining the business' liquidity, net worth, and overall profitability, as discussed at length in Part 4. Naturally, with each individual contractor the surety has special and unique concerns as to the particular financial risks it accepts. There are, however, concerns of sureties that extend past the individual client to the wider industry. Such can be termed macroeconomic variables.

As a result of macroeconomic factors such as inflation, trends and concerns emerge within the surety industry at multiple levels and in multiple guises. These factors change over time, and to understand any part of the industry it is crucial to determine the health of the industry as a whole and how certain developments are affecting such things as underwriting and claim management. Understanding whether the surety industry is in a profitable or struggling cycle, for example, yields information about the chances of certain kinds of firms securing bonding. In more profitable climates for the surety, the chances of a moderately capitalized builder finding a surety to underwrite its performance bonds are generally greater than when the market is suffering from a lack of capacity. Another example would be with proactive claim management. A contractor should expect a surety to become involved much earlier when presented with a claim by an owner or claimant under a payment bond than would have been the case even 10 years ago because of a change in the corporate climate (Klinger 1997).

Understanding the status of the surety industry today requires a brief analysis of where it was yesterday. The surety industry in the mid-1980s appeared to lose sight of its purpose, and as a result suffered significant financial losses. Instead of focusing efforts on underwriting, sureties wanted to maximize their premium income so they could invest more premium dollars in the high-interest financial market. The logic was that any defaults would be offset by the increased profit from investments. However, as losses piled up in excess of the investment income, sureties became increasingly concerned with sound underwriting, often tightening standards and making it increasingly difficult for under-capitalized or small contractors to secure bonding. This strict underwriting climate survived well into the 1990s.

Recently, however, the surety industry has again proven profitable. In this new wave of profitability, the industry has witnessed an incredible boom in the number of sureties underwriting construction contract bonds, making for some extremely high levels of competition. In view of this intense competition, many industry experts are pointing to a slackening of underwriting standards as sureties attempt to attract new contractor clients while holding on to their current ones.

Upwards of 600 surety companies or divisions are writing bonds, and many contractors who would have been denied bonding in the mid-1980s today have their choice of sureties. For the contractor wishing to pursue bonded work and for the owner wondering how requiring bonds will affect competition for bids, the surety market itself must be understood.

What follows is a discussion of industry-wide concerns and trends. The earlier sections of this book, while in no way ignoring the wider industry, have focused on how an individual contractor, owner, or surety goes about bonded construction and settling the claims or defaults that arise. Part 6 is a detailed exploration of how the surety industry has responded to macroeconomic factors and how it is currently evolving. Included is a chapter on alternatives to surety bonding, such as self-surety and letters of credit, and an exploration of guaranty in an international context. A chapter each will discuss aggregate industry performance, sources of information for the surety, and common disputes arising from surety contract bonds. To conclude, the final chapters will explore special surety concerns and industry trends, offering the author's analysis of where the industry is and is not heading. A summary follows.

Chapter 21

ALTERNATIVES TO SURETY BONDING

Through the Miller Act, federal law requires surety on government public works. Many states have similar laws on their books, the so-called "Little Miller Acts," and the private sector is increasingly using surety. Furthermore, the industry is in a highly profitable cycle where sureties vie to underwrite healthy contractors, lowering premiums and sometimes, underwriting standards. However, for various reasons, there are still many contractors who have and will continue to encounter difficulty securing bonding.

In a congressionally mandated study of the surety industry, the U.S. General Accounting Office (GAO) found that of small contractors who had received bonding in 1990-1993, over 20% were denied a bond during the same period. While the surety industry rightfully points out that 80% of those requesting bonds were not denied access to this product, there will always be contractors wishing to bid bonded jobs that are unable to do so. In addition, owners need to be aware of alternatives to surety bonding to be more informed consumers and to better understand the specific type of coverage a surety bond provides. These alternatives include: 1) waiver of bonds, 2) individual sureties, 3) self-sureties, 4) cash equivalent, and 5) letters of credit.

WAIVER OF BONDS

On a private project, the owner decides unconditionally whether or not to require bonds. If after calling for surety in project specifications the low bidder cannot secure bonding, the owner can collect under the bid bond or choose to allow the low bidder to provide some other form of guaranty. Additionally, the owner always has the right to waive any security requirements as it sees fit. Therefore, if any situation should arise where a contractor has trouble securing the required surety bonds, one option to consider would be presenting the owner with an alternative form of security discussed below. Chances are good, however, that in such a case the owner will go to the second-lowest bidder and call the lowest bidder's bid bond.

For public projects, this alternative to surety includes two options: 1) waiving bonds on construction projects or 2) increasing the project size threshold at which bonds are required. The waiving of construction project bonds for projects involving public monies does not appear to be a viable option given current statutory laws that make surety bonding a requirement. Federal law has been passed allowing the federal official in charge of a construction project to waive bonds on occasion, accepting in its stead some other guarantee, but it does not appear as if any other form of guaranty will anytime soon displace surety bonds.

However, increasing the construction size threshold has been and will continue to be a viable option to avoid surety. An amendment to the Miller Act and many state statutes was made in 1991, increasing the minimum project requiring surety from $25,000 to $100,000. In the future, if it is found that requiring surety on smaller projects does not best serve the public interest, then this threshold could be raised again.

For private owners, the decision to use surety bonds is of course entirely discretionary. There is no law compelling owners to purchase the protection surety bonds provide. For a company that is unsure of whether to use surety bonds or whether to waive the requirement for a low bidder who cannot secure the specified bonds, this chapter should provide an overview of why bonds should at least be considered. In addition, the following alternatives to surety bonding could be considered as well.

INDIVIDUAL SURETIES

Originally, suretyship was personal, but like Antonio in Shakespeare's *Merchant of Venice*, problems with the terms of contracts were rife. Due to a lack of security, collection problems with personal sureties led to the development of corporate surety in the 19th century. Today, individual sureties are still used on occasion.

An individual surety pledges personal assets to back the performance of second person to a third party guaranteed by a bond, much like corporate surety. The individual surety determines the terms and conditions of bonds written, and although there is a perceived lower cost margin, premiums generally exceed those of corporate sureties. Remember that a premium is used to in part cover inevitable losses. An individual surety, with more personal risk in the case of loss, must further hedge any risk by increasing premiums to cover expenses when a contractor fails.

This alternative has had more than its share of problems ("FBI Probes" 1990; "Surety Firm" 1990; "Bad Bonds Bring" 1990). In 1989, the federal government made official the regulation of individual sureties on federal construction. Published in the *Federal Register*, the rule specifies the types of assets that can be pledged, which cannot include accounts receivables, foreign securities, foreign real estate, jewelry, corporate assets, or "speculative assets" (e.g., mineral rights). The surety must present the government officer with "objective evidence" of ownership. The government contracting officer must first determine whether the individual surety is acceptable, meaning that it possesses proper assets in an amount sufficient to cover the bond. If it is determined that the individual possesses the necessary funds, they should be placed in an escrow account. If domestic property is used instead, then the government agency should be given priority on any liens ("Government Reins In" 1989).

The Associated General Contractors of America (AGCA) and the American Subcontractors Association (ASA), among others, have voiced concern that this rule undermines reasonable standards of governing sureties. Individual sureties often lack the administrative support staff and in-house expertise for handling bond claims effectively, meaning that in the case of default, there is the chance the public, as well as any subcontractors on the job, will not be well-served by such a guarantee. Individual sureties are not regulated or subject to the rules and requirements pertaining to the insurance business, such as corporate sureties are with the Treasury List. Possibly because of this, there have been numerous instances of fraudulent individual sureties in the last 10 years alone.

Any private owner considering accepting a surety bond issued by an individual should make every effort to authenticate the bond and the individual's assets. For smaller firms denied access to corporate surety bonds, individual surety may be a viable option for them to still bid bonded construction work. The private owner should make every effort to determine the financial position of the surety. The private owner and the contractor should also be very cautious of the authenticity of all bonds, not just individually backed ones. Fraudulent corporate bonds are not uncommon. If there is any question about a certain bond, contact the surety directly, as a short phone call can often establish without question whether a bond has indeed been issued.

SELF-SURETY

Self-surety is a practice similar to individual surety whereby a public entity may provide its own bonds without engaging the services of a surety company. The public entity reallocates the money that would have gone toward the bond premium to their own resources. The public entity's surety program then essentially performs the same screening and prequalification process, evaluating the potential contractor's ability to complete the work successfully, much as a surety company would. Once the bonding program is in the hands of the public entity, the coverage can be as broad or as specific as the entity wishes. However, most jurisdictions have legislation that restricts the powers and activities of public entities in this regard. This means that a public entity cannot provide its own surety unless new legislation passes that allows the entity to do so.

To effectively implement a self-surety program, the entity must establish a surety management program consisting of an actuarially sound plan supported by an aggressive loss control program and a catastrophe fund. The entity must forecast losses over an extended period of time based on past performances and then discount them to present values and funds contributed to an expected loss or retention fund. Each year, if the actual losses do not exceed total losses, the entity may retain any surplus in the fund for use in future years. In addition to loss forecasting, the surety management program would also perform the other functions that the surety company typically handles. The program would include contractor prequalification, cost estimating, record keeping, quality control, and claims management.

If the entity lacks these resources, or is unable to reallocate them, the entity would need to buy or develop the experience needed to perform the underwriting activities (Shiatte 1989). Since the success of the self-surety program heavily depends upon constant data gathering and updating information, any staffing shortages may cause serious setbacks and possibly program failure depending on the scope of work in progress.

CASH EQUIVALENT

Cashiers checks, certificates of deposit, or cash are a means by which contractors can demonstrate financial security and responsibility. This method is sometimes used in place of a bid bond, but can also be used to guarantee performance and/or payment on a project. Concerns as to what amount is actually adequate to cover the potential losses exist with this method. Usually, an owner will require 5% to 10% of the contract, but depending on the project, some may require more security, i.e., more cash.

The contractor will not have access to these funds for the duration of the contract. Especially if the contractor is small, it is important for the owner to realize that this may put undue stress on the contractor's finances. This may increase incidence of default, but there is no statistical evidence to back up this assumption. A retention clause, whereby a portion of the funds are withheld from each payment to the contractor, might alleviate some of these problems.

LETTERS OF CREDIT

The letter of credit (LOC) is simply a cash guarantee that the owner will be satisfied. Owners "cash in" this type of security when the contractor fails to perform or deliver a product that meets the plans and specifications of the contract. Upon being called, the LOC converts to a payment to the obligee and an interest-bearing loan for the contractor. Unlike a performance bond, the letter of credit has no completion cause. Whereas the surety always has the option of completion, the guarantor of the LOC, generally a bank, has no interest in fixing or completing a project if and when a contractor fails to perform. If the owner becomes dissatisfied with the contractor for any reason, he or she usually has the unconditional right to immediately demand the funds per the letter of credit. If the owner has some burden of proof regarding the alleged contractor failure or poor performance, the letter of credit can then be termed a "conditional guarantee" or a "conditional letter of credit." This is a very important distinction, especially for the international contractor, and as such, the proper structure of an international LOC is discussed below and analyzed in more depth in Chapter 22.

A good part of the confusion surrounding the LOC is that these on-demand instruments, while sometimes called "bonds," are in actuality not surety bonds. Most LOCs used to secure contracts are issued by banks and are thus subject to banking and not surety law. As there exits no burden to investigate, most banks will be forthright in paying the owner upon demand. Therefore, it is imperative that the contractor completely understands the terms and conditions that can lead to the calling of a letter of credit before he or she provides one to guarantee performance or payment on a contract.

The nearly automatic payment by the bank upon notification of the owner or the owner's bank may be avoided by the structuring of the LOC, making it a so-called conditional letter of credit. However, even protective clauses worked into the language of the letter do not preclude unfair or unnecessary calling and consequent cashing of letters of credit. Once an owner has called in a letter of credit, payment can usually only be stayed with a court order, and such orders are not easy to obtain, especially not with a foreign owner or bank. In addition, banks, unlike surety companies, are generally unfamiliar with contractor operations and do not often have the resources to adequately assess project risks. Therefore, the element of risk for a bank without the indemnification provided the surety is considered to be greater. As such, premium rates for on-demand guarantees are often much higher than those charged for surety bonds. Typically, a contractor pays 1% or more of the contract for the conditional letter of credit. The total premium for a letter of credit may sometimes be less than that paid for a 100% surety bond because a lower percentage of the contract is usually being guaranteed. Whereas a performance bond is generally for 100% of the contract value, a corresponding letter of credit guaranteeing performance is usually only for 5% to 15% of the contract. However, if the letter is called, the amount of the payment will then be owed the bank by the contractor. Depending on the contractor and his or her relationship with the bank, interest on this loan will accrue at or above prime rates.

Table 21.1 Comparison of Forms of Guaranty
Source: Surety Information Office

Construction assurances (1)	Surety bond (2)	Self insurance (3)	Letter of credit (4)
Prequalification of contractor	Yes	Yes	No
Continuous independent project monitoring	Yes	No	No
Extraordinary protection/ indemnity to avoid loss	Yes	No	No
Limited financial protection	No	No	Yes
Full financial protection	Yes	No	No
Unqualified forfeiture	No	No	Yes
Required to investigate complaints	Yes	No	No
Ability to respond to contract problems by promising to:			
Finance contractor	Yes	Yes	No
Assume construction responsibility	Yes	No	No
Tender offer to a completing contractor	Yes	Yes	No
Assist existing contractor to complete contract assuming responsibility	Yes	Yes	No
Forfeit financial limit	Yes	No	Yes

The amount of a letter of credit and what it costs generally depends on where it is issued, both within the United States and internationally. The financial institution that issues the LOC takes into consideration the financial condition of the contractor, local economic conditions, and possibly macroeconomic conditions external to the project area. As with a surety bond, the knowledgeable contractor passes the costs of the letter along to the owner in the bid price and should therefore have established a relationship with the issuing lending institution prior to bidding.

One issue pertinent to both the owner and contractor is whether to obtain a LOC or utilize a retention clause in the bid. The retention clause typically states that a certain percentage (typically 5% or 10%) of all payments made during the course of the work may be withheld by the client to ensure that the contractor rectifies defects identified during the work or within any specified warranty or maintenance period subsequent to completion of the work. This dispute between these two types of security cannot be easily resolved because the project owner and the contractor may disagree on which is more appropriate to the project. Essentially, however, both the LOC and retained funds amount to the same type of protection for the owner and the same predicament for the contractor should he or she be defaulted or experience cash flow problems.

In the last several years the use of the letter of credit by American contractors has increased slightly in part due to two factors: 1) changes to Federal law and 2) the use of LOCs in some foreign countries. In many parts of the world and for many small and disadvantaged contractors, the use of letters of credit could even be termed prevalent. For a new contractor unable to secure surety bonding, the LOC allows the firm to perform guaranteed work it might otherwise be unable to perform. For these contractors, the LOC can be viewed as an instrument of growth. For many foreign owners unfamiliar or opposed to surety bonds, the LOC is the only type of guarantee accepted. Table 21.1 helps clarify some of the differences between surety bonds and letters of credit.

Recently the U.S. Office of Federal Procurement Policy has resurrected a rule allowing contractors to use an irrevocable letter of credit in lieu of payment and performance bonds on federal construction projects requiring guaranty, which is to say on projects of $100,000 or more. For such projects, the LOC should be in an amount equal to the penal value of a comparable bond, which is to say in the full amount of any payment and performance guarantees. This policy change was designed to give more flexibility and contracting opportunities to smaller and disadvantaged firms unable to obtain bonds. In actuality, however, it is questionable how much flexibility this amendment provides. To perform even a $100,000 contract, a contractor would need to post a letter of credit worth $150,000 to cover performance and payment guarantees required under the Miller Act ($100,000 for performance and 50% of the contract value for payment, or $50,000). A contractor qualifying for a $150,000 letter of credit should also be able to secure a surety bond, which from a contractor's perspective is generally preferable for a variety of reasons. For starters, the cost of a surety bond may be less than the cost of the LOC. In addition, the letter of credit is an on-demand instrument subjecting the contractor to the possibility of an unfair call and should a borrowing need arise, the LOC will diminish the amount of credit available to the contractor. Furthermore, the contractor electing the letter of credit does not have the benefit of the surety company opinion as to whether a particular call against the surety bond has merit. For small or disadvantaged contractors interested in acquiring bonding, the Small Business Association oversees several surety programs discussed further in Chapter 26.

The second factor stimulating an increased use of the letter of credit is that contractors operating overseas have found many foreign owners unwilling to accept surety bonds, in part due to their unfamiliarity with the product. To these foreign owners, a bank's guarantee of the contractor's performance is more desirable. However, this desire puts U.S. banks in a tough position because Federal regulations limit their contingent liability on construction projects as well as legal lending limits whereby they may extend credit to one customer of no more than a specified percentage of the bank's capital and surplus. Therefore, it is possible that a large foreign contract could exhaust a bank's legal lending limit and strain its ability to guarantee a construction project. The current merger of large insurers and banks might force some of these regulations to change, but it remains to be seen exactly how or when this might occur.

The prevalence of LOCs internationally is in part due to the past practice of surety companies being unwilling or unable to accept the risks of bonding a contractor performing work in foreign countries. However, this trend is slowly changing. Sureties are now more willing than ever to entertain the possibility of guaranteeing foreign work as more contractors are looking overseas for projects. In addition, foreign owners are recognizing the merit of the surety product—especially the clause allowing the surety to complete a defaulted project. In the following chapter, special attention will be paid to how LOCs can be structured to avoid many of the dangers of an unfair or unjustified call by a foreign owner, but as with any international endeavor, the level and kinds of risk change dramatically.

Chapter 22

INTERNATIONAL GUARANTY

In an era in which globalization is being realized and internal competition is fierce, construction outside the United States can be a very attractive option for mid-to large-size contractors. As such, many U.S. contractors are pursuing overseas projects or traveling across borders to capitalize on their experience and expertise. International construction is highly complex and dynamic and offers many opportunities, but it also exposes the contractor to some additional risks. For the perspective or current international contractor, understanding and carefully analyzing these new risks is essential. In addition, the contractor must also know how these various risks are allocated in other nations in order to better serve foreign owners and protect oneself.

Although surety bonds are widely used in the United States and Canada to secure the performance of a construction contract, much of the rest of the world is accustomed to using letters of credit. It is the purpose of this chapter to offer an overview of the complex world of risk allocation in international construction, exploring how these two major security tools are generally used.

Several resources may be of assistance to the contractor looking to explore international markets. However, as international construction is rapidly evolving, the responsible contractor should thoroughly investigate local laws and customs before entering into any contracts, as the penalties for mistakes in bid estimation may be much more costly than they are in the United States or Canada.

The first sources to turn to for information on construction guarantees in a particular nation are *The Uniform Customs and Practice for Documentary Credits* (UCP) and the *Uniform Rules for Contract Guarantees* (URCG), both published by the International Chamber of Commerce (ICC) based in Paris. As the name implies, the UCP provides normal contractual and business procedures and practices for the some 110 member countries the ICC represents, while the URCG provides guidelines for obligations triggered by contractual default (Tiong 1992). Other international monitoring organizations include the Federation Internationale Des Ingenieurs Conseils (FIDIC) and the Institute of Civil Engineers, United Kingdom (ICE). The FIDIC has created the International Conditions of Contract, while the ICE has a similar Conditions of Contract, both of which detail how international contracts can and should be structured. Even so, it is important to note that these organizations do not have the power of law in U.S. courts and may not have any power in many foreign courts either. The information they provide is only a guideline, and should the need for arbitration or litigation arise, even with a well-structured contract there is no guarantee as to the outcome. These references, though extremely helpful, are therefore not a substitute for a thorough investigation into the local peculiarities of a perspective foreign market. Still, these documents can be very useful in gauging the general lay of the land, so to speak, and in avoiding the more common contracting pitfalls endemic to certain markets.

Table 22.1 Comparison of Forms of Guaranty
Sources: The Miller Act; Canadian Construction Documents Committee (1994); Tiong (1992).

Category of guaranty (1)	United States surety bond (2)	Canadian surety bond (3)	International letter of credit (4)
Bid	Federal: 20% State: 10% Private: 5 to 20%	10% or less for all contracts Private: certified check sometimes used	Generally 1%, but up to 5%
Performance	Federal: 100% for all projects over $25,000 State: 100% for most projects Private: up to 100%	50% for most public and private projects[1] Sometimes as high as 100%	5% to15% of contract
Payment/ labor and material payment	Federal: 50% if contract < $1 million 40% if contract = $1 to 5 million $2.5 million if contract > $5 million State: similar guidelines Private: up to 100%; commonly same value as performance bond	Public: 100%[2] Private: Up to 100%	Rare; a retention guarantee of 5 to 10% of contract might be used instead
Maintenance	Varies	Varies	Around 10%
Mobilization/ advance payment	Rare Only in extreme climates--ex. Pipeline	Rare	Up to 100% of advance payment, which varies

[1] Two year expiration.

[2] Claims must be initiated 90 days after claimant last performed work and filed in appropriate court within one year after general contractor completes project.

SURETY BONDS AND LETTERS OF CREDIT IN AN INTERNATIONAL CONTEXT

With surety bonds, only when the contractor has been defaulted, whether by the obligee or by the contractor itself, can the surety be called upon to step in to protect itself or fulfill its obligation. This basic principal is the same for all forms of surety bonds, including bid, payment, performance, supply, and maintenance bonds. Upon notification of a contractor default, the surety will immediately begin an investigation to determine whether the contractor is indeed in default and if so, determine what it should do to address the situation. Surety bonds are used extensively in North and South America, Central and South Africa, Southeast Asia, Europe, Great Britain, and Australia.

The other class of guarantees, known as "on-demand guarantees" or letters of credit, are used in many countries exclusively, including the Middle East and parts of Africa, or in addition to on-default bonds to guarantee construction contracts. In addition, they are commonly used by private owners in many nations, including those where surety bonds are available. The most common reason stipulated for a prejudice toward letters of credit is an unfamiliarity with the surety product line, but as mentioned above, such attitudes are slowly changing.

Closest to the United States in terms of building culture and proximity is Canada. Bonding practices in both countries are similar in many respects, including the common use of bid, performance, and payment bonds. While irrevocable letters of credit and personal sureties might occasionally be used in the U.S. in lieu of performance or payment bonds, such practices are technically illegal in Canada. In practice, however, it is often the case that a letter of credit or cash equivalent may be used in lieu of a bid bond in Canada and in the U.S. Where the two markets essentially differ is over the amount of coverage required by law for public projects and the premium paid for such coverage. Table 22.1 shows the difference in the amount of the contract covered by U.S. and Canadian surety bonds and by letters of credit.

Avoiding Unjustified Calls

With a surety bond, the surety must find that the contractor is indeed in default to begin discharging its duties under the bond. In the case f a disagreement between the owner and the contractor, the surety might act as a mediator. Under many letters of credit, it takes only the word of the owner to translate the guarantee into cash. As such, the majority of the chapter will focus on advantageous ways to use letters of credit since, for all practical purposes, a default in a foreign country will follow the same steps as outlined in the Part 5 of this book.

It should be pointed out that the particularities of on-demand letters of credit are complicated and idiosyncratic. Regions in which owners, including most governments, require such measures include Northern Africa, the Middle East, and parts of Asia—all potentially lucrative markets. Unless properly structured, agreements with owners from these regions can entail a much higher degree of risk and expenditure for the global contractor.

Table 22.2 Threats to Letters of Credit
Source: Adapted from Tiong (1992)

Type of threat (1)	Explanation (2)
Unfair calling of guarantees	Basically, the owner can demand payment at any point in the construction process for whatever reason or for no reason at all.
Contractor bound to original price	Like a bid bond, a bid or tender guarantee can be cashed in if the contractor refuses to enter into contract, even if during the interim period between the submission of the bid and the award date the prices for labor, materials, or machinery greatly increase. Refusal to do so my result in the "proceed-or-call ultimatum," where the contractor has to choose between forfeiting the amount in the letter of credit or possibly losing money on the contract.
Extend-or-call threat	While an owner is considering the bids on a project, contractors may be told to extend the expiry date of the bid guarantee, even if it is the owner that is causing the award to be delayed. If it does not extend the date, the contractor risks losing the cash value of the tender guarantee. If it does extend the date, the contractor can only hope the owner will allow changes either before or after the contract has been accepted.
Expired guarantees not being expired, or exte or-else	Until returned to the contractor, maintenance guarantees, even though expired, may still be effective in many nations. It is common for an owner to issue an extend-or-else threat, meaning that the contractor may have to increase the warranty period or risk losing the cash value of the letter. Also, in some countries, like Jordan for example, some banks may continue to charge premiums until the letter is returned, even if well after the expiry date. In Iraq, interest accrues over time on letters of credit, increasing the contractor's potential financial exposure and premiums.
Drain on contractor's financial resources	Many situations arising from the use of letters of credit, including the four threats named above, may reduce the contractor's working capital, often making it very difficult to perform existing projects and/or to bid on new contracts.

Caution and familiarity with the region one wishes to work in is a must. In some countries, such Saudi Arabia, owners require letters of credit issued by a local bank in conjunction with the contractor's domestic bank payment (Stevens and Al-Dulaijan 1989). This counter-guarantee makes it even easier for the obligee to cash in a letter of credit should it become dissatisfied with the contractor's performance. It also opens the door for all sorts of other misunderstandings. One example that occurred in the Middle East is almost humorous. After a successful meeting with an owner, one contractor was notified that he had to extend the period of his maintenance letter of credit or it would be called. Upon appeal to the owner, the contractor learned that the local bank issued the threat without the knowledge of the owner because a clerk noted the expiration date was approaching and as a matter of course, sent out the ultimatum (Trodden 1986).

Robert L. K. Tiong has identified five general risks that accompany the letter of credit: 1) unfair calling of guarantees; 2) contractor bound to original price; 3) extend-or-call threat; 4) expired guarantees not being expired, or extend-or-else; and 5) drain on contractor's financial resources. The international contractor should be well aware of these five threats, which are described in Table 22.2.

Tiong identifies several strategies to deal with these possible owner threats if the letter of credit must be used. The contractor should first try to get the obligee to accept a surety bond instead of a letter of credit. Even though the contractor might have to pay a higher premium for this form of guarantee, since it is usually for 100% of the contract price instead of a lesser amount, the protection a surety provides for the contractor against an unfair calling is immeasurable. If this is not an option, a contractor should try to garner a conditional guarantee. This essentially means that to receive payment, the owner must satisfy some requirement proving that the contractor has defaulted. Usually a court order or arbitration certification will suffice, and such is the burden of proof required under the URCG of the ICC, for example.

Even though a surety bond or conditional guarantee is desirable for the contractor, it may not be so for the owner. If the owner demands a letter of credit, the contractor should attempt to negotiate the inclusion of several key clauses, although with many owners, such efforts will be futile. These clauses include:

1. Try to lower the total guarantee.
2. To avoid premature calls by the owner, the guarantee should only become effective once the contract is in full effect and only after the obligee has fulfilled its obligations. This way, the obligee cannot use such strategies as the extend-or-call threat.
3. The amount of the guarantee should be reduced proportionately with the amount of work completed, thus decreasing contractor liability and premium rates.
4. The expiry date should be clearly stated and should become effective regardless of who physically possesses the document. The expiry date can be a specific date, or after a certain event, such as completion. The former is preferable for the contractor as it makes definite the period of warranty.
5. The procedure for calling should clearly outline what steps the owner must take to demonstrate that the contractor is in default, as should the definition of performance. Should "performance" mean completion, or satisfactory, economic, or timely completion, or all three?
6. Include an arbitration clause making mandatory a stage of negotiation before default can be declared.

None of these strategies may be acceptable to some owners. In the case of an unfair call, the contractor should immediately notify the bank underwriting the letter and challenge the bank's commitment in the appropriate courts. Even so, the contractor has little protection under domestic law for unfair calls. Therefore, it is often advisable for the international contractor to acquire some sort of insurance if letters of credit are to be used, increasing the contractor's cost, but transferring the bulk of the financial burden in the case of an unfair or unjustified defaulting. Such forms of insurance can include using more than one bank to guarantee the letter of credit. The contractor can also acquire insurance from the U.S. government or from the insurance industry that indemnifies the contractor against unfair calling. In the United States, the Overseas Private Investment Corporation, a U.S. government agency, will indemnify contractors and prospectors for up to 90% of the dispute amount. If and when such a claim arises, this agency will investigate the nature of the claim, taking it off the contractor's balance sheet.

With a departure from familiar territory into the possibly muddy waters of foreign markets and laws, the international contractor should be cautious and take every opportunity to understand how guaranty works in each country in which he or she wishes to build. International construction can be profitable and exciting but carries with it a fair amount of risk. In his book *Building International Construction Alliances: Successful Partnering for Construction Firms*, Roberto Pietroforte (1997) has noted that international partnering can be a good way to allocate risk for the international contractor. By teaming up with a complementary firm (i.e., a firm with compatible skills, experience, and corporate culture), the international contractor can become familiar with local customs and guaranty practices in the complementary firm's native country. Also, risk on every shared contract would be divided between both firms.

Whether individually or team-oriented, however, international ventures offer the contractor a great deal, including profits and the opportunity to acquire new techniques and insights. Before entering a foreign market, however, the international contractor should be aware of the potential risks, as there are many.

Chapter 23

AGGREGATE INDUSTRY PERFORMANCE

As has been alluded to throughout this book, the surety industry tends to be cyclical in nature with availability and pricing to a large extent dependent upon past aggregate claim history cost ("Bonding, Insurance Falling Short" 1988). For example, many surety companies incurred large losses from 1984 to 1987. In response to these incurred losses, surety companies tightened their underwriting criteria, severely affecting bond cost as well as the contractor's ability to obtain bonds. In the late 1990s, despite minor losses in 1996, the surety industry has been profitable. In response, large numbers of new companies have begun underwriting contractors. This has driven the price of premiums down to levels commonly lower than The Surety Association of America's (SAA) standard rates adopted in 1987. Although there will always be contractors unqualified to receive surety bonds, in this current cycle there have been numerous instances of sureties courting healthy contractors to underwrite, often offering low premiums and superior service.

Measure of aggregate performance of the surety industry requires the collection of data on companies offering contract bonds. Using statistics that the SAA publishes annually, Table 23.1 presents the aggregate performance of the surety industry from 1967 to 1997 regarding contractor surety bonds. Contract surety represents roughly two-thirds of the total surety premium written in any given year, and virtually all of this premium is from bonds written to protect the owners of construction projects (i.e., through Bid, Performance, Payment, and Maintenance bonds). The remainder of the contract premium is derived from bonds written in connection with the other types of contracts included in the contract section of the SAA "Manual of Rules, Classifications, and Loss Costs" (1987).

It is worth noting that when talking about the surety industry, aggregate figures can vary depending on the source. SAA numbers are typically derived from those companies using the SAA statistical service and constitute the majority of surety companies, but not all. SAA purchases some data from A.M. Best because the Best data base includes a larger universe of reporting companies than does the SAA data base. This explains why industry totals appearing in exhibits based on Best's data are larger than the industry totals based on SAA data. In addition, only SAA breaks the surety premium down into various sub lines. Thus, whenever one sees "contract" surety figures, they are most likely from the SAA data base.

Table 23.1 lists direct premiums earned, losses incurred, and loss ratios on contract bonds for the surety industry over the 30-year period between 1967 and 1997, and Figures 23-1 and 23-2 plot this data. Assumed for the industry is a 60% expense ratio, which seems to be a reasonably accurate average. This ratio includes cost elements such as salaries, rent, heat, duplication equipment, required

Table 23.1 Data Regarding Surety Contract Bonds 1967-1997[1]
Source: The Surety Association of America

Year (1)	Direct premiums earned ($) (2)	Direct losses incurred ($) (3)	Loss ratio current year (%) (4)	Combined ratio (%) (5)	Profit (loss) (%) (6)
1967	194,811,000	55,324,000	28.4	88.4	11.6
1968	207,859,000	69,276,000	33.3	93.3	6.7
1969	217,833,000	70,452,000	32.3	92.3	7.7
1970	231,761,000	93,582,000	40.4	100.4	(0.4)
1971	247,547,000	99,260,000	40.1	100.1	(0.1)
1972	269,562,000	79,959,000	29.7	89.7	10.3
1973	289,381,000	123,072,000	42.5	102.5	(2.5)
1974	323,151,000	274,064,000	84.8	144.8	(44.8)
1975	337,550,000	269,729,000	79.9	139.9	(39.9)
1976	334,755,000	187,167,000	55.9	115.9	(15.9)
1977	352,440,000	176,893,000	50.2	110.2	(10.2)
1978	452,414,000	229,293,000	50.7	110.7	(10.7)
1979	504,099,000	209,226,000	41.5	101.5	(1.5)
1980	554,434,000	367,816,000	66.3	126.3	(26.3)
1981	597,031,000	203,482,000	34.1	94.1	5.9
1982	598,201,000	211,584,000	35.9	95.9	4.1
1983	616,385,000	245,625,000	39.8	99.8	0.2
1984	656,029,000	375,791,000	57.3	117.3	(17.3)
1985	710,160,000	670,149,000	94.4	154.4	(54.4)
1986	764,884,000	788,734,000	103.1	163.1	(63.1)
1987	848,522,000	564,826,000	66.6	126.6	(26.6)
1988	973,785,000	327,325,000	33.6	93.6	6.4
1989	1,082,476,000	394,334,000	36.4	96.4	3.6
1990	1,152,614,000	350,819,000	30.4	90.4	9.6
1991	1,156,689,000	308,517,000	26.7	86.7	13.3
1992	1,143,568,000	374,350,000	32.7	92.7	7.3
1993	1,171,950,000	262,860,000	22.4	82.4	17.6
1994	1,219,360,000	428,870,000	35.2	95.2	4.8
1995	1,208,899,000	393,587,000	32.6	92.6	7.4
1996	1,286,927,000	558,232,000	43.4	103.4	(3.4)
1997	1,429,395,000	403,120,000	28.2	88.2	11.8
Total	21,134,472,000	9,167,318,000	43.3	103.3	(3.3)

[1] The Expense factor must be added to the loss cost in order to develop a combined ratio. Expenses vary from company to company but are estimated to average about 60%; a combined ratio in excess of 100% indicates an underwriting loss; for example, in 1984 the underwriting loss would be 17.3% assuming a 0% expense factor.

communication services (e.g., telephone, postage, fax machine), taxes, and the agents' commission (approximately 25% to 28% of the bond premium). The remainder of the premium, as discussed in the first part of the book, is used for profit and to cover losses.

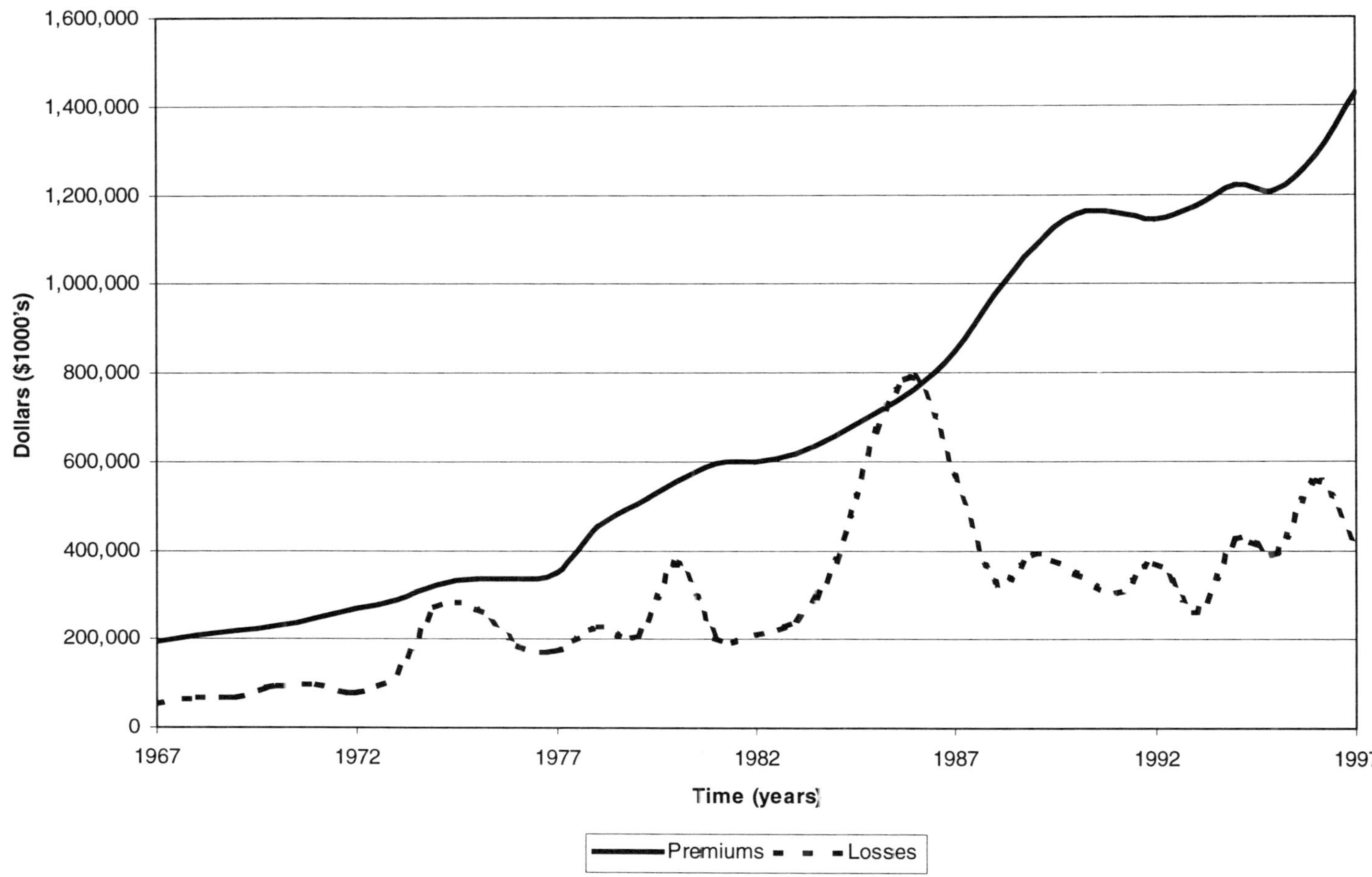

Figure 23-1 Surety Industry Results, 1957-1997
Source: Surety Association of America

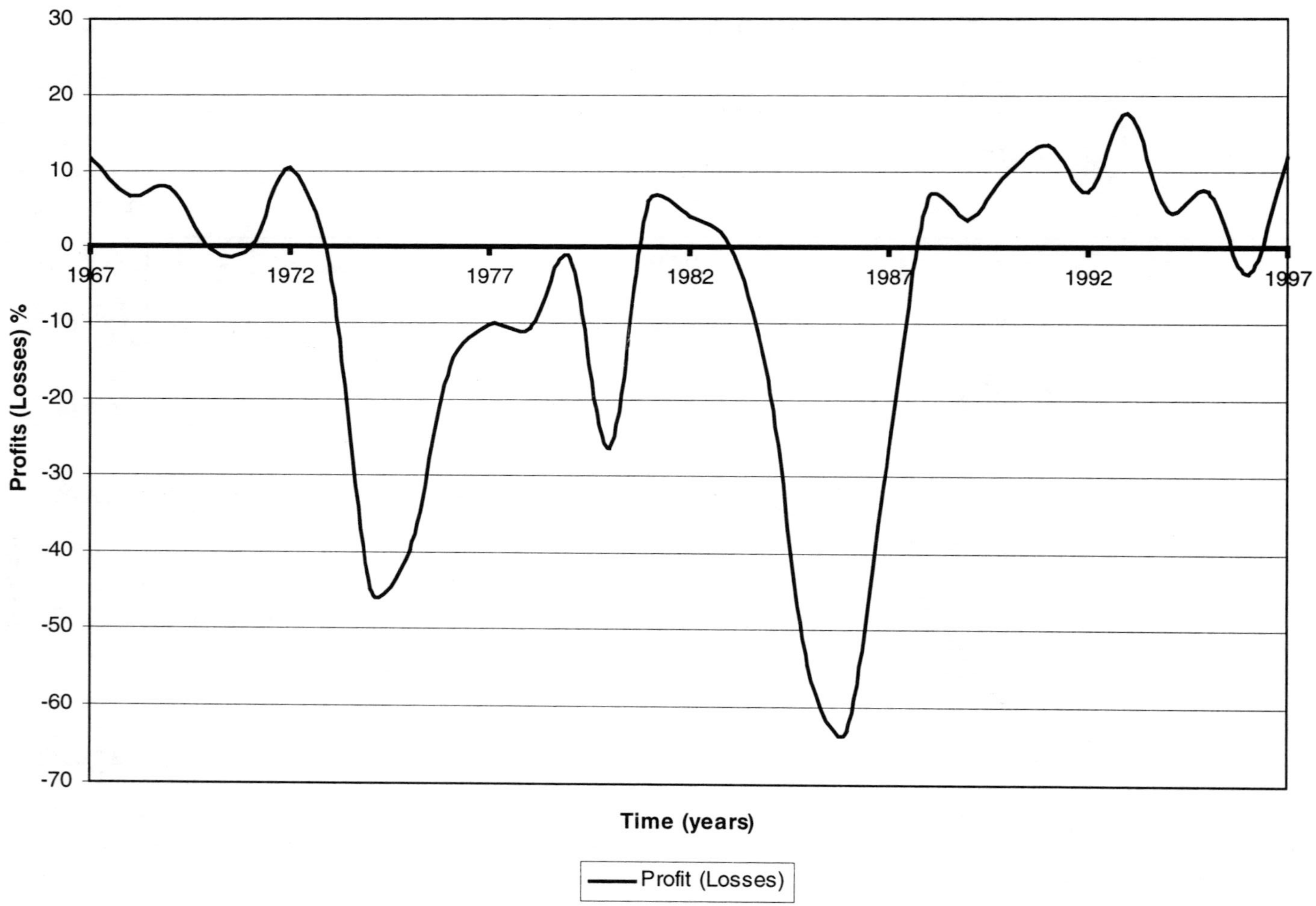

Figure 23-2 Surety Industry Results, 1967-1997
Source: Surety Association of America

The combined ratio derives from adding the loss ratio and the 60% expense ratio together. For example, in 1994, a 35.2% loss ratio added to a 60.0% expense ratio yields a 95.2% combined ratio. A combined expense ratio less than 100% represents a profit (see 1988-1995 and 1997), whereas a combined expense ratio greater than 100% reflects a loss (see 1973-1980, 1984-1987, and 1996). There have been more years of underwriting profit than years of loss, but assuming a 60% expense ratio, the industry has lost over 3% between 1967 and 1997.

It is important to remember that in any given loss year, it is often the case that a handful of underwriters suffer severe losses, working to bring industry averages down. Even in the worst years, however, the majority of sureties manage to remain profitable. In 1987, for example, the industry loss ratio was 66%, but if the sureties with the four largest losses are discarded from the data pool, the total loss is halved to about 34%. Coupled with a standard 60% expense ratio, in 1987 the rest of the industry actually made close to a 6% profit (34% + 60% = 94% combined ratio).

When looking at Figure 23-1, it is important to keep in mind at least three reasons why the dollar value of the direct premiums earned has increased. The first reason is that there were rate increases on the price of bonds of 12%, 16%, 20%, and 20% in 1976, 1977, 1985, and 1987, respectively. The second reason is that the table has not been adjusted for inflation. The third reason is that both the total volume of projects and the volume of projects requiring surety bonds from 1967 to 1997 saw growth. Figure 23-2 helps illustrate the relative loss or profit of the surety industry over time.

Underwriting losses were minimal from 1967 through 1972, with small losses in 1970 (0.4%) and 1971 (0.1%), and with profits ranging from 0.6% to 11.6%. Sureties experienced losses in the following eight years, reaching a "peak loss" of 44.8% in 1974 and 39.9% in 1975. The industry didn't show a profit again until 1981. Sureties saw marginal profits through 1983, when yet another four years of losses set in. In 1985 direct premiums earned were just above direct losses incurred, while in 1986 direct losses incurred exceeded direct premiums earned for the only time in this 30-year period. The combined losses for these two years amounted to over $1.45 billion. However, as noted above, losses experienced by surety companies have not been uniformly distributed over the industry participants as a whole.

Regardless of aggregate performance, every year some sureties lose money while the majority of sureties manage to make a profit. Specific profit makers and losers within the industry vary from year to year, with a small number of companies generally accounting for a great portion of the surety loss. As we saw in 1987, severe losses incurred by a few underwriters can more than offset modest profits earned by all other industry participants. This explains the reason why most sureties remain in business despite the fact that the industry as a whole has been slightly unprofitable since 1967. Furthermore, there is current debate as to whether the 60% expense ratio may be too high. If costs can be lowered to even 57% of premiums earned through corporate efficiency, then significant profits can and have been made. Nonetheless, it still appears as though the surety industry was unprofitable in 1996 (103.4 combined ratio).

Tables 23.2, 23.3, 23.4, and 23.5 list the 15 largest surety writers for 1968, 1978, 1988, and 1996, respectively, while Table 23.6 presents the top two underwriters in each state in 1996. Note how companies such as Fidelity and

Deposit Group and Fireman's Fund Insurance Company are perennially leading underwriters. Nonetheless, Table 23.7 lists those surety companies in receivership or liquidation from the turbulent years of 1984-1991. Note the cluster of companies leaving the industry in 1989, probably due to the timelag between writing bonds and discharging obligations brought about by contractor default.

The data reported in these tables includes most all types of surety bonds such as contract, licensee, courts, coal reclamation, and custom, among others. These tables also present the direct premiums written and earned for these writers. The difference between direct premiums written and earned is that direct premiums earned depends upon how the company has elected to earn the premium over the life of the obligation. The following hypothetical example illustrates this difference.

A bond premium is $1,200, and the duration of the obligation is 12 months. The bond is written on May 1, 1998 and the accounting year ends on December 31, 1998. Therefore, the premium is earned at a rate of 1/12 per month. On December 31, 1998 the direct premium written will be $1,200 and the direct premium earned will be $800 (8/12 of $1,200). On December 31, 1999 the direct premium written will be zero and the direct premium earned will be $400 (4/12 of $1,200).

When direct premiums earned exceeds the direct premiums written, the underwriting volume is shrinking (see Travelers Insurance Group in Table 23.3). The converse of this scenario—when direct premiums written exceeds direct premiums earned—occurs when the underwriting volume is increasing (see Seaboard Surety Company in Table 23.3).

In comparing the 15 largest surety writers over time, certain trends manifest themselves. The amounts of the direct premiums written and direct premiums earned have increased due to the previously discussed reasons of rate increases, inflation, and increasing volume. Moreover, the top 15 spots are consistently filled by most of the same companies, providing ample evidence that over time the surety industry has proven profitable—if it wasn't, one would expect to see sufficient turnover. However, during the same period, the market share of the fifteen largest writers has generally decreased. In particular the market share of the top three surety writers decreased slightly from 24.1% in 1968 to 21.9% in 1997.

In the mid-1980s the performance of the surety product line was for the most part unprofitable. Many surety operations got caught up in the insurance industry's emphasis on "cash flow" underwriting in the hopes that the investment income would more than offset underwriting losses. It has been said that the smell of profits were too strong to ignore. Sureties, eyeing large rates of return, underwrote contract bonds in order to generate premiums for investment at high rates of interest. As a result, the prequalification process suffered, some very large contractors failed, and the industry suffered. Some large reinsurers backing the market pulled out and capacity shrank. Thereafter, sureties tightened their belts and the entire industry witnessed a dearth in the availability of bonds.

Table 23.2 Fifteen Largest Surety Writers for Calendar Year 1968[1]
Source: The Surety Association of America

Rank (1)	Company or group (2)	Direct premiums written ($) (3)	Percentage of all companies (%) (4)	Direct premiums earned ($) (5)	Direct losses incurred ($) (6)	Loss ratio (%) (7)	Combined ratio (%) (8)	Profit (loss) (%) (9)
1	United States F&G Group	28,074,198	8.2	25,823,545	10,452,049	40.5	100.5	(0.5)
2	Fidelity & Deposit of Maryland	27,581,424	8.0	26,082,895	5,288,469	20.3	80.3	19.7
3	Aetna Life and Casualty Group	26,968,529	7.9	23,983,213	4,009,975	16.7	76.7	23.3
4	Travelers Group	22,187,816	6.5	19,452,931	2,155,150	11.1	71.1	28.9
5	Fireman's Fund American Ins.	16,311,269	4.8	15,049,387	3,700,165	24.6	84.6	15.4
6	Maryland American General Ins.	15,118,894	4.4	13,907,824	2,323,179	16.7	76.7	23.3
7	St Paul Companies	14,338,480	4.2	13,600,944	7,826,117	57.5	117.5	(17.5)
8	Hartford Insurance Company Group	13,005,337	3.8	13,213,563	849,600	6.4	66.4	33.6
9	Chubb & Son, Inc. Group	12,739,169	3.7	12,089,192	1,016,592	8.4	68.4	31.6
10	Seaboard Surety Company	11,545,397	3.4	10,487,925	766,777	7.3	67.3	32.7
11	Safeco Insurance Group	11,282,420	3.3	9,597,130	-1,165,255	-12.1	47.9	52.1
12	Reliance Insurance Companies	10,976,203	3.2	11,484,040	5,654,665	49.2	109.2	(9.2)
13	Continental Insurance Companies	9,971,940	2.9	10,994,219	4,548,251	41.4	101.4	(1.4)
14	Insurance Company of North America	9,681,491	2.8	9,117,621	1,684,104	18.5	78.5	21.5
15	United Pacific Insurance Group	9,204,699	2.7	9,145,640	1,594,600	17.4	77.4	22.6
	Total other writers[2]	103,870,934	30.2	100,016,831	34,240,313	34.2	94.2	5.8
	Total all companies	342,858,200	100.0	324,046,900	84,944,751	26.2	86.2	13.8

[1]Total surety line.

[2]Approximately 200 companies constitute other surety writers.

Table 23.3 Fifteen Largest Surety Writers for Calendar Year 1978[1]
Source: The Surety Association of America

Rank (1)	Company or group (2)	Direct premiums written ($) (3)	Percentage of all companies (%) (4)	Direct premiums earned ($) (5)	Direct losses incurred ($) (6)	Loss ratio (%) (7)	Combined ratio (%) (8)	Profit (loss) (%) (9)
1	Aetna Life and Casualty Group	68,169,528	8.2	52,560,751	20,092,372	38.2	98.2	1.8
2	Fidelity and Deposit Group	65,374,434	7.8	60,264,823	45,028,081	74.7	134.7	(34.7)
3	United States F&G Group	62,264,958	7.5	56,918,827	12,899,592	22.7	82.7	17.3
4	Travelers Insurance Group	50,772,555	6.1	52,234,848	61,888,009	118.5	178.5	(78.5)
5	Fireman's Fund Group	46,883,866	5.6	40,644,922	8,143,288	20.0	80.0	20.0
6	Reliance Insurance Group	39,136,051	4.7	31,959,385	4,041,089	12.6	72.6	27.4
7	Chubb Group Insurance Company	35,745,244	4.3	31,190,462	2,795,309	9.0	69.0	31.0
8	St Paul Companies	33,759,931	4.0	27,435,808	3,426,782	12.5	72.5	27.5
9	INA Corporation Group	29,119,822	3.5	25,674,935	21,045,369	82.0	142.0	(42.0)
10	Safeco Insurance Group	25,348,334	3.0	24,619,181	7,441,465	30.2	90.2	9.8
11	Seaboard Surety Company	24,068,927	2.9	22,325,891	3,005,460	13.5	73.5	26.5
12	Hartford Insurance Group	23,604,288	2.8	23,809,406	3,776,137	15.9	75.9	24.1
13	Continental Corporation	20,104,636	2.4	18,047,964	1,282,222	7.1	67.1	32.9
14	CNA Insurance Group	17,310,436	2.1	15,368,727	1,788,743	11.6	71.6	28.4
15	Ohio Casualty Company	13,857,745	1.7	11,573,209	1,823,593	15.8	75.8	24.2
	Total other writers[2]	280,316,369	33.4	249,269,346	123,730,395	49.6	109.6	(9.6)
	Total all companies	835,837,124	100.0	743,898,485	322,207,906	43.3	103.3	(3.3)

[1] Contract bonds represent approximately 60% of the surety business.
[2] Approximately 300 companies constitute other surety writers.

Table 23.4 Fifteen Largest Surety Writers for Calendar Year 1988[1]
Source: The Surety Association of America

Rank (1)	Company or group (2)	Direct premiums written (3)	Percentage of all companies (4)	Direct premiums earned (5)	Direct losses incurred (6)	Loss ratio (7)	Combined ratio (8)	Profit loss (9)
1	Fidelity and Deposit Group	126,313,324	6.8	131,375,188	12,017,937	9.2	69.2	30.8
2	Aetna Life and Casuality Group	114,704,349	6.1	107,383,653	48,668,979	45.3	105.3	(5.3)
3	United States F&G Group	109,542,818	5.9	108,151.278	29,236,275	27.0	87.0	13.0
4	Reliance Insurance Group	102,567,272	5.5	93,724,705	10,890,049	11.6	71.6	28.4
5	Fireman's Fund Company	86,143,474	4.6	90,351,892	-8,148,995	-9.0	51.0	49.0
6	Seaboard Surety Company	84,852,232	4.5	72,920,274	15,144,407	20.8	80.8	19.2
7	Chubb Group Insurance Comp.	75,943,155	4.1	77,081,744	78,588,994	102.0	162.0	(62.0)
8	Safeco Insurance Companies	66,831,517	3.6	60,449,637	43,030,532	71.2	131.2	(31.2)
9	Hartford Insurance Group	60,282,797	3.2	59,547,017	22,127,787	37.2	97.2	2.8
10	St Paul Group	55,144,705	3.0	52,603,798	8,580,335	16.3	76.3	23.7
11	Travelers Insurance Group	53,372,612	2.9	52,415,139	15,425,589	29.4	89.4	10.6
12	CNA Insurance Companies	52,569,896	2.8	49,932,179	16,813,067	33.7	93.7	6.3
13	Continental Insurance Company	46,019,151	2.5	41,581,174	13,956,144	33.6	93.6	6.4
14	Transamerica Insurance Group	44,576,453	2.4	45,964,184	26,392,774	57.4	117.4	(17.4)
15	Cigna Group	43,557,516	2.3	55,934,524	-32,730,279	-58.5	1.5	98.5
	Total other writers[2]	744,337,888	39.8	706,849,413	183,482,504	26.0	86.0	14.0
	Total all companies	1,866,759,159	100.0	1,806,265.799	483,476,099	26.3	86.8	13.2

[1] Contract bonds represent approximately 60% of the surety business.

[2] Approximately 435 companies constitute other surety writers.

Table 23.5 Fifteen Largest Surety Writers for Calendar Year 1997[1]
Source: The Surety Association of America

Rank (1)	Company or group (2)	Direct premiums written ($) (3)	Percentage of all companies (%) (4)	Direct Premiums earned ($) (5)	Direct losses incurred ($) (6)	Loss ratio (%) (7)	Combined ratio (%) (8)	Profit (loss) (%) (9)
1	CNA Insurance Companies	232,611,026	8.8	161,629,926	39,662,481	24.5	84.5	15.5
2	Reliance Insurance Companies	179,418,027	6.8	126,899,372	14,673,021	11.6	71.6	28.4
3	U.S. Fidelity &Guaranty Group	167,530,825	6.3	138,259,016	41,143,343	29.8	89.8	10.2
4	Fidelity & Deposit Group	158,084,929	5.9	112,530,612	13,411,354	11.9	71.9	28.1
5	Travelers Property Casualty Corporation	153,913,514	5.8	126,587,926	32,976,414	26.1	86.1	13.9
6	St Paul / Seaboard Group	145,018,503	5.5	112,793,702	12,075,108	10.7	70.7	29.3
7	American International Group	118,268,306	4.5	86,847,058	22,806,418	26.3	86.3	13.7
8	Safeco Insurance Companies	101,552,673	3.8	50,571,750	13,461,959	26.6	86.6	13.4
9	Fireman's Fund Insurance Companies	97,808,523	3.7	70,639,107	23,357,064	33.1	93.1	6.9
10	Chubb Group of Insurance Companies	78,526,909	3.0	52,656,804	4,907,422	9.3	69.3	30.7
11	Hartford Insurance Group (The)	71,983,323	2.7	63,120,495	14,733,547	23.3	83.3	16.7
12	Amwest Insurance Group	68,176,279	2.6	63,441,540	17,558,086	27.7	87.7	12.3
13	Frontier Insurance Group	63,672,885	2.4	50,444,514	22,218,255	44.0	104.0	(4.0)
14	Liberty Mutual Group	53,421,102	2.0	40,753,831	14,363,058	35.2	95.2	4.8
15	Western Surety Company	44,122,123	1.7	39,482,244	834,859	2.1	62.1	37.9
	Total other writers[2]	922,041,966	34.5	964,425,804	210,639,045	—	—	—
	Total all companies	2,656,150,913	100.0	2,261,083,701	498,821,434	33.0	—	—

[1] Total surety line.
[2] Approximately 600 companies constitute other surety writers.

Table 23.6 Premiums and Leading Surety Writers by State, 1996
Source: Patrick M. Sweeney/A.M. Best 1997

State (1)	Rank (2)	Number of companies (3)	Dir. premiums written[1] ($) (4)	Portion of U.S. total (%) (5)	Premium change (%) (6)	Adjusted loss ratio[2] (%) (7)	Leading writer (8)	Portion of state market (%) (9)	Second leader (10)	Market share (11)
AL	25	192	38,014	14	15.9	33.4	Zurich Ins Grp US	16.0	US F & G	13.6
AK	42	129	14,702	0.5	9.7	19.1	Reliance Ins Grp	15.8	SafeCo Ins Co	12.8
AZ	10	185	69,355	2.5	-65.0	31.7	Amer Express Grp	29.4	Merchants Bond (Mut)	5.7
AR	35	173	19,417	0.7	1.4	109.9	St. Paul Companies	9.2	US F & G	9.1
CA	1	222	331,870	12.1	-0.1	19.9	Reliance Ins Grp	7.8	Firemans Fund Cas.	7.2
CO	17	203	49,848	1.8	7.3	16.1	Pioneer General Ins	12.4	Travelers PC Grp	9.1
CN	27	142	34,991	1.3	7.2	90.8	Travelers PC Grp	16.9	US F & G	7.7
DE	49	136	6,853	0.3	-5.8	-9.5	Clarengdon Ins Grp	18.0	Firemans Fund Cas.	6.8
DC	28	152	30,867	1.1	3.1	38.3	Frontier Ins Grp Inc	14.3	CNA Ins Grp	10.4
FL	5	228	173,932	6.4	4.6	19.6	Accredited Sur & Cas	8.4	Reliance Ins Grp	6.1
GA	9	206	70,101	2.6	-0.5	23.5	St. Paul Companies	7.4	Zurich Ins Grp US	6.8
HI	38	106	18,542	0.7	-0.8	14.3	Reliance Ins Grp	14.7	Firemans Fund Cas	13.8
ID	43	157	12,222	0.4	12.2	-26.5	Reliance Ins Grp	21.9	SafeCo Ins Co	12.0
IL	4	235	131,241	6.6	84.1	50.7	CNA Ins Group	46.9	US F & G	4.7
IN	21	201	44,733	1.6	14.2	11.1	US F & G	16.6	CNA Ins Grp	10.8
IA	37	176	19,014	0.7	-5.1	20.9	Merchants Bond (Mut)	20.3	US F & G	16.3
KS	34	185	29,945	0.9	8.7	12.4	Zurich Ins Grp US	12.7	Firemans Fund Cas	6.6
KY	23	213	39,085	1.4	3.6	7.7	Cumberland Surety Co.	10.7	US F & G	7.9
LA	20	186	45,747	1.7	-4.1	28.7	Zurich Ins Grp US	18.8	CNA Ins Grp	8.5
ME	46	111	7,941	0.3	-3.2	-9.8	Zurich Ins Grp US	11.3	Travelers PC Grp	10.8
MD	15	196	55,134	2.0	-1.0	1.8	Zurich Ins Grp US	11.1	Reliance Ins Grp	9.7

Table 23.6 Premiums and Leading Surety Writers by State, 1996 (Continued)
Source: Patrick M. Sweeney/A.M. Best 1997

State (1)	Rank (2)	Number of companies (3)	Dir. premiums written[1] ($) (4)	Portion of U.S. total (%) (5)	Premium change (%) (6)	Adjusted loss ratio[2] (%) (7)	Leading writer (8)	Portion of state market (%) (9)	Second leader (10)	Market share (11)
MA	12	149	67,325	2.5	-7.8	23.3	St. Paul Companies	11.9	US F & G	10.9
MI	16	186	50,144	1.8	5.7	45.3	CNA Ins Group	8.8	Hartford Ins Grp	5.9
MN	26	180	36,546	1.3	0.0	11.0	St. Paul Companies	11.5	US F & G	9.6
MS	31	183	27,198	1.0	3.4	15.6	US F & G	25.9	Zuric h Ins Grp US	11.4
MO	19	209	46,091	1.7	1.1	19.9	Zurich Ins Grp US	8.8	US F & G	8.0
MT	44	155	40,439	0.4	16.8	56.3	St. Paul Companies	15.0	Reliance Ins Grp	14.7
NE	41	181	15,280	0.6	-1.1	-0.8	Universal Surety Grp	17.9	US F & G	8.2
NV	24	163	38,178	1.4	16.7	8.4	Meadowbrook Ins Grp	9.4	Frontier Ins Grp, Inc.	7.8
NH	45	124	10,150	0.4	16.2	9.1	Ing Group	19.4	US F & G	13.0
NJ	8	175	89,155	3.3	0.2	37.6	First Indemn of Amer	8.1	Frontier Ins Grp, Inc	7.7
NM	39	174	18,051	0.7	-1.5	15.8	CNA Ins Grp	13.9	Nat. American Ins	6.8
NY	3	195	191,677	7.0	5.9	60.1	Amer Intern Grp	9.4	St Paul Comapnies	8.1
NC	14	183	57,283	2.1	13.9	38.1	Reliance Ins Grp	11.1	Travelers PC Grp	9.4
ND	50	152	6,318	0.2	-4.2	14.2	CNA Ins Grp	18.6	St Paul Companies	8.1
OH	7	209	95,295	3.5	134	37.7	Westfield Co.	9.4	Cincinnati Fin Corp	6.1
OK	33	176	25,971	0.9	0.8	4.8	Amer Financial Grp	16.0	Nat. American Ins	8.2
OR	22	173	39,699	1.5	13.3	17.3	Reliance Ins Grp	13.0	SafeCo Ins Co	8.4
PA	6	210	102,952	3.8	3.6	19.7	Zurich Ins Grp US	9.2	Reliance Ins Grp	7.1

Table 23.6 Premiums and Leading Surety Writers by State, 1996 (Continued)
Source: Patrick M. Sweeney/A.M. Best 1997

State (1)	Rank (2)	Number of companies (3)	Dir. premiums written[1] ($) (4)	Portion of U.S. total (%) (5)	Premium change (%) (6)	Adjusted loss ratio[2] (%) (7)	Leading writer (8)	Portion of state market (%) (9)	Second leader (10)	Market share (11)
RI	47	131	7,935	0.3	-2.5	3.4	Travelers PC Grp	14.0	CNA Ins Grp	10.7
SC	29	157	29,495	1.1	4.6	35.1	US F & G	9.9	Reliance Ins Grp	7.4
SD	30	155	27,219	1.0	11.0	8.9	Hica Holding Group	62.3	St Paul Companies	17.4
TN	18	205	49,335	1.8	4.4	30.2	Reliance Ins Grp	9.9	US F & G	9.8
TX	2	258	199,922	7.3	9.0	5.5	Capsure Group	7.7	CNA InsGrp	6.8
UT	36	155	19,121	0.7	2.4	28.6	CNA Ins Grp	10.8	Reliance Ins Grp	10.7
VT	51	116	3,486	0.1	3.0	80.0	US F &G	9.2	Ing Grp	16.1
VA	13	194	65,689	2.4	3.6	6.8	Zurich Ins Grp US	6.8	Travelers PC Grp	10.9
WA	11	177	68,301	2.5	-2.1	10.0	Reliance Ins Group	19.4	SafeCo Ins Co	13.4
WV	40	140	15,629	0.6	9.6	12.1	US F & G	16.9	Westfield Co	10.5
WI	32	163	26,924	1.0	-3.8	4.0	Nationwide Group	12.9	Zurich Ins Grp US	12.0
WY	48	139	7,708	0.1	-15.8	7.3	United Fire & Cas Grp	11.3	Amer. Intern Group	9.7
US Total	—	**556**	**2,737,035**	**100.0**	**1.7**	**26.6**	**CNA Ins Group**	**8.4**	**Reliance Ins Grp**	**6.5**

[1] Premiums are in thousands (000 omitted) and before reinsurance.
[2] Adjusted loss ratio includes pure losses and policyholder's dividends but not loss adjustment or underwriting expenses.

Table 23.7 Surety Companies in Receivership or Liquidation 1978-1991
Source: The Surety Association of America

Company or group (1)	Location (2)	Type of failure (3)	Date (4)
Allied Surety Company	Portland, ME	Liquidated	01/78
Reserve Insurance Company	Chicago, IL	Liquidated	05/79
Cosmopolitan Mutual Insurance Company	New York, NY	Liquidated	10/80
Proprietors' Insurance Company	Delaware, OH	Liquidated	08/81
Kenilworth Insurance Company	Chicago, IL	Liquidated	04/82
Cotton Belt Insurance Company	Memphis, TN	Liquidated	07/82
Beacon Insurance Company	Richardson, TX	Rehabilitation	02/84
Dome Insurance Company	Christiansted, Virgin Islands	Rehabilitation	04/84
Northwestern Fire Insurance Company of PA	King of Prussia, PA	Liquidated	06/84
Nassau Insurance Company	Jamaica, NY	Liquidated	06/84
Surety Insurance Company of California	LaHabra, CA	Liquidated	08/84
Excaliber Insurance Company	Carrollton, TX	Liquidated	09/84
Aspen Indemnity Corporation	Englewood, CO	Liquidated	09/84
Universal Casualty Insurance Company	Coral Gables, FL	Rehabilitation	09/84
Ideal Mutual Insurance Company	New York, NY	Rehabilitation	12/84
Eastern Indemnity Company of MD	Rockville, MD	Liquidated	01/85
Iowa National Mutual Insurance Company	Cedar Rapids, IA	Liquidated	10/85
Union Indemnity Insurance Company of NY	New York, NY	Liquidated	10/85
Union International Insurance Company of Delaware	New York, NY	Liquidated	10/85
Guard Casualty & Surety Insurance Company	Indianapolis, IN	Liquidated	12/85
RGAF Underwriters	Ocala, FL	Liquidated	03/86
Midland Insurance Company	New York, NY	Liquidated	04/86
American Druggists' Insurance Company	Cincinnati, OH	Liquidated	04/86
Texas Fire & Casualty Company	Indianapolis, IN	Liquidated	06/86

Table 23.7 Surety Companies in Receivership or Liquidation 1978-1991 (Continued)
Source: The Surety Association of America

Company or group (1)	Location (2)	Type of failure (3)	Date (4)
Allied Fidelity Insurance Company	Indianapolis, IN	Liquidated	06/86
Integrity Insurance Company	East Orange, NJ	Rehabilitation	12/86
Mission Insurance Company	Los Angeles, CA	Liquidated	09/87
American Mutual Insurance Company of Boston	Wakefield, MA	Liquidated	03/89
American Mutual Liability Company	Wakefield, MA	Liquidated	03/89
American Policyholders' Insurance Company	Wakefield, MA	Liquidated	03/89
Pacific Marine Insurance Company	Milwaukee, Wi	Liquidated	08/89
Southeastern Casualty & Indemnity Insurance Company	Plantation, FL	Receivership	08/89
Southeastern Casualty & Indemnity Company of NJ	Livingston, NJ	Receivership	08/89
Southeastern Reinsurance Company	Plantation, FL	Receivership	08/89
Louisiana Underwriters Insurance Company	Lafayette, LA	Rehabilitation	10/89
Oil and Gas Insurance Company	Columbus, OH	Liquidated	01/91
Protective Casualty Insurance Company	Baton Rouge, LA	Liquidated	05/91

The net effect was an inevitable return to underwriting basics in order to make a profit. Other responses by the industry include trimming operating costs, increasing premiums, and better defining their market niche (e.g., size of contractor, type of contractor). Contractors of all sizes felt the pinch, but small to medium-sized firms found acquiring bonding especially difficult. As late as 1992, the situation persisted even though the surety industry had been in the black for several years. In a poll of independent surety agents performed at the end of 1991, 38% responded that surety bonding would be difficult to get in 1992 ("Small Shops Face" 1991).

Eventually, as sound underwriting became the industry norm once again, the confidence of reinsurers grew, and there was a concomitant reemergence of industry capacity. Since these changes have taken effect, the industry has seen almost 10 profitable years—one of the longest profitable stretches ever. Even though some money was lost in 1996, the industry had a combined profit of 11.8% in 1997. As profits continue, more companies are entering the market. Sureties are competing to bond successful contractors, often offering reduced rates, and it has become easier for small- and medium-sized firms to acquire bonds. However, this competitive trend has some industry experts fearing that the pendulum is again swinging toward an era of decreased underwriting acumen, which will mean inevitable losses. The trend towards increased competition in the surety industry is explored in more depth in Chapter 26.

REASONS POSTULATED FOR LOSS

The performance of the surety industry, as is the case with any single surety or contractor, is difficult to consistently predict. An examination of some of the risks facing sureties may better explain why the industry is inherently cyclical and risky. This stems from the process surety companies use to evaluate risks associated with bond underwriting for a construction company project. Problems with the underwriting process derive from three primary issues, and each issue receives attention below:

1. *Subjectivity*—Evaluation of contractors involves parameters that are qualitative, subjective, and imprecise; hence, the decision-making process requires subjective judgment, which is bound to be imperfect.
2. *Macroeconomic conditions*—Changes in national or world-wide economy can occur, and over a short period of time create volatile conditions resulting in problems for both the surety industry and construction companies.
3. *Idiosyncratic nature of projects*—Characteristics and circumstances vary from project to project, and multiple unknowns, such as the presence of Native American remains and unstable subterranean conditions, perpetually exist.

Subjectivity

The evaluation process remains largely an art that requires subjective judgment based on the underwriter's experience. A summary of parameters evaluated and previously described include:

1. Financial stability—helpful for evaluating the company's financial longevity and its ability to meet financial obligations, both short and long term, as well as the contractor's financial reporting practices
2. Experience record—used to measure the current level of expertise that the company's principals possess
3. Management capabilities—reveals the level of competence and capability of the contractor's staff
4. Past performance track record—reveals the success of prior completed projects
5. Current work load relative to total capacity—reveals whether the contractor has the capacity to add an additional project
6. Organizational structure—used to assess the company's organization and identify key relationships among its personnel and departments
7. Trust and confidence in contractor-surety relationship—reveals the nature of the contractor-surety relationship
8. Continuity of organization—used to evaluate the company's longevity and ability to continue operating in the event a key person dies or is disabled
9. Business plan—reveals the planning and forecasting the contractor performs.

With the exception of financial stability, each parameter is qualitative and therefore difficult to evaluate with relevant and precise data. Furthermore, the parameters are time-dependent. The evaluation performed reflects the contractor at *that particular point in time*. For example, this picture can change radically with the occurrence of such unforeseen events as the departure of a firm's principal owner, planned or unplanned; death of key personnel; receipt of a new, larger contract; and litigation on a project. Understanding the simultaneous interaction of these parameters further complicates the extremely subjective evaluation process and resulting decision.

As a result of the risk associated with bonds and the evaluation process involved, many surety companies base their decision predominantly on the credit worthiness of the applicant. In many instances, a construction company (especially small to medium-sized firms) does not have adequate financial reserves to meet the surety's credit requirements and consequently may have trouble obtaining bonds. This has forced contractors to invest more resources into their operation or find a bonding company that has less stringent requirements. Another alternative for the small contractor is to seek assistance through the Small Business Administration's Surety Bond Guarantee Program, described in more detail below.

Even with a stringent financial analysis, the decision whether to accept a contractor as an account is ultimately a subjective decision. In the current profitable economic cycle, industry experts predicting a downturn are pointing to the growing number of emerging companies writing surety bonds and the inevitable dearth of experienced personnel. With so many new players in the surety field, it would be impossible for there to exist experienced underwriting staffs at every enterprise, meaning that many unqualified or untested underwriters are issuing bonds. Although inexperience does not guarantee a loss, it is believed to increase the incidence of bonding unworthy contractors, leading to more defaults. No one has found a way to underwrite contract bonds without losses, and some say it is only a matter of time before the current boom cycle collapses. Experienced underwriters make mistakes as well, but the feeling in the industry is that the greatest number of losses will come from the newer ranks of underwriters.

Macroeconomic Conditions

A huge factor in the health of the surety industry and what may precipitate a downturn as well as an upturn, is the macroeconomic climate. Several factors have resulted in the industry's inability to control its underwriting outcome such as:

1. Late 1970s inflation
2. High interest rates of 1979-1981
3. Severe recession of 1981
4. Major restructuring of mature industries and rapid growth of new technological industries in 1980s
5. Regional boom and bust of economies in 1980s
6. Severe federal budget deficits in 1980-1987
7. General deregulation of major industries in 1980-1987
8. Low inflation in 1982-1987
9. New tax laws in 1986
10. Growth of U.S. and world economy in the 1990s
11. Asian financial crisis 1997-1999
12. Slowing of U.S. GDP in the late 1990s.

In illustration, the oil embargo resulted in the cost per square yard of bituminous asphalt to rise from $5 per square yard to $14 per square yard. Numerous contractors went into default as a result of this totally unexpected escalation in material cost. Neither contractors nor underwriters had any way to forecast the occurrence of such an event, and thus did not anticipate its consequences. Likewise, it is unclear at the time of publication how exactly the slowing of U.S. and Asian economies will affect the rate of new construction and the incidence of contractor default. However, as we approach the 21st century, changes stemming from this recession will no doubt be felt. There may be a spate of defaults and surety bankruptcies as there were in the mid-1980s, or the effects could be worse or not as severe. Still, it is certain that the surety market will be affected and that this effect will show up in industry results.

Additional problems influence the underwriting process as a direct result of the factors listed above (Surety Seminar summary 1987). Some include:

1. Excess capacity in the construction industry and not enough concentrated capital
2. Geographic dispersion of construction projects and companies
3. Increased diversification in project type and size

4. Tighter cash management
5. A slow-down in payment of accounts receivable
6. Rapid increase in litigation
7. Slim profit margins with little or no contingency for losses
8. Excessive leverage in debt and size of work programs
9. Over-investment in fixed assets
10. Investment in outside ventures (e.g., land development).

Many of these factors are direct contributors to the increase in contractor failure, but are difficult to predict, much less quantify.

Idiosyncratic Nature of Projects

In addition to surety underwriting and macroeconomic events, each construction project has unique characteristics and peculiarities that must factor into the underwriting process. Items such as the type and size, owner characteristics, geographic location, regulatory requirements, and contractual type play a role in the risk of surety in light of their potential impact on a construction company's performance.

Although a surety doesn't typically approve every project a contractor bids individually, but rather the contractor's work program as a whole, deviations from the established norm are accounted for. In the agreement that a surety has with the independent surety agent representing the contractor, if the contractor wishes to pursue a new type of construction, perform work in a new area, or bid an especially large job, the surety has to give prior approval. Therefore, it is unlikely that a surety will become involved in a project that is much outside the parameters it annually or semi-annually determines for a contractor.

However, sureties sometimes approve projects anomalous to the contractor's normal work program. If a surety does approve a job larger than the single-job limit, a job using new techniques or skills, or a job in a new geographic area, the surety will be more interested than normal in the characteristics and qualifications of not only the contractor, but the project owner. Prospective owners should cooperate with surety personnel, as in the end, a cordial relationship will help resolve potential difficulties. In particular, sureties typically evaluate private owners on the following criteria:

1. Financial Commitment for the Construction Funding—including lender's certification of availability, escrow of cash, and other funding sources
2. General Credit Worthiness—may include trade clearances
3. Adequate Property Insurance and Liability Insurance—particularly if the owner is engaged in a hazardous business
4. Credit Insurance—especially if a foreign construction site is involved

Based on this information, if the quality of the project owner is inappropriate, the surety may decline to provide a bond even on behalf of a qualified contractor.

Chapter 24

SOURCES OF INFORMATION PERTAINING TO SURETIES

Surety companies provide important protection for owners as well as contractors. Since the obligee is the beneficiary of the bond, the principal must use a surety that is acceptable to the obligee. To obtain information regarding companies that issue contract surety bonds, sources are available that provide financial as well as other relevant data. These sources include state insurance departments, the *Federal Register*, A.M. Best Company, telephone directories, and insurance agents and brokers. Appendix XV lists contact information for several organizations mentioned in this chapter and in other parts of the book.

Before writing surety bonds in the United States, a company should be licensed by the insurance department of one or more states. Generally, licensing is required by the state in which the surety is doing business and/or the state where the guaranteed bond obligation is being performed (some exceptions apply). State insurance departments are responsible for performing periodic examinations and anyone may request information on a particular surety company.

In addition, the U.S. Treasury Department prequalifies surety companies and maintains a listing published each July first in the *Federal Register*. The publication includes such information as business name and address, underwriting limitation(s), the states where the company is licensed, and in what state the business is currently incorporated (see Figure 24-1). Companies that desire inclusion in this listing must provide financial and other relevant data to the Treasury Department for evaluation. Hard copies of the Treasury List are available from the Government Printing Office for a nominal charge.

The Treasury Department analyzes this data, determines each company's capital (net worth), and establishes the maximum amount for a single bond. This limitation is generally 10% of capital and surplus. However, it is important to understand that no limit is set on the number of bonds that surety companies can underwrite under this amount. For example, a company could underwrite 40 bonds below their given threshold. Consequently, if these bonds were called in full, the surety might have insufficient capital available to settle all the claims, which could reach as high as four times its net worth (40 * 10% = 400%).

The A.M. Best Company performs annual analysis and ratings of almost all insurance companies operating in the United States. Their results appear in the annual publication entitled *Best's Insurance Reports, Property-Casualty*, which lists the companies by alphabetic rating and financial size category (*Best's Insurance Reports* 1997). The report is a good investment for companies that frequently accept surety bonds, although public and financial libraries often maintain copies of this publication. Best's also publishes a shortened version of

The Cincinnati Insurance Company. BUSINESS ADDRESS: Post Office Box 145496, Cincinnati, OH 45250-5496. UNDERWRITING LIMITATION b/: $34,662,000. SURETY LICENSES c/: All except CT, GU, HI, LA, ME, NH, SD, VT, VI.
INCORPORATED IN: Ohio.

Commercial Insurance Company of Newark, New Jersey. BUSINESS ADDRESS: 180 Maiden Lane, New York, NY 10038. UNDERWRITING LIMITATION b/: $6,378,000. SURETY LICENSES c/: All except GU, PR, VI.
INCORPORATED IN: New Jersey

Commercial Union Insurance Company. BUSINESS ADDRESS: One Beacon Street, Boston, MA 02108. UNDERWRITING LIMITATION b/: $21,789,000. SURETY LICENSES c/: All except GU.
INCORPORATED IN: Massachusetts.

Consolidated Insurance Company. BUSINESS ADDRESS: 115 North Pennsylvania Street, Indianapolis, IN 46204. UNDERWRITING LIMITATION b/: $1,855,000. SURETY LICENSES c/: FL, IL, IN, IA, KY, MI, OH, OR, TN, WA, WI.
INCORPORATED IN: Indiana.

Figure 24-1 Excerpt of Federal Register

Bond number (if any)

Name and address of principal (including all of the names of the parties if the principal is a joint venture)

Name and address of obligee

Amount of performance bond

Amount of payment bond

Date bond executed

Name of person signing bond for the surety

Brief description of project

Contract price

Directions as to whom the confirmation should be sent

Figure 24-2 Items To Check Authenticity of Bond
Source: Users Information, The Surety Association of America 1991

1. Is the surety licensed to conduct business in the geographical area of the project?

2. Is the contract reasonably described? Does it contain the minimum information indicated by the following wording:

   Entered into a contract dated ________ *day* of ________, *(Year), for the construction of a* ____
   at ____?

3. Has the performance bond been properly signed by the contractor and a surety?

4. Is the performance bond penalty correct?

5. Is Power of Attorney attached to the bong? If so, does it coincide with the person who executed the bond and is it properly dated?

6. Has a bond been included if requested? Has it been properly executed and is the penalty correct?

7. Has the bonding of subcontractors been considered?

8. Are the bond wordings acceptable?

Figure 24-3 Pre/Bid Bond Checklist
Source: Bonds $ Insurance for Contractors 1989

this guide, which sells for approximately 20% of the price, yet contains alphabetic ratings and information regarding financial size.

Two other sources include the telephone directories and insurance agents and brokers. A telephone call to the place of business can help establish the legitimacy of the firm. Contacting local insurance agents and brokers can also serve as a means of establishing the authenticity of the company claiming to provide contract bonds. Any locally established agency knowledgeable of the insurance field should be able to assist in verifying a firm. In addition, the National Association of Surety Bond Producers in Washington, D.C., will provide the names of insurance agencies whose specialty is surety bonding.

The above suggestions address investigating the authenticity of a company providing surety bonds. A similar investigation can help authenticate a bond. Figure 24-2 presents the steps that are useful in checking the authenticity of a bond. The Surety Association of America provides a list of surety companies that voluntarily help verify the authenticity of surety bonds.

In addition, Figure 24-3 illustrates a pre-bid-bond checklist ("Bonds & Insurance for Contractors" 1989). In the format shown it is useful for verifying that the surety is legitimate and that the bond(s) are both complete and correct.

Chapter 25

SURETY AND DESIGN/BUILD: THE FUTURE

Without question, the most important development facing the surety industry in the 21st century, as well as the construction industry as a whole, is Design/Build. From 1987 to 1997, the annual domestic volume of Design/Build (D/B) construction has jumped from $6 billion to over $56 billion (Port-Hull 1997). It has been estimated that by the year 2000, 50% of domestic construction will be D/B, and that by 2010, 50% of all construction is going to be D/B. Until recently, however, lack of adequate insurance and bond availability has been a huge obstacle for contractors and owners interested in both this delivery system and the protection of surety bonds. As D/B has gained in popularity, however, the surety industry has been forced to respond.

In D/B, a single entity is liable for not only the physical structure of a building, as is the case with traditional contract bonds with an architect and a contractor, but the underlying design as well. The exact manner in which a surety company underwrites this increased risk differs substantially, but the growing popularity of one-stop construction made it imperative that the surety industry find a way to underwrite D/B ventures. A recent federal procurement act was signed into law allowing agencies and departments to individually select fast-track construction for public projects. Many states have or are in the process of adopting similar statutes. The surety industry has responded to this growing interest and use of D/B by revising contract bond language and offering more comprehensive contractor insurance polices, but currently there is no industry standard in D/B underwriting.

One thing is certain: no one in the construction industry can afford to ignore this trend. Today, D/B is a growing sector of the surety industry, and no longer feared or even something to be avoided by most sureties. Sureties continue to advance underwriting procedures, and insurance companies continue to look for better ways to provide the necessary coverage on projects without lapses or "gaps" in coverage. However, there is little consensus as to how best to underwrite the increased risk to the surety. The Design-Build Institute of America (DBIA) has promulgated information and statistics for the growing industry and has worked closely with the SAA, the AIA, and the American Insurance Association, along with many individual sureties, to help develop means of underwriting D/B projects. However, the issues surrounding D/B guaranty are complicated and daily being negotiated. The following attempts to outline the surety industry's response to D/B.

HISTORY

D/B is a delivery system in which the owner has one point of contact from the design to the construction of a building. It is similar to the "master-builder" approach popular in this country during the first half of the century in which an owner hired a single firm to design and construct a building. However, after W.W.II, competition became more intense in the construction industry, architects became increasingly professionalized, and competitive bidding became the norm. As this happened, the master-builder style went out of favor in the United States. As construction has advanced technologically and in complexity, a single point of contact has become more desirable for many owners.

D/B is attractive for those who during every stage of the project want to know exactly who to turn to. In addition, D/B projects are fast-tracked, meaning ground is often broken much earlier than with traditional design-bid-build delivery where plans must be finalized and approved before construction can begin. Once a general plan is adopted in the D/B model, construction starts immediately, saving the owner significant amounts of time and very often, money.

D/B can be attractive for the contractor and design professional as well. It generally allows them more control in the project, as constructability becomes a concern not only for the contractor, but also for the designer. At the same time that this saves the owner money, D/B can translate into a higher margin of profit for the contractor and design team.

As attractive as D/B might seem with its win-win structure, until recently, the insurance and surety industry have considered D/B ventures much more risky than traditionally delivered projects fitting the old template. Instead of simply purchasing construction services, as an owner does in the traditional design-bid-build model, under the D/B model, the owner is purchasing an end product. Ten years ago, as D/B was emerging as a viable alternative to the standard design-bid-build delivery model, the prospect of underwriting these increased risks was daunting to both interested contractors and sureties alike. The old model of underwriting and insurance didn't allow for adequate coverage for this new complete product—not only for flaws in workmanship, but in design and functionality. Under the traditional performance and payment bonds and the architect/engineers errors and omissions (E&O) insurance, issues of liability in the case of poor structural performance were seen as blurred between contractors and designers. As such, the pricing structure of bonds and premiums were not suited to absorb the increased exposure. Thus, many sureties responded conservatively by refusing to underwrite D/B ventures, or selecting only the most capitalized and experienced builders to underwrite. This resulted in a difficulty for legitimate D/B firms to secure bonding.

Even in the case where a D/B venture was able to secure bonds, there were often significant "gaps" in coverage. Claims often involved expensive and time-consuming litigation to determine who was at fault and which insurance policy or bond would cover any defects in the performance of the structure. In the traditional design-bid-build model, if a problem arises due to an error in design, it is the E&O insurer who compensates the owner and/or rectifies the situation. If the lack of performance is due to an error in construction, however, then the contractor or its surety is responsible for any modifications to properly outfit the structure under the performance or maintenance bond agreement.

For example, if the HVAC system in a building on a bonded job fails to function properly, the owner would investigate and/or notify the surety. After an investigation occurs, if it is determined that the system is failing because the vent-size specified in the engineer's plans were too small, any repair would be paid under the engineer's E&O policy. On the other hand, if the system doesn't work because the contractor used an inferior compressor, then the contractor and its surety would be liable for bringing the structure up to the specifications of the contract. Under the E&O and performance bond obligations, the owner would also most likely be able to recoup any loss of income directly traceable to improper functioning of the HVAC, making it extremely important—and difficult—for the interested parties to determine what and who is at fault.

In addition, E&O insurance is only available on a claims-made basis, meaning that when a claim is made, the policy must be active for there to be coverage. Unlike a surety bond that is for a single project, an E&O policy is for a specified amount of money *for all projects the design professional works on.* If the engineer in the above example no longer carried E&O insurance, or if the policy has been depleted by claims from other projects, the owner typically must absorb the cost of a design error or attempt to extract it from the surety. Under a D/B arrangement, the owner would have a good case to recoup loss from the contractor if the designer's E&O policy was depleted and/or nonexistent. Thus, in the case of a design error, the owner would have a case against the contractor's surety as well, often after extensive litigation. For this reason, many sureties have been reluctant to bond D/B operations.

In the past two years, however, there have been numerous developments within the surety and insurance industry to change the ambiguity between what is defined as a performance and what is defined as a design error. Changes have been reflected in the language of contract bonds themselves and with companion insurance policies similar to E&O insurance for contractors discussed below. The above example demonstrates how intricately related E&O insurance and surety bonds are for the D/B firm. Thus, D/B trends in the surety industry dovetail with insurance concerns and will be discussed in tandem.

A MORE THOROUGH UNDERWRITING

The first response to the growing use of D/B has been increased specificity in the language of contract bonds. While many sureties are still reluctant to accept the increased risk of design, many others, through careful underwriting, are willing to provide the necessary bonds. What follows is an outline of the underwriting process that serves to supplement the information gathering discussed in Part 4, allowing sureties to confidently underwrite D/B ventures.

First of all, the underwriter of a proposed D/B venture evaluates the legal climate in the proposed construction area. Some state and local statutes forbid contractors from participating in design, while in other states, a design professional may not be tantamount to a contractor. Knowing the particular details are important to avoid litigation, and as some public entities require competitive bids, the surety must be comfortable with any and all legal ramifications regarding the use of D/B. Once satisfied that the D/B venture is legal and conforms to local statues, the surety underwriter proceeds to review the contractor and design team or firm.

Next, the underwriter performs a standard review using the criteria set forth earlier in this book. In addition to this review, the underwriter will want to determine what type of D/B venture it is. The contractor can be subcontracting design, the design firm can be subcontracting a contractor, the two can be joint partners, or they can be a single firm. When one considers that a construction manager may be used to subcontract a contractor and a designer or both, the necessity to understand the peculiarities of each venture becomes obvious. The surety may request documents supporting whatever claims are made about the length and scope of a partnership, and the D/B team should be forthcoming about producing any documents the surety requests.

The underwriter will want to be comfortable with the communication channels set up between the contractor and designer, no matter what the arrangement. The underwriter will want to know the capital resource of all entities, as well as what insurance the designer carries. The underwriter will also note any D/B experience and any projects on which the designer and the contractor have collaborated in the past, even if under the traditional design-bid-build model. A mission statement and documents detailing the scope, goals, and legal and financial agreements in place for the collaboration are all useful for the underwriter and may be required by some sureties. The surety will want to know the exact nature of the relationship, how long of a commitment has been agreed to, and how long any commitment has been in place. Individual underwriters may interview key personnel and even require a personnel chart listing the chain-of-command and channels of communication. All of this will allow the surety to evaluate the likelihood of problems arising on a specific project and therefore determine the level of risk this operation represents.

Once satisfied with state of the contractor and the designer, the underwriter will alter the traditional design-bid-build performance and payment bonds to reflect the added risks associated with any D/B venture, and this one in particular. At one time sureties charged a 10% to 50% surcharge for D/B construction, but such is no longer the case. In the current soft market, D/B contractors can expect to pay the same rates that traditional builders do. However, a D/B firm should inquire with its surety to know exactly what charges will be incurred, so as to include that cost in the bid.

In addition, it is common in D/B construction for the owner to require a warranty of more than the standard one year included with most performance bonds. This is to insure that the building is not only constructed in a timely and appropriate manner, but that it performs to the owner's specifications. The D/B team should determine at the outset who will be responsible for any maintenance or upkeep required in the project and factor in these costs in the bid. Generally this expanded coverage does increase the premium; therefore, a surety underwriter must be apprised of any warranties stipulated in contract documents. The increased use of warranties and maintenance bonds qualifies as a trend in and of itself and is discussed below.

The owner and/or the surety may require that an independent third-party monitor the disbursement of contract funds and construction progress, as in the D/B model the independent status of the design professional is somewhat complicated. The designer has a professional obligation to make sure that the plans are followed, the funds are paid out according to the stipulated payment terms of the contract, and work progresses at a reasonable pace. There is no reason to think that the D/B design professional would be lax in performing these duties, but just in case, an independent monitor might be called for in the language of either the contract or the surety bond to protect the surety and the owner from overpayment.

Finally, the surety will want to make sure that the coverage under the bond fits snugly with the contractor's insurance and design professional's E&O policy. The underwriter will want to make sure there are no gaps in coverage to avoid the timely and costly process of determining fault in case an error is found to be due to both design and construction. The increased task of underwriting D/B contracts may increase the bond price somewhat, but such an increase shouldn't be significant. In addition, there are several options when acquiring the necessary insurance that in the unified arrangement of D/B may actually save the contractor, design professional, and owner money.

CONTRACTOR'S LIABILITY INSURANCE

In addition to a more specialized underwriting detailed above, other insurance products have been used to provide complete coverage to the owner interested in acquiring a secured form of D/B. This insurance has several names and comes in two forms, one acquired by the contractor and another by the owner. Each policy should be written with the design professional's E&O policy in mind.

It might be assumed that under a contractor's general liability insurance policy there would be some clause for property damage or bodily injury claims arising from errors in project design. In fact, many policies systematically exclude such coverage. By stipulating changes at the outset of a policy, D/B contractors and construction managers can maintain limited coverage, but a serious gap exists in the amount and scope of financial damages that can be paid out in the case of loss of life or of poor performance due to a design error when the contractor is responsible for that aspect of construction.

The insurance industry has responded by offering professional liability insurance for contractors, alternatively called contractor's E&O. These policies, purchased in addition to a contractor's general liability policy, protect the contractor, and thus the owner, in the case of any damage to property or persons due to negligence or design flaw. Unlike designer's E&O, all of which are claims-made policies, contractor's or construction manager's liability insurance can be sold on an annual practice or single-project basis. There are upper-end limits to both policies, and advantages to each, but both provide protection not found in the performance bond or the designer's E&O policy.

The annual policy is considered favorable by some because it minimizes administrative procedures and provides continuous coverage on projects completed. In addition, it is often more affordable. On a project basis, if a policy expires before a statute of repose, or the time during which a contractor is liable for projects, the contractor may be without coverage, especially if the designer's E&O policy is not in force. However, if the contractor has an annual policy, this would cover projects completed in the past.

On the other hand, a single-project policy can be advantageous, especially for designers. Any losses on the project would not be reflected in the designer's E&O policy, and thus premiums would not increase. The cost of project policies might be more if purchased by the contractor, but the exact coverage needed will be acquired. Alternatively, if the insurance is included as part of a package purchased by the project owner, then the costs can be significantly less.

This second type of coverage is known as an owner-controlled insurance program (OCIP), alternatively called “wrap-up” insurance or an umbrella policy. Because it is such a large policy, administrative costs are reduced, and thus a premium discount is granted, lowering the cost for the owner. An OCIP has many advantages, especially for a large project owner, as in addition to lowering costs it can assure that no lapses in coverage exist. Also, an owner can acquire an excess liability policy that, as the name implies, would provide coverage after the contractor’s liability and designer’s E&O policies were depleted. Such coverage typically costs less than a project policy and may be preferable for some owners (“Gaps or Going Bare?” 1998)

Whether the owner or the D/B team purchases a policy, and whether it is for a single job or an entire work program, it is recommended that the standard coverage of the design professional’s E&O policy, the contractor’s general liability policy, and the performance bond be supplemented with contractor’s liability insurance. The added coverage will help clear the waters that can be seriously muddied as the layers of design and construction become increasingly mixed. If an owner wishes to purchase the assurance that surety bonds provide, he or she will most likely be interested in covering any gaps that might exist with the traditional surety and insurance package. While the exact products vary a great deal with individual surety and insurance companies and states, the industry is converging on a model in which the insurance products are completely integrated with surety bonds. As of yet, there are still many models out there and no one seems to be dominating any other. The savvy owner should make sure that the surety company is aware of any potential concerns, and that every effort is made to provide complete coverage.

Chapter 26

SPECIAL SURETY CONCERNS AND INDUSTRY TRENDS

The future of the surety industry will always be difficult to predict due to the reasons mentioned in the previous chapter. Despite recent improved financial performance, the surety business remains as risky and uncertain as construction itself. Current competition between sureties and contractors may drive up the incidence of contractor default as profit margins shrink and every default weighs more heavily on a surety's resources. On the other hand, this spate of competition could as easily be a motive force in causing the surety industry to evolve more quickly with the construction industry by streamlining its expenses and altering its underwriting to better serve owners and contractors. This is already happening with respect to remediation and more recently with Design/Build.

By nature, the surety industry is highly conservative. It is through prudent judgment that an individual company and the industry as a whole can remain profitable. As such, a good indicator of change in the industry are areas in which current underwriting practice seems insufficient or unnecessarily cumbersome, as it was with Superfund work and as it still is for Design/Build. Locating where the surety industry is attempting to assimilate new developments in the construction industry reveals how surety is evolving as well as its mechanisms for change. To accomplish this, the final chapter will explore several trends in the surety industry. Included as well are predictions from industry experts as to how these trends will shape and be shaped by surety practice.

Of course, not every problem with the surety industry is novel, nor every hot spot an indicator of change; several surety concerns have been around for awhile. One perennial concern is small and medium-sized firm, another is surety "red flags." These concerns, addressed alongside surety trends, reveal general information about the climate of the surety industry and are as follows: 1) warranties and maintenance bonds, 2) competition among sureties and contractors, 3) private market, 4) hazardous waste and Superfund legislation, 5) small to medium-sized and disadvantaged firms, 6) Miller Act, and 7) transfer of ownership.

WARRANTIES AND MAINTENANCE BONDS

Maintenance bonds, alternatively called warranty bonds, are increasingly being used to guarantee the quality and wearability of completed construction projects. The duration of coverage as well as the ways in which they are being used is changing, leaving many surety experts to predict some sizable changes in the cost and underwriting of these bonds.

Typically, a one-year warranty is provided with a performance bond. Sometimes this period increases to two years. Since it is extremely difficult to predict how long a contractor will remain in business to repair any problems that arise, the surety underwriting such a long guaranty may be faced with huge costs. Therefore, it is difficult to determine how well the current pricing of this product will provide for inevitable loss due to default.

To make matters more complicated, many owners have recently begun purchasing 5 to 10-year warranties that greatly extend the scope and use of the maintenance bond. In effect, the owner is requiring the contractor to not only build, but guarantee performance for a significant amount of time after a project is completed. Since parts of even well-constructed buildings may fail over time, this coverage can become quite costly over time. And since it is impossible to determine what will fail and how much it will cost to fix it, in effect, both the contractor and the surety are hedging. It remains to be seen how losses will affect pricing.

Some states are experimenting with maintenance bonds and highway construction. Contractors are asked to provide complete guaranties for up to 10 years that the asphalt will continue to perform, even in the case of severe weather such as flooding. The contractor is responsible for maintaining the highway, and in some cases, removing debris and snow. In effect, the state contracts out the up-keep of its road to a contracting firm.

COMPETITION AMONG SURETIES AND CONTRACTORS

In today's market there are over 600 sureties and surety divisions competing to write bonds whereas the number was half that 15 years ago. Capacity is high and losses have dropped from over $788 million in 1986 to just over $403 million in 1997. In addition, markets are full of willing contractors—by last estimate there were over 1 million—but especially in the public sector, construction starts are static. The current market is being described by industry monitors and surety personnel as having "heavy competition," "severe competition," and more specifically as "the most competitive market I've seen during my 25 years in business." The surety market has been relatively stable with little new business, and the flooding of new players has made it easier for many contractors to get surety bonds as both prices and underwriting standards have dropped. According to John Savage, Co-CEO & President of AMWEST Surety Insurance Company, this extreme competition "is exerting a downward force on pricing," while according to Richard Farnsworth, Senior Vice President of the Fireman's Fund's Surety Division, "in addition to driving down prices, competition is also driving down underwriting problems." Furthermore, he adds, "it's a serious problem" (Hirsch 1998).

The consensus of contractors and surety personnel is that acquiring a bond is much easier than it was 10 years ago. According to J. Martin Huber, executive vice president of the National Association of Surety Bond Producers, "You have more [surety] companies chasing the same accounts, especially the medium-to-large 'cream-of-the-crop' accounts," which in turn is making surety bonds "a commodity with an emphasis on price" ("Surety Firm Competition" 1996). At the same time, the increased surety competition allowing many contractors to perform bonded work has lowered profit margins for both sureties and contractors as everyone attempts to remain profitable. The situation is at best tenuous as buffer margins slowly and steadily erode.

In the current state of affairs where the emphasis for sureties has moved from prequalification and claims management toward pricing and service, a possible widespread lack of sound underwriting may have serious financial consequences. This has led many to point to a situation on the horizon not dissimilar to that which occurred in the early 1980s. It has been reported that some sureties are waiving indemnification, while others are more relaxed about geographic restriction—both of which were unheard of in the wake of the losses from 1984-1987 ("Surety Firm Competition" 1996; "No Surety or Fidelity" 1997; Hirsh 1998). As one surety expert has said, "This industry cannot but help shooting itself in the foot." As sureties compete with each other to underwrite contractors, lowering their premiums and underwriting criteria in the process, the industry may be setting itself up for huge losses.

Furthermore, competition and its effects create a problem for owners. The less discriminate the surety becomes in selecting contractors to bond, the less prequalification is actually achieved. Prequalification is fundamental to what the surety line ostensibly provides owners. By selecting only the contractors fit to perform on a given job, the owner gets not only the surety's guarantee that the project will be completed, but the assurance that the chances of contractor failure are slim. While it is difficult to see the effect of loosening underwriting and heightened competition, since the effects will not present themselves for several years down the road, there are indications that the industry is slumping.

Data shows that contractor failures raised 6% in 1996. While the amount of these liabilities is lower than 1995 and about four times as low as 1992, this could mark the beginning of a period of scarcity. In 1997, D&B reports indicate that both liabilities and contractor failures increased pointedly. Contractor failures were up over 11% to almost 11,000, the most since 1992. Is this the end of an 11-year string of good surety performance? Probably not, says Lynn Schubert, SAA President. While profits are being squeezed out of the market, Schubert reports that "most of the companies I have spoken with have retained their strong underwriting standards. They tell me they are looking for creative ways to price the product" (Hirsh 1998). Alternatively, this spate of competition may lead to better communication between sureties and contractors, better claims handling, and a more rapid acceptance of D/B, bonding in the global marketplace, and environmental clean-up, three areas where intrepid and sound sureties are making healthy profits.

As sureties collect fewer premium dollars, they are forced to avoid contractor default and forced to search out profitable sectors of the industry to remain profitable. This is achieved with solid market research and a more thorough underwriting. Also, if a contractor has problems, the surety can help it finish troubled jobs, whether by financing with "look-see" money or with the reinsurer's consent. The lower margin of profit may drive sureties to become more active in responding to any owner complaints and force them to new levels of efficiency. Many have already observed a more proactive trend in claims handling, and such could be a sign that competition is actually honing underwriting and claims skills (Klinger 1997). While many fear poor underwriting, there is no definitive method of determining just how lax underwriters have become other than by noting liability dollars and number of contractor failures. Although both of these numbers may be increasing, the industry can easily remain profitable. Although for five years people have predicted a surety slide, doomsday has so far been staved.

Either way, the trend toward increased competition will most probably continue as long as the surety industry remains profitable. Eventually, if construction tapers off, the construction and surety markets will no doubt shrink as players leave the industry. But before this happens, or perhaps simultaneously with it, there may be several large defaults—as was the case in the mid-1980s—or perhaps numerous smaller ones. This might clear out some of the less-capitalized sureties and builders, as well as reinsurers. If this were to happen, eventually premiums would no doubt be increased to a level allowing sureties an acceptable margin of profit. So far, however, there appear to be only downward trends in pricing, and it remains to be seen how long the current profitable period will last.

PRIVATE MARKET

One perennial interest of the surety market is expanding the surety line into the private market. Roughly 18% of the surety's book is with private owners, while the majority of private projects are unbonded (see Figure 1-1). This means that there is significant room for growth into this market. In addition, as competition makes it more difficult for sureties to remain profitable, there is more incentive than ever for the surety to actively expand into the private sector rather than compete in an ever fiercer realm for the existing and static public-dominated market.

In 1993, The Surety Information Office (SIO) was founded by the SAA and the NASBP to help promote surety bonds. Its mission is "to increase the use of corporate surety bonds in the private sector and disseminate positive information on the important role of corporate suretyship in public and private construction." The office freely dispenses surety speeches, articles, and presentations to the interested public. The information is also readily available on their internet site, http://www.sio.org (see Appendix XV). In addition to general information about surety bonds, the SIO promulgates the benefits of surety bonds over letters of credit, how surety claims works, and how a private owner might be better served with surety bonds than without.

The existence of this organization is evidence that surety companies are looking to expand into the private market. In the current competitive climate, such an expansion would benefit the surety market tremendously; however, it is difficult to gauge how successful this effort has been or will continue to be. Many surety companies report a public awareness effort, but so far, few report large numbers of private owners converting to the use of surety bonds. While premium dollars keep increasing, they are barely staying abreast of inflation, meaning that growth is minimal. Perhaps as the construction industry shifts to D/B—with all of their eggs in one basket, so to speak—owners may want assurance that plans are followed and projects are successfully completed in a timely fashion. In this case, the surety industry has an historic opportunity to begin linking their services to D/B delivery in an effort to promote their product and increase the satisfaction of owners. The DBIA, SAA, and NASBP are working closely together to create surety bonding for D/B, and it is only a matter of time before it is commonplace for surety firms to bond D/B as well as other delivery systems. In the near future, the surety industry has an opportunity to capitalize on this trend.

HAZARDOUS WASTE AND SUPERFUND LEGISLATION

The 1935 Miller Act established the requirement of performance bonds on federal construction jobs—including hazardous waste and toxic remediation. However, it was not until the passage of the 1980 Comprehensive Environmental Response, Compensation and Liability Act, commonly referred to as the Superfund law, that efforts to clean up the nation's most hazardous sites was initiated. Under the Environmental Protection Agency (EPA), Superfund was to support the cleaning of what was initially identified as 400 sites. In 1986, the Superfund Amendments and Reauthorization Act renewed Superfund, establishing funding through 1990. Funding was again extended in 1995, and most recently in 1997. However, a May 15, 1998 deadline was not reached to renew funding for the program, and currently, it is unclear how Superfund will evolve.

From day one, Superfund has been mired in litigation. Although it appears as if the remediation industry is slowly moving ahead with necessary surety bonds, sureties were until recently increasingly reluctant to bond hazardous work because of "deep pockets" fears. Sureties, remembering what happened to writers of liability insurance for asbestos manufacturers, feared that long after the insurance was written, they would be forced to pay liabilities to claimants. In 1989, bonding Superfund work was named the biggest issue facing the surety industry by the SAA president, Lloyd Provost ("Sureties in Black" 1989). In 1990, the U.S. Congress assuaged some of these fears by passing a bill removing some liability from sureties and insurers for long-term tort claims in Federal courts (Provost 1991). However, this amendment didn't clear the way for an uncomplicated bonding climate, and to this day many projects have been stalled and many sites lay dormant.

So far, 500 sites have been cleaned up for a cost over $30 billion, a third of which has been spent on legal fees alone, under current Superfund law. Part of the reason for so much litigation stems from the fact that the Superfund law is retroactive, meaning companies performing acts illegal under current Superfund law before it was passed are still liable. In other words, there were no grandfather clauses. Also, each company involved in a site is jointly and severally liable, which means that even small contributors to a toxic site can be forced to foot the entire clean-up bill if they are the only surviving entities (Gastel 1998).

Because Superfund work is public, even if a private company is found to be responsible, proceeding with remediation requires a surety bond. It has been estimated by A.M. Best that to clean up the nation's toxic sites will eventually cost insurers and sureties over $55 billion. Since 1991, they have already paid out over $9 billion. However, in lieu of the murky legislative climate, most sureties have shunned hazardous waste clean-up resulting in delayed schedules, reduced competition, and ultimately an increase in the price of waste clean-up, not to mention further litigation. Congress has attempted to revamp Superfund several times, clarifying to what extent parties are liable, but there has been little consensus and currently the entire program is in question.

Recently, as sureties search for new markets in the competitive climate, a few companies have become experts in assessing risk on Superfund and environmental remediation work. There are about 8,000 contractors who currently perform environmental construction and consulting, but a study conducted in 1993 and 1994 found that over 50% were uninsured. Although there are still issues of open-ended liability, sureties such as Reliance/ECS Underwriting Inc., AIG/C & I, Zurich, United Coastal, and GenStar are taking advantage of this market potential for not only surety bonds, but professional liability insurance with pollution coverage (Gilbert 1995). There is room for sureties to grow into this market, but specialized and technically proficient underwriting is a must.

In addition to individual surety's initiative, legislation continues to push Superfund forward in Congress and the court system. In the summer of 1998, the U.S. 4th Circuit Court of Appeals backed the Institute of Scrap Recycling Industries' (ISRI) friend-of-the-court brief that Superfund liability applies to the disposer of waste at a contaminated site and not to vendors of secondary materials. In the past, ISRI members have been declared "potentially responsible parties" for using or selling contaminated products, but under this ruling and amendments to state Superfund and EPA laws in 1997 and 1998, the parties actually producing waste, and not those selling or transporting products that produced the waste, are solely responsible for environmental degradation ("Scrap Companies Win" 1998). Such clarification of Superfund law is making it easier for a surety to assess its total risk, and thus will eventually help the United States dispose of an estimated 35,000 toxic and hazardous sites.

Another change in Superfund law may come from a refocusing of the goals of the program. There have been movements toward remediating the so-called "brownfields." Brownfields are abandoned manufacturing and industrial properties that are generally much easier to remediate than other Superfund sites. There are currently 1,300 identified Superfund sites, and over 500,000 brownfields in the United States. While Superfund sites still appear daunting from both legal and technical angles, according to George Kanapel, a brownfield specialist, "brownfields are a potential asset, a chance to turn a property around, maybe make a profit" (Wright and Roe 1997). Since initiating a pilot brownfield redevelopment program in 1995, the EPA has dispensed over $13 million to 78 projects. On July 15, 1998, Columbus, Ohio, received a federal grant for $2,000,000 to administer the Brownfield Assessment Demonstration Pilot. The funds will help the Columbus Brownfields Task Force assess eight Columbus-area brownfields, as well as a public outreach program. The sites are to be turned over for future development (Pryce 1998).

Such a reengineering of Superfund legislation to focus on less-toxic sites may help move legislation to take care of all the nation's toxic sites. Another possible solution to the current legislative and legal logjam may be to relax the bonding requirements on Superfund jobs. This could be achieved with an amendment to the Federal Acquisition Regulation to allow letters of credit or some other form of guaranty in lieu of bonding. Overall, however, clarification is needed in the language of the Superfund law, especially over third-party liability. It needs to be decided exactly how long and to what extent sureties and insurers of remediation projects and industrialists are liable. Until this happens, it appears as if progress in detoxifying the environment will be at best plodding.

SMALL- TO MEDIUM-SIZED AND DISADVANTAGED FIRMS

Sureties have historically been reluctant to bond small, emerging contractors, whether minority, disadvantaged, or otherwise. In fact, access to adequate bonding constitutes the major obstacle faced by the minority contractor and any small contractor trying to enter the business by performing public work or bonded private projects ("Opponents Find Common Ground" 1990).

With few exceptions, most contractors find themselves forced to start out performing private, unbonded projects. If they can make a go of it—build a track record, accumulate some working capital and net worth—then they have a chance of establishing a bonding program. Thus, while small, emerging, minority contractors may not be able to take bonded projects, they can compete for unbonded work. With the total construction market in the United States estimated at close to $500 billion annually, of which at most 25% is bonded, there is plenty of opportunity for start-up contractors. Bonding is not a barrier to those who wish to enter the construction business. It is only a barrier to those who want to perform work that requires bonds. For the small or minority contractor wishing to perform bonded work, however, there are several paths to achieve this goal.

Construction companies with adequate capital and experience will almost always be able to obtain surety bonds, while those without these adequate resources will have to pursue unbonded work. Just what exactly is defined as "adequate," however, changes constantly with wider economic concerns. Today, sureties are much more willing to bond small and minority contractors than they were even five years ago. In general, however, small to medium-sized firms ($1 to $25 million in sales) tend to experience more difficulty in securing necessary bonds than larger or more well-capitalized builders. The major reason for this is, of course, their insufficient working capital and overall net worth, both of which factor heavily into a surety's underwriting decision. Other variables are important as well, and for the small to medium-sized contractor looking to perform bonded work, the following might serve as a road-map toward that goal.

According to a National Association of Minority Contractor's (NAMC) survey taken in 1985-86, about three-fourths of those surveyed had won contracts through competitive bidding but had to forego the contract due to the inability to secure bonds ("Surety Bonds and Minority Contractors" 1988). This was an especially difficult time for sureties, and at least part of the high rate of denial to minority-owned firms was directly related to an industry-wide tightening of the belt. However, more recent evidence suggests that such firms receive higher degrees of scrutiny and less success in securing bonds.

According to a U.S. General Accounting Office (GAO) survey of small construction firms that had received bonds between 1990 and 1993, minority-owned firms were almost twice as likely to be denied a bond as non-minority firms (36% to 20%). In addition, the GAO found that sureties set more conditions more often when considering a minority firm than they did when considering a non-minority firm. The study found that women- and minority-owned firms were "more likely to be asked for certain types of documentation," and that minority firms were "more likely to be asked to provide collateral and meet other conditions than the firms not owned by minorities." Sam Carradine, executive director of the NAMC in 1995, concluded the survey showed "in a very clear and quantitative way that there is a great disparity in the opportunity for getting bonds between small minority firms and their counterpart-sized majority firms" ("GAO Suggests Bias" 1995; "Minority-Owned Firms" 1995).

The surety industry pointed to a higher availability of bonds to smaller contractors in general. In addition, the NASBP reported that 18% of the 6000 firms represented by its members are minority-owned. However, it is clear that on some level disadvantaged firms are kept from acquiring bonds more often than are white-owned contractors of the same size. For the minority contractor, it is especially important to have his or her finances in order and to keep abreast of any developments in construction and surety that might affect the availability of bonds.

Although in the current profitable cycle many small to medium-sized firms of all ownership report an easier time securing bonding, for the surety, no matter what the overarching economic climate, concerns revolve around the ability of such firms to succeed under increases in project size, changes in the types of construction undertaken or geographical area, replacement of key personnel, and overall contractor profitability. If anything unusual should happen—possibly even something outside the contractor's field of influence (e.g., an owner goes bankrupt)—the surety will want reassurance that the contractor will not fold.

Typically, small- to medium-sized firms lack financial and technical sophistication necessary to remain consistently profitable. Job cost information and financial record keeping are often poor or nonexistent; hence, these firms generally do not implement project management and control in the field. This can lead to financial difficulties because the contractor does not know the cost status of on-going projects. In addition, as work becomes more scarce, these firms are more apt to pursue projects that are beyond their demonstrated experience. If they are unfamiliar with the type of work or experience problems with an increased project scope, difficulties arise that can trigger the failure of the firm. Therefore, sureties must be convinced that a small contractor does not represent more than a tolerable level of risk.

As mentioned earlier, prequalification is crucial in the surety's assessment of risk. Consequently, firms with little or no construction background make a risk analysis more difficult, if not impossible. Lowering underwriting standards may mean financial losses. Furthermore, minority firms, as do all small contractors, tend to acquire relatively small projects. In addition, the surety expends approximately the same amount of time and effort evaluating contractors for small projects as it does for contractors bidding for larger projects whose return (premium) is far greater.

No matter what their size or financial reserves, contractors who have been unprofitable or have an erratic profit pattern over the past few years are unattractive to sureties. Since business cycles tend to be more abrupt for smaller contractors, this is of special concern for sureties considering bonding smaller contractors. Consequently, the contractor's demonstrated ability to make a profit in a competitive environment is often a critical factor in whether he or she gets a bond. Sureties exist that bond small contractors, and they sometimes offer considerably higher premiums to compensate for the additional risk. Still, there are ways for even the smallest of contractors to perform bonded work.

Generally, a well-articulated business plan, complete with succession papers in the event of loss or incapacitation of key personnel, a codified set of bid preparations, and reliable accounting procedures will cast a favorable light on the contractor seeking bonding. This is especially true for a disadvantaged firm. The more prepared the contractor is, the more attractive the firm will be to a prospective surety. The contractor might consider using an individual surety or an alternative form of guaranty to get their foot in the door and develop a track record. In addition, the Small Business Association (SBA) offers two programs for small or disadvantaged contractors to acquire corporate surety bonds and has the authority to waive surety bonding requirements for "small disadvantaged businesses" contracting with the Federal Government.

The SBA and Surety Bond Programs

The SBA Bond Guarantee Program and Preferred Surety Bond Guarantee Program provide one of the best opportunities for small and minority contractors to demonstrate their abilities without being restrained by their lack of experience. The first program, known as Plan A, was established in 1968 by the Federal Government. The program's basic objectives were and still are: 1) to provide bonds for small and minority contractors who may have difficulty obtaining bonds and 2) to increase their viability so they can eventually make the transition to the regular bonding system (Johnston 1981; McGreevy 1991).

Plan A permits the SBA to insure up to a certain percentage of any federally listed surety company's losses as a result of a small contractor's breach of the contract terms, usually 80%. If contract losses are less then $100,000 and the owner of the firm is economically or socially disadvantaged, then the SBA will guarantee up to 90% of the surety's losses. To qualify, the contractor must demonstrate revenue below a specified amount and be able to show that it is bidding on a project that requires a contract bond that cannot otherwise be obtained without an SBA guarantee.

Under Plan A, sureties must have prior approval of SBA before issuing bonds. This has been viewed as an impediment to furthering the guarantee program, as the paperwork and waits can become excessive. In 1988, to help further efforts to bond small and minority contractors, the SBA introduced the Preferred Surety Bond Guarantee Program, known as Plan B. This program, unlike Plan A, allows approved sureties to issue bonds *without prior approval* to qualifying contractors. A surety must be on the U.S. Treasury List, have an approved underwriting capacity of at least $1.25 million, and have completed an SBA review. Currently 14 surety companies participate in this program. Contractors must also be qualified by the SBA and must have an average earning over the past three years that does not exceed $5 million annually. In addition, the contractor must have a good business reputation and track record and must be willing to purchase the computer hardware and software necessary to transmit economic information to SBA (Hannum 1997).

Plan B insures losses up to 70%. For the increased control sureties have over underwriting—under Plan A the SBA had final say—the guaranteed amount is slightly decreased from Plan A's 80% to 90%. The surety cannot charge more than the SAA advisory rates, while the contractor is charged $6.00 per $1,000 of contract price by SBA. Therefore, the fees for this form of bonding will be higher for the contractor, but the access of surety bonds allows a small or disadvantaged company to bid bonded work. A surety cannot participate in both SBA programs.

For any SBA programs, contractors must fill out the following SBA forms (Adapted from Hannum 1997):

912—Statement of Personal History: completed by proprietor; each partner in a partnership; each officer, director and/or holder of 20% or more voting stock in corporation; and any other person who has authority to speak for management

991—Guarantee Agreement Addendum: puts SBA backing in force as soon as contractor begins work on a project

994F—Schedule of Uncompleted Work: lists all uncompleted work, both bonded and unbonded

994—Application for Surety Bond Guarantee Assistance: contains specific questions relating to the contractor's business size, type, ability to obtain bonding, minority status, and relationship to SBA

1261—Statements Required by Law and Executive Orders: gives applicants notices of various federal government and executive orders that affect SBA's business

1624—Certification Regarding Debarment: certifies contractor is not personally debarred, suspended, proposed for debarment, declared ineligible, or voluntarily excluded from transactions with any federal agency

Interested contractors can contact a surety agent, or the SIO or SBA directly. Their addresses, telephone numbers, and web sites are available in Appendix XV.

MILLER ACT

The 1991 revision to increase the minimum federal job requiring bonds to $100,000 was the latest amendment to the 1935 Miller Act. While this piece of legislation has proven highly resilient to time, the current levels may need to be adjusted again, especially for payment bonds. If for no other reason, the severe amount of inflation between 1935 and the 21st century suggests some adjusting of the limits.

On all federal jobs over $5 million, payment bonds are capped at $2.5 million. In addition, claims can only be made by first- and second-tier subcontractors and their direct suppliers. For large, multi-million dollar jobs with lots of subcontracted work, significant amounts of money in excess of the $2.5 million is tied up, often with third-tier subcontractors. If a primary contractor defaults on a large federal job, the chances of recovery might be slim to none, especially for small contractors. *ENR* reported in 1997 that an electrical contractor was owed over $3 million on a $103 million federal prison where the primary contractor defaulted. Other subcontractors were owed money as well, but since the limit for a payment bond is $2.5 million, it is doubtful that the U.S. government or the surety will foot the bill. The subcontractor has to take the primary to court or receive only a portion of the money it is owed ("Miller Act Frustrates Subs" 1997).

ENR recently reported as well activity in Congress to amend the Miller Act, but as of publication there has been no major indication that such legislation is forthcoming. Rep. Carolyn B. Maloney (D-N.Y.) has been outspoken in the House, calling the Miller Act "wholly inadequate today when major construction contracts cost hundreds of million [of dollars]." Maloney is the author of a proposed bill calling for the payment bond amount to equal the subcontracted amount, no matter what the size of the contract. Such a change would afford all subcontractors equal protection under the bond. The White House agrees that changes need to be made, but does not currently support Maloney's proposal. So far, no companion legislation has been introduced in the Senate, but many feel that in 1999 some sort of overhauling will take place (Winston 1998).

The author would not be surprised if in the near future payment bonds will by law be issued for the same amount as performance bonds on federal and state projects, much as they are for private contracts where a 100% payment and 100% performance bond costs the same amount as a 100% performance bond. This increased coverage may raise premiums, but since the performance and payment bond is sold as a unit, any increase should be minimum.

TRANSFER OF OWNERSHIP

Since most sureties bond the company, not its owners, it is entirely possible in the case of sole proprietorships or partnerships that the surety in essence bonds the sole proprietor or partners. In this case, a surety must be concerned if a construction firm voluntarily transfers ownership to a new construction firm or entity, as what belongs to the owner(s) and what belongs to the company become highly important. Therefore, the surety must first examine the organization of the construction firm in order to determine whether the bond will carry over in the event of a transfer of ownership and what effect such a transfer would have on the bond.

If the sale does not affect the status of the bond for the firm, the surety will still want to examine the new owners and determine their strengths and weaknesses. The surety wants to know the assets and liabilities of the new owner as well as their experience in the construction field. The surety should also consider such other factors that may be relevant, including those they considered before initially issuing the bond. The surety should consider these factors on a case-by-case basis and within the context of the ownership transfer. In some cases the transfer of ownership will void the construction bonds. In such a case the surety must re-issue new bonds according to the criteria it considers relevant, such as assets and liabilities, construction experience, credit rating, current construction workload, the management structure, biographical references, and others.

In certain cases of ownership transfer, such as by will or trust from a father to his children, there may be no need for an interruption in the bond. In such scenarios, the surety agent will most probably have some knowledge of the succeeding owners. It is when the transfer is made to an unrelated third party that the surety may become concerned. In all such matters, sureties determine action on a case-by-case basis. Because the transfer of ownership is an important event, the surety will most often request information from the potential or new owner. The new owner should furnish this information willingly.

Not all transfers are due to business transactions. Sometimes, tragedy may befall a contractor and necessitate immediate transference. A recent survey conducted by the CPA firm Grant Thornton found that while many contractors are planning for such contingencies, many more need to do so. Especially for family-owned business, which account for roughly 75% to 85% of contractors, the need for a strategic succession plan is important not only for the company's well-being, but for securing bonds. While almost 75% of contractors with annual revenues of $50 million or more have a succession plan, only 29% of those earning less than

$10 million annually report that they have one ("Survey Examines 6 Issues" 1998). While a succession plan takes time, it provides contracting firm owners, investors, and relatives the peace of mind that in the case of the loss or incapacitation of a key employee, the business will continue to be viable. The surety looks for a succession plan as a sign of the company's vision and management acumen, and many underwriters insist on some kind of transition plan. Small contractors looking to bond should take special note of this fact. Simply taking the time to provide an articulated plan of succession, even if informal, can help tremendously in securing bonding.

Chapter 27

SUMMARY

Fundamentally, surety bonding involves offering a prequalification service through the extension of credit where no loss is anticipated. Therefore, suretyship as it is currently performed is an arrangement whereby underwriters rely heavily on the integrity and financial resources of the construction companies they bond, with only minimal provision for loss in their premium. However, this theory of "no expected losses" has been proven incorrect. The increases in premiums during the 1980s demonstrate the surety's need to account for inevitable and unpredictable losses.

The surety industry and the process though which its members arrive at an underwriting decision are not well-understood by outsiders. Factors evaluated in the underwriting process are often unfamiliar to construction industry professionals, and thus have been identified and described as much as possible. For example, the factors affecting premium computation can be difficult to fully comprehend. Armed with the information this book presents, however, contractors are better prepared to meet the requirements for establishing and maintaining a proper relationship with their bonding agent and surety company. This helps to ensure that important information is not overlooked, which could result in the rejection of a bond or delays in the bonding decision. With the information presented in mind, contractors can make a more informed decision as to whether to pursue projects requiring bonds. Contractors choosing to pursue bonded work should establish a relationship with a surety agent *as soon as possible*. As remarked on in several places, delays in establishing such a relationship can result in denials of requested bonds. In the event that a bond is denied, the contractor should seek the reason(s) for the decision and what corrective action may prevent future rejections. Finally, it is important that contractors keep their surety and independent surety agent abreast of their operations by continually providing information on the status of current projects and future anticipated work program. Such information enables the agent and surety to better assist the contractor with its bonding needs.

Considerations in handling a surety bond claim are outlined above. As discussed, alternatives to surety bonding, such as 1) waiver of bonds, 2) individual or personal sureties, 3) letters of credit, and 4) cash equivalents are possible solutions to the difficulty in obtaining surety bonding. There are also the SBA's two bonding programs for small or disadvantaged firms. The reader also learned about certain special concerns of sureties and how macroeconomic variables can affect the availability of bonds, the most striking cases being the relative lack of capacity in the mid-1980s and the large capacity of the mid-1990s.

The owner/contract administrator should check the authenticity of the company providing the contract bonds as well as the information presented on each of the bonds. Sources of information regarding the surety companies include state insurance departments, *Federal Register*, A.M. Best Company, telephone directories, and insurance agents and brokers. Fostering an understanding of contract default and disputes arising from surety bonds are vital to comprehending the surety business, and clarifying these subjects, along with trends in the surety industry, were the goals of this book.

Until a solution to the bonding dilemma develops, bonding recipients can be more adaptable by having an enhanced knowledge of the surety process. The author set out to enable project owners, design professionals, and contractors to better understand the surety industry and the construction contract bond underwriting process. As the construction industry grows increasingly uncertain and complex, surety bonds will remain an integral part of the mix. An increased awareness of the surety industry and underwriting process should enhance the bonding recipient's capacity for functioning in the complex world of surety bonding at the same time that the surety continues to improve its product and better serve owners.

Over the long term, better business management with regard to capacity and capital will evolve in the construction industry due to surety bonding. Because the typical construction company is undercapitalized and yet desires to increase its volume of work, intense competition, low-profit margins, and high failure rates are more common in this industry than many others. Astute surety companies will only bond construction firms with the capacity and financial capability to complete all the projects that it undertakes. If a contractor wants to acquire more work than it is capable of handling, it will be unable to obtain bonds for the work; the inevitable result will be that contractors will be more cautious expending time and resources pursuing work for which they do not have the capacity to complete profitably. Into the 21st century, surety bonding will clearly have an impact on the structure of the construction industry.

Looming over this project is the current financial situation in Asia and the United States. It is unclear exactly how the slowing of the fastest-growing economies in the 1970s and 1980s will affect the surety and construction industries domestically, but rest assured that they will. If the surety market experiences large or numerous losses, there will be a mass exodus of the smaller and less-experienced sureties. The current level of competition is fierce, and something will surely upset the balance sooner or later. Perhaps the competition itself will be the trigger. If so, expect a tightening of underwriting criteria and a move back to sound underwriting. In addition, there may be an increase in premium rates, and smaller contractors will have difficulty bidding bonded work as they will have a difficult time securing bonding. Even in the worst of times, however, there will always be private and federal construction jobs requiring surety bonds, and it is up to the individual contractor to prove him- or herself worthy to perform bonded work. If this is the goal, the principles set forth in this book should help a contractor achieve a bonded program no matter what the business climate, and for those interested, the author wishes them best of luck.

GLOSSARY

Agent—One who solicits, negotiates, or sells surety bonds on behalf of a surety company. An agent's right to exercise various functions, authority, and obligations, and the obligations of the surety company to the agent are subject to the terms of the agency contract with the surety, to statutory law, and to common law. Contrast with "broker" and "consultant."

Bank Guarantee—See "letter of credit."

Bid Bond—Guarantees that the bidder (contractor) has passed a surety's prequalification process and is capable of entering into the contract.

Broker—Individual or organization representing a contractor in soliciting, negotiating, or buying a surety bond and rendering services incidental to these functions. By law, a broker may also be allied with the surety company for certain purposes, such as delivery of a policy or collection of payment.

Claim Personnel—Employees of the surety company who manage and control surety exposure in case of contractor default. Contrast with "Surety Underwriter."

Consequential Loss—Loss which arises as a result of direct property damage. See "direct loss."

Consortium—In surety bonding, an organization formed to spread risk.

Constructability—The concept of using construction knowledge and experience in planning, designing, procuring, and building to achieve overall project objectives more efficiently and synergistically.

Consultant—An individual or organization that offers, for a fee, advice to technical assistance on matters pertaining to surety bonding.

Contractor (principal)—The construction contractor who agrees to carry out the terms of a contract for the obligee (project owner).

Contractual Liability—Liability assumed under a written contract.

Design-Bid-Build—Traditional construction delivery system in which the owner (obligee) first commissions a design, then offers the project up for bidding to interested contractors. The owner then selects the lowest qualified contractor to construct the building according to the plans.

Design/Build (D/B)—Construction delivery system in which the owner (obligee) hires a single firm (principal) to design and build a building. Similar to "master builder" approach popular in early 20th century. Contrast with "Design-Bid-Build."

Direct Loss—Loss that is a direct consequence of a particular peril. Fire damage to a refrigerator would be a direct loss; spoilage of food in the refrigerator as a result of the fire would be an indirect or consequential loss. See "consequential loss."

Discovery Period—Under certain surety bonds, the time allowed to discover, after termination of the bond, that a loss has occurred. The terminated bond covers the loss if it is discovered during the specified time period.

Environmental Remediation Contract—A broad class of construction contracts covering remediation of environmental damage.

Expense Ratio—Ratio obtained by dividing written premiums by incurred expenses over a given period. This ratio measures the percentage of premium used to pay all the costs of acquiring, writing, and servicing surety bonds, and is generally thought to be equal to around 60% of the premium.

Fiduciary—A corporation, person, or other entity that occupies a position of trust; especially one who manages the affairs of another.

General Indemnity Agreement (GIA)—Agreement by which one party agrees to indemnify a second party for losses suffered by the second party.

Heard Act—An 1894 federal act requiring surety bonding of federal public works programs; later expanded by the Miller Act of 1935.

Insurance—A two-party risk-transfer mechanism through which one party pays to have another party protect it from certain well-defined risks. Suretyship, in contrast, is a three-party relationship akin to that of a credit transaction in which no money changes hands until a default occurs or a problem arises.

Letter of Credit—There are many types of letters of credit. The type used to guarantee a contractor's performance is a "standby" letter of credit in which a bank stands ready to pay over the amount of the letter to the owner of the project (obligee) in the event of a declared default. A letter of credit differs significantly from a surety bond, and one is not a substitute for the other.

Loss Ratio—Fraction calculated by dividing the amount of surety losses by the amount of surety premiums, expressed as a percentage of the premiums. Various bases are used in calculating the loss ratio.

Miller Act—A 1935 statute mandating surety bonds on all federal public works contracts in excess of $100,000. State and local public works projects are protected by "Little Miller Acts."

NASBP—The National Association of Surety Bond Producers is an international organization of insurance agencies and brokerage firms that represent surety companies and specialize in arranging corporate surety bonding and insurance for construction contractors.

Non-Compliance—Failure to obtain surety bonding for a public project. Board members of public entities have been held personally responsible for losses resulting from defaults by contractors who should have been covered by surety bonds.

Obligee—Usually the construction project owner. Surety bonds are guarantees to the obligee, the primary beneficiary of the surety bond, that the surety company will fulfill the contractor's performance of the contract and payment of certain labor and material bills.

Payment Bond—Guarantees payment for labor and materials associated with the project.

Performance Bond—Protects the owner (obligee) from financial loss caused by contractor failure to complete the project.

Preferred Surety Bond Program (PSB)—A program administered by the Small Business Administration. Provides incentives to encourage underwriters to provide surety coverage for small, minority-owned and woman-owned contractors. Also known as "Plan B."

Premium—Payment required for surety bonds.

Prequalification—A rigorous review performed by the surety or the owner to certify that a contractor is capable of performing the work in accordance with the terms and conditions of the contract.

Producer—The agent, generally an independent surety agent, responsible for writing the surety bond.

Reinsurance—Insurance that involves acceptance by one insurer, called the reinsurer, of all or a part of the exposures covered by another insurer.

Reserves—Funds set aside for meeting obligations as they fall due.

Risk—Possibility of loss or exposure to loss; probability or chance of loss; peril which may cause loss.

Risk Control—All methods of reducing the frequency and/or severity of losses including exposure avoidance, loss prevention, and loss reduction.

SAA—The Surety Association of America, established in 1908, is an association of licensed companies engaged in the business of suretyship.

SIO—Surety Information Office, disseminates useful information about the benefits of corporate surety bonding in both public and private construction.

Surety—A company licensed in one or more states to write surety bonds. The U.S. Treasury Department maintains a list of surety companies that it has qualified to write surety bonds for U.S. Government projects. The surety stands ready to facilitate completion by providing trained personnel to consult and solve problems, bringing in a completion contractor, paying subcontractors and suppliers, and arranging financial assistance for the contractor.

Surety Agencies—Insurance agencies and brokerage firms that arrange corporate surety bonding and insurance for contractors.

Surety Bond—An agreement guaranteeing that a principal will carry out the contractual obligations the principal has agreed to perform or, alternatively, to compensate the other parties to the contract for losses resulting from the principal's failure to perform. Under many surety bonds, the principal is a contractor.

Surety Bonding—The process of prequalifying the contractor (principal) and guaranteeing to the project owner (obligee) that the contractor will fulfill the contract's terms and conditions and pay subcontractors and suppliers.

Surety Credit—Contractors qualify for surety credit following an analysis of their financial statements, integrity, and abilities by the surety underwriter.

Surety Underwriter—An employee of the surety company who evaluates applications for surety bonds and determines the terms under which the applicant will be bonded. Contrast with "claim personnel."

Suretyship—An extension of credit without funds being dispersed (unless there is a default by the contractor principal on the bonds.) The surety is the third party to whom an owner (obligee) can turn if a contractor fails or cannot perform its obligations.

APPENDICES

Appendix I. Sample General Indemnification Agreement: Old Republic Surety

Source: Old Republic Surety Company

GENERAL AGREEMENT OF INDEMNITY

"Company" means as interest(s) may appear, the Old Republic Surety Company, any and/or all subsidiary(ies), and/all parent company(ies) of Old Republic Surety Company, and/or any affiliated company(ies) within the Old Republic International General Insurance Group, as well as/or any and/all reinsuring surety(ies), co-surety(ies) and any surety(ies) which have been procured to execute the bond(s), their successors or assigns.

I. DEFINITIONS

The following definitions apply to this Agreement:

BOND(S) "Bond" means any and/or all contracts of suretyship, guaranty or indemnity, undertaking or other writings obligatory in nature, including continuations, extensions, alterations or substitutions thereof, the Surety has or may execute or procure the execution of, for and/or on behalf of any and/or all Indemnitors in the capacity of Principal.

CONTRACT "Contract" means all documents comprising the written contract including general and special conditions, specifications and drawings.

EXPENSE "Expense" means any and all expense, including but not limited to investigative charges, accounting charges, engineering charges, fees and disbursements of counsel (whether on salary, retainer or otherwise.)

INDEMNITOR "Indemnitor" means the Subscriber to this Agreement or addendum hereto, whether an individual, organization, partnership, corporation, unincorporated association, joint venture or other entity, for themselves, their heirs, executors, administrators, successors and assigns. Subscription to this Agreement shall include the Named Indemnitor and any and all presently existing and future subsidiaries, and all other existing and future corporations, partnerships, organizations or other entity controlled by the Subscriber. The word Indemnitor or any personal pronoun shall apply regardless of number or gender.

OBLIGEE "Obligee" means the party in whose favor a bond runs.

PRINCIPAL "Principal" means an Indemnitor (a) by himself, or (b) a joint venture not named herein for which the Surety executes bond(s) and which has the primary responsibility to the Obligee.

SURETY "Surety" means as their interest may appear, any named company in Old Republic International General Insurance Group or all reinsuring surety(ies), co-surety(ies) and any surety(ies) which have been procured to execute the bond(s), their successors or assigns.

INTEREST The Surety shall be paid interest at the prime rate plus 2% as determined by First National Bank of Chicago, Chicago, Illinois (unless prohibited by law, then at the maximum rate allowed).

II. WITNESSETH:

WHEREAS, Principal has, is, or may desire or be required to execute and deliver or procure the execution and delivery of bond(s) by the Surety; and

WHEREAS, the Indemnitor understands the Surety expressly requires the delivery of this Agreement as part of the consideration for the bond requested; and

WHEREAS, the Indemnitor has a substantial, material and beneficial interest in the obtaining of bond(s) by the Principal(s) or in the Surety's refraining from cancelling such bond(s);

NOW, THEREFORE, in consideration of the execution and delivery by the Surety of the bond(s), the Indemnitor jointly and severally convenant and agree with the Surety as follows:

III. AGREEMENTS

A. **Premium** – The Indemnitor will pay to the Surety premiums and charges at the rates in the Surety's schedule of rates at the times specified in respect to each such bond or as may be agreed between the Principal and the Surety with any additions or amendments thereto, by reference made a part hereof, and will continue to pay the same until the Surety shall be discharged and released from any and all liability and responsibility upon and from each such bond or matters arising therefrom.

B. **Indemnification** – The Indemnitor shall continually indemnify and save the Surety harmless from and against every claim, demand, liability, loss, cost, and expense which the Surety may pay, sustain or incur in consequence of having executed or procured the execution of bond(s), or the failure of the Indemnitor, to perform or comply with the convenants and conditions of this Agreement to enforce the right of the Surety to any collateral taken specifically or otherwise, to enforce any and/or all obligations of the Indemnitor under this Agreement, or to defend any action(s) against the Surety arising out of the execution of any bond(s) on behalf of the Principal or the Surety's exercise of any rights under this Agreement.

The Surety shall have the exclusive right for itself, the Principal and the other Indemnitors to determine whether any claim, demand, suit or judgement brought against the Principal and/or Surety upon a bond executed by the Surety shall on the basis of liability, expediency or otherwise, be paid, settled, defended or appealed and its determination shall be conclusive upon the Principal and/or Indemnitor. Payment shall be made by the Indemnitor to the Surety for 1. all loss, cost and expenses as soon as the Surety becomes liable therefor whether actually paid by the Surety in whole, any part thereof, or not, and; 2. Interest on any payments made by Surety referred to in 1. above from the date of payment by Surety until Surety is reimbursed by Indemnitor. The Indemnitor agrees to accept the voucher or other evidence of such payment as prima facie evidence of the propriety thereof, and of the Indemnitor's liability therefor to the Surety.

C. **Right of Surety to Have Funds Deposited** — If the Surety shall establish or increase a reserve to cover any claim, suit, judgment, or other possible liability under any bond(s) which is the subject of this Agreement, the Indemnitor will immediately upon demand deposit with the Surety a sum of money equal to such reserve or increase. Such sum will be held by the Surety as collateral security on such bond, which with any other money or property on deposit or hereafter deposited or pledged as collateral security on any such bond shall, unless otherwise agreed in writing by the Surety, be available, at the discretion of the Surety, as collateral security on any other or all bonds coming within the scope of this Agreement. Surety shall have the right to use the deposit in payment or settlement of any liability, loss or expense for which the Indemnitor would be obligated to indemnify the Surety under this Agreement. The Indemnitor shall be entitled to the return of the unused portion of the deposit upon the termination of the liability of the Surety and the performance of the Indemnitor under the terms of this Agreement. The Surety's demand shall be sufficient if sent by Certified Mail to the Indemnitor at the addresses stated herein or at the addresses of the Indemnitor last known to the Surety. Return of collateral so deposited by the Surety shall not prejudice the Surety's present or future rights, remedies or defenses against the Indemnitor.

D. **Loans** — The Surety may make loans, advances or guarantee loans to be used for the account of the Principal in the performance of obligations covered by bond(s) without the necessity of seeing to the application of the proceeds thereof. The Indemnitor shall be obligated to indemnify the Surety in accordance with the terms of this Agreement for the amount of all such advances and loans, including Interest and all attendant charges and costs, notwithstanding that the proceeds of any part thereof have not been so used by the Principal. Such monies shall be considered a loss under the terms of this Agreement and not a voluntary payment.

E. **Trust Funds** – The Principal and Indemnitor expressly declare that all funds due or to become due under the contracts or obligations covered by a bond are trust funds, whether in the possession of the Principal or another, for the benefit and payment of all persons to whom the Principal incurs obligations in the performance of such contract or obligations for which the Surety would be liable under the bond. If the Surety discharges any such obligation it shall be entitled to assert the claim of such person to the trust funds. This declaration shall constitute notice of such trust. The Principal shall, upon demand of the Surety and in implementation of the trust or trusts hereby created, open an account(s) with a bank or similar depository designated by the Principal and approved by the Surety which account(s) shall be designated as trust account(s) for the deposit of such trust funds and shall deposit therein all monies received pursuant to said contract(s) or obligation(s). Withdrawals shall be by check or similar instrument signed by the Principal and at the Surety's option, countersigned by a representative of the Surety. Said trust(s) shall terminate on the payment by the Principal of all contractual obligations for which the trust was created or the expiration of the applicable statute of limitations, whichever shall first occur.

F. **Assignment of Indemnitor's Property** – If any bond is executed or issued by the Surety, the Indemnitor, subject only to the trust herein created, does hereby assign, transfer and convey to the Surety all of their right, title, interest and estate in and to all of their property, whether real, personal or mixed, tangible or intangible, wherever situated or of whatever nature and all interests and rights now owned or hereafter acquired and if sold the proceeds therefrom. Personal or mixed property shall include but not be limited to:

1. All goods, consumer or otherwise.
2. All commercial paper, negotiable or otherwise.
3. All monies, bank accounts and deposits, checking, savings or otherwise.
4. All warehouse receipts, bills of lading and other documents of title.
5. All securites, investment or otherwise.
6. All accounts, receivable or otherwise.
7. All contract rights.
8. All equipment, machinery, tools and plant(s).
9. All furniture and fixtures.
10. All general intangibles
11. All farm products.
12. All inventory.
13. All property described in Article 9 of the Uniform Commerical Code in which the Indemnitor presently has or hereafter obtains an interest.
14. All monies due or to become due the Principal under or as a result of the contract covered by the bond herein or hereinafter applied for, including but not limited to, progress payments, deferred payments, retained percentages, compensation for extra work and proceeds of damage claims. To implement the foregoing, the Indemnitor hereby authorizes the Surety to endorse in the name of the Payee, and to attest any checks, drafts, warrants or other instruments made or issued in payment of any such sums and to disburse the proceeds thereof.
15. All right, title and interest of the Principal in and to all supplies, tools, plant, equipment and materials of every nature and description that may now or hereafter be in, on or around the site of, or the work under, the contract covered by the bond and materials purchased or ordered for the performance of said contract whether in process of construction, in transit to the site, or in storage elsewhere.
16. All right, title and interest of the Principal in and to all subcontracts covered by bond, any extension, or alteration thereto as well as all right, title and interest of the Principal in and to all surety bonds covering such sub-contracts or purchase orders.
17. All actions, causes of action, claims and demands whatsoever which the Principal has against any sub-contractor, laborer, or any person furnishing or agreeing to furnish or supply labor, material, supplies, machinery, tools or other equipment in connection with or on account of the contract covered by the bond and against any surety(ies) of any such sub-contractor, laborer or materialman. The Surety is authorized to assert and prosecute any right or claim assigned, transferred or conveyed in the name of the Indemnitor on such terms as it considers reasonable under the circumstances. The Surety may sell any property assigned to it pursuant to this Agreement at Public or Private Sale without notice at any time or place without incurring any liability of any kind. The Surety shall not be obligated to perform or discharge the terms, convenants or conditions on the part of the Indemnitor in connection with any matter or thing assigned.

Such assignment is effective as of the effective date hereof unless there is no abandonment, no breach, no delay, or no default by Indemnitor, or no declaration of default by any Obligee, or by other third party beneficiaries of the bond in the performance of any of the following:

1. Obligations contained in or covered by any such bond,
2. Provisions of this Agreement or any other agreement between the Surety and any Indemnitor, or
3. Payment, satisfaction or discharge in any and all of the obligations which might constitute a possible claim under any such bond.

G. **Right of Surety to Record and File** — The Surety and its attorneys are hereby authorized, without notice to the Indemnitor, to record or file this document as is, or at the Surety's option to make schedules of property assigned under this Agreement, real and/or personal and/or mixed, and to attach the same hereto at any time and the same when attached shall by this reference be incorporated herein and made a part of this assignment and when made and attached the Surety and its attorneys are authorized without notice to the Indemnitor to file or record a copy of same under any provisions of law governing such. The Surety may, in conjunction with the aforementioned filing or recording or in lieu thereof, perfect its interest in such property at any time without notice to the Indemnitor by exercising its power of attorney granted in this document and by executing and recording a Quitclaim deed or filing and/or recording a Financing Statement or any other similar document in its favor. In exercising these rights the Surety shall not, in any way, abrogate, restrict or limit its rights under this Agreement or under any law, or in equity. This Agreement constitutes a security agreement to the Surety and a Financing Statement in accordance with the provisions of the Uniform Commercial Code. The Surety may make such additions as necessary to permit its filing as a Financing Statement under such Code. The Indemnitor will, on demand, deliver such future instruments as required by the Surety. The Indemnitor specifically empowers and authorizes the Surety to file and record a copy of this instrument certified by an authorized representative of the Surety that it is a true and correct copy.

H. **Right of Surety to Demand Conveyance** — The Indemnitor expressly acknowledges and agrees that one of the primary purposes of this Agreement is to secure fully, completely and adequately, the Surety against any and all possible losses and expenses, and to that end, jointly and severally agrees that upon demand by the Surety, he will immediately provide the Surety with a deed, deed of trust, or other appropriate instrument of conveyance, to all of his property, both real and personal, mixed, tangible and/or intangible, with a power of immediate sale, authorizing the trustee or the Surety to sell any and all of said property so that the same may be converted into money and deposited and used as provided in Paragraph C above, time being of the absolute essence. The Indemnitor further expressly acknowledges that the purpose of this paragraph is to save the Surety the time and expense of having to realize its interest in the property assigned to the Surety under Paragraph F through the more cumbersome and time-consuming coventional process of foreclosure. Thus, the Indemnitor expressly acknowledges that this obligation is specifically enforceable against him in a court of law, and that his property may be sold at a judicial sale without right of redemption and without any homestead exemption.

I. **Waiver of Exemptions** – The Indemnitor does hereby release and abandon, as to the Surety, all right to claim any property, including his homestead, as exempt from levy, execution, sale or other legal process under the law of any state, province or other government. The Indemnitor also specifically waives the provisions of any statute, ordinance or law governing the limit of the Indemnitor liability.

J. **Right of Surety to Complete Contracts** — Whenever the Surety determines it is necessary or advisable to complete the contracts (and its decision in this regard shall be conclusive on the Indemnitor) or in the event of any Event of Default, hereafter stated, the Surety shall have the absolute right, at its option and in its sole discretion and is hereby authorized to, with or without exercising any of the rights or options conferred by this Agreement, law or equity, to enter upon and take possession of all equipment, machinery, tools, plant and materials described in F of this Agreement and enforce, use and enjoy the possession thereof, consuming, if necessary, the same in performance of the contract so that same may be used in completing the contract work and to take possession of any part or all of the work under any contract covered by any such bond or bonds and at the expense of the Indemnitor to complete or arrange for the completion of the same and the Indemnitor shall promptly upon demand pay to the Surety all expenses so incurred plus interest thereon unless the Indemnitor can prove the Surety acted in bad faith. The Surety shall have no duty to exercise the rights herein conferred and the Indemnitor agrees the Surety shall have no liability to the Indemnitor for exercising or not exercising the rights herein conferred. Event of Default shall include the following:

1. Any abandonment, forfeiture or inability of or failure, refusal, or inability to perform any contract covered by a bond or any bond liability.
2. The failure, delay, refusal, or inability of the Principal to pay bills or other indebtedness incurred in, or in connection with the performance of any contract covered by a bond.
3. The failure to perform or comply with, any of the terms, covenants and obligations to this Agreement.
4. The failure to pay and discharge when due any other indebtedness of the Principal to the Surety.
5. Any assignment for the benefit of creditors or the appointment of a receiver or trustee, voluntarily or involuntarily under the terms of any state law.
6. The commencement or contraction of any proceeding which deprives the Principal of, or interferes with, his use of any of the supplies, tools, plant, machinery, equipment or materials in, on, or around the site of the work under the contract covered by any said bond.
7. If the Principal is an individual, the Principal's dying, absconding, disappearing, incompetency, being convicted of a felony or imprisioned, and if the Principal is any other type of entity, any change or threat of change in the character, identity, control, management, beneficial ownership or existence of the Principal.

K. **Right to Control Performance of Principal** – The Surety shall be entirely within its rights and remedies under the terms of this Agreement and as Surety if it becomes necessary or advisable in its judgment to control, administer, operate or manage any and all matters connected with the performance of any obligation undertaken in connection with a defined bond issued by the Surety for the purpose to minimize the ultimate loss to the Indemnitor or the Surety, or to enable the Surety to discharge its obligation of suretyship.

L. **Right to Information** – The Indemnitor will furnish to the Surety such information as it may request concerning the financial condition of the Indemnitor, the status of work under any contract covered by a bond, the condition of the performance of any such contract and the payment of obligations incurred in connection therewith. The Surety may at reasonable times and places examine and copy the books, records and accounts of the Indemnitor. The Surety may obtain information concerning the affairs and operations of the Indemnitor and any transaction between or among the Indemnitor from any banks, depositories, obligees of the bonds, materialmen, supply houses, credit reporting agencies or other persons, including governmental entities, who are hereby expressly authorized to furnish such information to the Surety. The Indemnitor waives any confidentiality requirements whether internal or imposed by statute, regulation, etc.

M. **Notice of Claim** – If the Indemnitor becomes aware of any action, demand, notice, suit or proceeding which may result in liability to the Surety under any bond, the Indemnitor shall notify the Surety immediately thereof in writing.

N. **Discharge from Suretyship** — The Indemnitor will, upon the request of the Surety, procure and discharge the Surety from any bond and from all liablity by reason thereof. The Surety may take such action as it deems necessary or proper to obtain its release from any and all liability.

O. **Waiver of Notice** – The Indemnitor hereby waives notice of the execution of any bond and of any act, fact or information concerning or affecting the rights or liabilities of the Surety or the Indemnitor.

P. **Consent to Change** – The Surety is authorized, without notice to or knowledge of the Indemnitor, to assent to any change whatsoever in any bond and any contract covered by a bond including, but not limited to, any change in the time for the completion of the contract and for payments or advances thereunder, to take any assignment or assignments, to execute or consent to the execution of any continuations, extensions, renewals, or alterations of any bond and to execute any substitute(s) thereof with the same or different conditions, provisions and obligees and with the same, larger or smaller penalties, and the Indemnitor shall remain bound under the terms of this Agreement even though any such assent by the Surety does or may substantially increase the liability of the Indemnitor.

Q. **Appointment of Surety as Attorney-In-Fact** – The Indemnitor hereby irrevocably nominates, appoints and designates the Surety or any person(s) designated by the Surety as his Attorney-In-Fact with the right to exercise all the rights assigned, transferred and set over to the Surety by this Agreement, and in his name to execute and deliver any and all additional or other assignments, instruments, or documents including but not limited to deeds, mortgages, subordination agreements and/or Financing Statements, deemed necessary or desirable by the Surety:

1. To vest in the Surety absolute title to any and all monies, property and rights hereby assigned and
2. To provide the protection and rights to the Surety contemplated by all of the provisions of this Agreement, the Surety being empowered, but not obligated, to do any thing which the Indemnitor might do, if personally present, if this instrument were not executed. All acts and actions of the Surety as Attorney-In-Fact are hereby ratified and confirmed. The Surety shall also have the right to fill in any blanks in this Agreement or otherwise complete same as behalf of the Indemnitor.

R. **Subordination of Indemnitor** — The indemnitor hereby waives any right of indemnity against the Principal or his property until the Principal's obligations to the Surety under this Agreement have been satisfied in full.

S. **Nature of Rights** – All rights and remedies of the Surety under this Agreement shall be cumulative, and the exercise of or the failture to exercise any right or remedy at any time shall not be an election of remedy or a waiver of any other right or remedy. Failure to the Surety to pursue any remedy against any one or more of the Indemnitors shall not release or waive any right against any other Indemnitor. The Surety is not required to exhaust its remedies or rights against the Principal or to await receipt of any dividends from the legal representatives of the Principal before asserting its rights under this Agreement against the Indemnitor.

The rights, powers and remedies given to the Surety by the Agreement shall be and are in addition to and not in lieu of, any and all other rights, powers and remedies which the Surety may have or acquire against the Indemnitor or others whatsoever by the terms of any other agreement, by operation of law or otherwise.

The Indemnitor shall continue to be bound under this Agreement even though the Surety may with or without notice to or knowledge of the Indemnitor, have heretofore or may accept or release, other agreements of indemnity or collateral in connection with the execution of the bonds from the Indemnitor or others. The Surety is authorized to settle or compromise with any one or more indemnitors, without reference to the other remaining Indemnitor continuing to be bound by this Agreement with the remaining Indemnitor(s) expressly authorizing said settlement or compromise.

The Indemnitor waives any defense that this instrument was executed subsequent to the date of any such bond admitting and convenanting that such bond was executed pursuant to Indemnitor request and/or reliance of the Indemnitor promise to execute this Agreement.

T. **Partial Invalidity or Execution** – If any of the persons or parties named herein as Indemnitor fails to execute this Agreement or if the execution hereof by any Indemnitor shall be defective or invalid for any reason, such failure, defect or invalidity shall not in any manner diminish or otherwise affect the obligation or liability hereunder of any other Indemnitor. Failure of the Principal to sign any bond shall not relieve the Indemnitor of liability under this Agreement.

If any provision(s) or sub-part of any provision of this Agreement is held to be void or unenforceable under the laws of the place governing its construction or enforcement, this Agreement shall not be void or unenforceable thereby but shall continue as though such provision or sub-part of such provisions were omitted.

U. **Right of Surety to Decline to Execute Bond** – The Surety may decline, without disclosing the reasons, to execute any bond applied for without incurring any liability whatever to the Principal or Indemnitor, including but not limited to "loss of profits". If the Surety shall execute a bid or proposal bond or any similar undertaking, it may nevertheless decline to execute any and all bonds that may be required in connection with any award made on the proposal for which the bond is given and the Principal shall have the right to procure from another Surety any bond(s) that may be required in connection with any award under the proposal for which the bond is given.

V. **Separate Action** – Separate suits may be brought on this Agreement against any Indemnitor and the bringing of a suit or the recovery of a judgment upon any cause of action shall not prejudice nor bar the bringing of other suits upon other causes of action, whether theretofore or thereafter arising.

W. **Service of Process** – If any proceeding is brought against the Surety in which the Surety desires to join any Indemnitor by reason of the undertakings in this Agreement, each Indemnitor will, upon written notice of the Surety to do so, voluntarily appear in such proceedings and accept service of process and other papers either personally or by an attorney of the Indemnitor's choice. If any Indemnitor fails upon such notice from the Surety to appear, such Indemnitor hereby designates the Secretary of State or equivalent of any Territory, Province, or Government in which proceedings are pending, as his agent, for the service of process in any such proceedings. With respect to any action brought by the Surety on this Agreement in a jurisdiction in which one or more Indemnitors resides, are domiciled, are doing business or are found, each Indemnitor in the jurisdiction hereby designates each other Indemnitor in such jurisdiction as his agent to receive on his behalf service of process in such action.

X. **Right of Surety to Exercise Rights Without Liability** – The Surety, or any of its authorized representatives shall not be liable to the Indemnitor for any damage or injuries that may be sustained by them of whatever kind or nature caused by or arising out of any action taken, or statements, verbal, written or otherwise, made in good faith by the Surety in exercising or attempting to exercise any of the rights or privileges under this Agreement or any other agreement between the Surety and any Indemnitor, under law, in equity, under or relating to bonds executed by the Surety.

Y. **Modifications** – The rights and remedies afforded to the Surety by the terms of this Agreement may not be waived or modified orally and no written change or modification shall be effective until signed by an officer of the Surety.

Z. **Termination** – This is a continuing Agreement which remains in full force and effect until terminated. This Agreement may be terminated by the Indemnitor or by one or more of them upon twenty (20) days written notice, from receipt by the Surety, sent by Certified Mail to the Surety but any such notice of termination shall not relieve the Indemnitor of any obligations under this Agreement with respect to bonds which are executed prior to such termination or with respect to bonds executed after the date of such termination:

1. Upon the award of a contract to the Principal or a bid or proposal with respect to which the Surety has executed a bid or proposal bond or similar undertaking prior to such date.
2. Which the Surety has become obligated, prior to such date to execute.
3. Any renewal, substitution or extension of a bond in force, and,
4. Any maintenance or guarantee bonds executed incidental to any other bond executed prior to such date and renewals, substitutions and extensions thereof.

Such notice of termination shall operate only with respect to the Indemnitor upon whose behalf such notice of termination shall have been given.

ZA. **Separate Application** – Any separate application shall be considered as supplemental to, and not in lieu, of this Agreement of Indemnity.

ZB. **Offset** – At the Surety's option monies due or to become due the Indemnitor from any Company in the Old Republic International General Insurance Group, through insurance proceeds or bonding payment may be utilized to pay or help pay obligations incurred under this Agreement as an "Offset".

ZC. **Applicable Law and Jurisdiction** – Indemnitor agrees that this Agreement shall be deemed to have been made in the State of Incorporation of the Surety. The Surety, at its option, shall be entitled to enforce this Agreement by an action, arbitration or proceeding brought in the said State of Incorporation or in the State where Surety's principal office is located, and Indemnitor agrees that in any such action, arbitration or proceeding he will be subject to the jurisdiction of any court of competent jurisdiction in the said State of Incorporation and in the State where Surety's principal office is located, and service of process may be made on Indemnitor in any manner then permitted, and that venue shall be proper and convenient in the County in which Surety's principal office may be located. Interpretation and enforcement of this Agreement, at the Option of Surety, shall be governed by the laws of the said State of Incorporation (excluding conflict of laws principles). For the purposes of this paragraph the principal office for the Surety shall be deemed to be located in Brookfield, Waukesha County, Wisconsin.

ZD. **Representations** – THE INDEMNITOR REPRESENTS TO THE SURETY THAT THEY HAVE CAREFULLY READ THE ENTIRE AGREEMENT AND THAT THERE ARE NO OTHER AGREEMENTS BETWEEN THE PARTIES WHICH IN ANY WAY LESSEN OR MODIFY THE OBLIGATIONS SET FORTH HEREIN.

AGREEMENT

FAIR CREDIT REPORTING ACT NOTICE This notice is given to comply with the Federal Fair Credit Reporting Act (Public law 91-508) and any similar state law which is applicable. As part of our underwriting procedure, a routine inquiry may be made which will provide information concerning character, general reputation, personal characteristics and mode of living. Upon request, additional information as to the nature and scope of the report, if one is made, will be provided.

IN WITNESS WHEREOF, The Indemnitor who is an individual hereunto sets his (her) hand and all other Indemnitor(s) have caused this Agreement to be duly executed by their duly authorized representatives, and acknowledge receipt of a copy thereof.

This Agreement effective this ________________ day of ________________, 19 ________.

ATTEST OR WITNESS

____

____

INDEMNITOR: CORPORATE

________________________________ (SEAL)

By: ________________________________

P.O. Address ________________________________
Street

City State Zip Code

INDEMNITOR: CORPORATE

________________________________ (SEAL)

By: ________________________________

P.O. Address ________________________________
Street

City State Zip Code

INDEMNITOR: FIRM OR PARTNERSHIP

____

By: ________________________________

P.O. Address ________________________________
Street

City State Zip Code

INDEMNITOR: INDIVIDUAL

____

By: ________________________________

P.O. Address ________________________________
Street

City State Zip Code

Social Security Number ________________________________

INDEMNITOR: INDIVIDUAL

____

By: ________________________________

P.O. Address ________________________________
Street

City State Zip Code

Social Security Number ________________________________

INDEMNITOR: INDIVIDUAL

____

By: ________________________________

P.O. Address ________________________________
Street

City State Zip Code

Social Security Number ________________________________

INDEMNITOR: INDIVIDUAL

____

By: ________________________________

P.O. Address ________________________________
Street

City State Zip Code

Social Security Number ________________________________

(over)

NOTARIAL ACKNOWLEDGEMENT – CORPORATE

STATE OF ______________________________)
) SS
COUNTY OF ______________________________)

On this ______________ day of ______________ 19______ , before me personally appeared ______________

____

____ ,

personally known to me (or proved to me on the basis of satisfactory evidence) to be the person who executed this instrument as ______________
Title of Officer

on behalf of the corporation therein named and acknowledged to me that the corporation executed it.

____
Notary Public

NOTARIAL ACKNOWLEDGEMENT – CORPORATE

STATE OF ______________________________)
) SS
COUNTY OF ______________________________)

On this ______________ day of ______________ 19______ , before me personally appeared ______________

____

____ ,

personally known to me (or proved to me on the basis of satisfactory evidence) to be the person who executed this instrument as ______________
Title of Officer

on behalf of the corporation therein named and acknowledged to me that the corporation executed it.

____
Notary Public

NOTARIAL ACKNOWLEDGEMENT – FIRM OR PARTNERSHIP

STATE OF ______________________________)
) SS
COUNTY OF ______________________________)

On this ______________ day of ______________ 19______ , before me personally appeared ______________

____

____ ,

personally known to me (or proved to me on the basis of satisfactory evidence) to be the person who executed this instrument, on behalf of the firm or partnership and acknowledged to me that the firm or partnership executed it.

____
Notary Public

NOTARIAL ACKNOWLEDGEMENT – INDIVIDUAL

STATE OF ______________________________)
) SS
COUNTY OF ______________________________)

On this ______________ day of ______________ 19______ , before me personally appeared ______________

____

____ ,

personally known to me (or proved to me on the basis of satisfactory evidence) to be the person whose name is subscribed to this instrument, and acknowledged that he (she or they) executed it.

____
Notary Public

NOTARIAL ACKNOWLEDGEMENT – INDIVIDUAL

STATE OF ______________________________)
) SS
COUNTY OF ______________________________)

On this ______________ day of ______________ 19______ , before me personally appeared ______________

____

____ ,

personally known to me (or proved to me on the basis of satisfactory evidence) to be the person whose name is subscribed to this instrument, and acknowledged that he (she or they) executed it.

____
Notary Public

Appendix II. Sample Bid Bond Form: AIA Document A310
Source: American Institute of Architects

THE AMERICAN INSTITUTE OF ARCHITECTS

AIA Document A310

Bid Bond

KNOW ALL MEN BY THESE PRESENTS, that we

(Here insert full name and address or legal title of Contractor)

as Principal, hereinafter called the Principal, and

(Here insert full name and address or legal title of Surety)

a corporation duly organized under the laws of the State of
as Surety, hereinafter called the Surety, are held and firmly bound unto

(Here insert full name and address or legal title of Owner)

as Obligee, hereinafter called the Obligee, in the sum of

Dollars ($),

for the payment of which sum well and truly to be made, the said Principal and the said Surety, bind ourselves, our heirs, executors, administrators, successors and assigns, jointly and severally, firmly by these presents.

WHEREAS, the Principal has submitted a bid for

(Here insert full name, address and description of project)

NOW, THEREFORE, if the Obligee shall accept the bid of the Principal and the Principal shall enter into a Contract with the Obligee in accordance with the terms of such bid, and give such bond or bonds as may be specified in the bidding or Contract Documents with good and sufficient surety for the faithful performance of such Contract and for the prompt payment of labor and material furnished in the prosecution thereof, or in the event of the failure of the Principal to enter such Contract and give such bond or bonds, if the Principal shall pay to the Obligee the difference not to exceed the penalty hereof between the amount specified in said bid and such larger amount for which the Obligee may in good faith contract with another party to perform the Work covered by said bid, then this obligation shall be null and void, otherwise to remain in full force and effect.

Signed and sealed this day of 19

	(Principal) *(Seal)*
(Witness)	
	(Title)
	(Surety) *(Seal)*
(Witness)	
	(Title)

Appendix III. Sample Performance Bond Form: AIA Document A312

Source: American Institute of Architects

THE AMERICAN INSTITUTE OF ARCHITECTS

AIA Document A312

Performance Bond

Any singular reference to Contractor, Surety, Owner or other party shall be considered plural where applicable.

CONTRACTOR (Name and Address):

SURETY (Name and Principal Place of Business):

OWNER (Name and Address):

CONSTRUCTION CONTRACT
Date:
Amount:
Description (Name and Location):

BOND
Date (Not earlier than Construction Contract Date):
Amount:
Modifications to this Bond: ☐ None ☐ See Page 3

CONTRACTOR AS PRINCIPAL
Company: (Corporate Seal)

Signature: ______________________
Name and Title:

SURETY
Company: (Corporate Seal)

Signature: ______________________
Name and Title:

(Any additional signatures appear on page 3)

(FOR INFORMATION ONLY—Name, Address and Telephone)
AGENT or BROKER:

OWNER'S REPRESENTATIVE (Architect, Engineer or other party):

1 The Contractor and the Surety, jointly and severally, bind themselves, their heirs, executors, administrators, successors and assigns to the Owner for the performance of the Construction Contract, which is incorporated herein by reference.

2 If the Contractor performs the Construction Contract, the Surety and the Contractor shall have no obligation under this Bond, except to participate in conferences as provided in Subparagraph 3.1.

3 If there is no Owner Default, the Surety's obligation under this Bond shall arise after:

3.1 The Owner has notified the Contractor and the Surety at its address described in Paragraph 10 below that the Owner is considering declaring a Contractor Default and has requested and attempted to arrange a conference with the Contractor and the Surety to be held not later than fifteen days after receipt of such notice to discuss methods of performing the Construction Contract. If the Owner, the Contractor and the Surety agree, the Contractor shall be allowed a reasonable time to perform the Construction Contract, but such an agreement shall not waive the Owner's right, if any, subsequently to declare a Contractor Default; and

3.2 The Owner has declared a Contractor Default and formally terminated the Contractor's right to complete the contract. Such Contractor Default shall not be declared earlier than twenty days after the Contractor and the Surety have received notice as provided in Subparagraph 3.1; and

3.3 The Owner has agreed to pay the Balance of the Contract Price to the Surety in accordance with the terms of the Construction Contract or to a contractor selected to perform the Construction Contract in accordance with the terms of the contract with the Owner.

4 When the Owner has satisfied the conditions of Paragraph 3, the Surety shall promptly and at the Surety's expense take one of the following actions:

4.1 Arrange for the Contractor, with consent of the Owner, to perform and complete the Construction Contract; or

4.2 Undertake to perform and complete the Construction Contract itself, through its agents or through independent contractors; or

4.3 Obtain bids or negotiated proposals from qualified contractors acceptable to the Owner for a contract for performance and completion of the Construction Contract, arrange for a contract to be prepared for execution by the Owner and the contractor selected with the Owner's concurrence, to be secured with performance and payment bonds executed by a qualified surety equivalent to the bonds issued on the Construction Contract, and pay to the Owner the amount of damages as described in Paragraph 6 in excess of the Balance of the Contract Price incurred by the Owner resulting from the Contractor's default; or

4.4 Waive its right to perform and complete, arrange for completion, or obtain a new contractor and with reasonable promptness under the circumstances:

.1 After investigation, determine the amount for which it may be liable to the Owner and, as soon as practicable after the amount is determined, tender payment therefor to the Owner; or

.2 Deny liability in whole or in part and notify the Owner citing reasons therefor.

5 If the Surety does not proceed as provided in Paragraph 4 with reasonable promptness, the Surety shall be deemed to be in default on this Bond fifteen days after receipt of an additional written notice from the Owner to the Surety demanding that the Surety perform its obligations under this Bond, and the Owner shall be entitled to enforce any remedy available to the Owner. If the Surety proceeds as provided in Subparagraph 4.4, and the Owner refuses the payment tendered or the Surety has denied liability, in whole or in part, without further notice the Owner shall be entitled to enforce any remedy available to the Owner.

6 After the Owner has terminated the Contractor's right to complete the Construction Contract, and if the Surety elects to act under Subparagraph 4.1, 4.2, or 4.3 above, then the responsibilities of the Surety to the Owner shall not be greater than those of the Contractor under the Construction Contract, and the responsibilities of the Owner to the Surety shall not be greater than those of the Owner under the Construction Contract. To the limit of the amount of this Bond, but subject to commitment by the Owner of the Balance of the Contract Price to mitigation of costs and damages on the Construction Contract, the Surety is obligated without duplication for:

6.1 The responsibilities of the Contractor for correction of defective work and completion of the Construction Contract;

6.2 Additional legal, design professional and delay costs resulting from the Contractor's Default, and resulting from the actions or failure to act of the Surety under Paragraph 4; and

6.3 Liquidated damages, or if no liquidated damages are specified in the Construction Contract, actual damages caused by delayed performance or non-performance of the Contractor.

7 The Surety shall not be liable to the Owner or others for obligations of the Contractor that are unrelated to the Construction Contract, and the Balance of the Contract Price shall not be reduced or set off on account of any such unrelated obligations. No right of action shall accrue on this Bond to any person or entity other than the Owner or its heirs, executors, administrators or successors.

8 The Surety hereby waives notice of any change, including changes of time, to the Construction Contract or to related subcontracts, purchase orders and other obligations.

9 Any proceeding, legal or equitable, under this Bond may be instituted in any court of competent jurisdiction in the location in which the work or part of the work is located and shall be instituted within two years after Contractor Default or within two years after the Contractor ceased working or within two years after the Surety refuses or fails to perform its obligations under this Bond, whichever occurs first. If the provisions of this Paragraph are void or prohibited by law, the minimum period of limitation available to sureties as a defense in the jurisdiction of the suit shall be applicable.

10 Notice to the Surety, the Owner or the Contractor shall be mailed or delivered to the address shown on the signature page.

11 When this Bond has been furnished to comply with a statutory or other legal requirement in the location where the construction was to be performed, any provision in this Bond conflicting with said statutory or legal requirement shall be deemed deleted herefrom and provisions conforming to such statutory or other legal requirement shall be deemed incorporated herein. The intent is that this Bond shall be construed as a statutory bond and not as a common law bond.

12 DEFINITIONS

12.1 Balance of the Contract Price: The total amount payable by the Owner to the Contractor under the Construction Contract after all proper adjustments have been made, including allowance to the Contractor of any amounts received or to be received by the Owner in settlement of insurance or other claims for damages to which the Contractor is entitled, reduced by all valid and proper payments made to or on behalf of the Contractor under the Construction Contract.

12.2 Construction Contract: The agreement between the Owner and the Contractor identified on the signature page, including all Contract Documents and changes thereto.

12.3 Contractor Default: Failure of the Contractor, which has neither been remedied nor waived, to perform or otherwise to comply with the terms of the Construction Contract.

12.4 Owner Default: Failure of the Owner, which has neither been remedied nor waived, to pay the Contractor as required by the Construction Contract or to perform and complete or comply with the other terms thereof.

MODIFICATIONS TO THIS BOND ARE AS FOLLOWS:

(Space is provided below for additional signatures of added parties, other than those appearing on the cover page.)

CONTRACTOR AS PRINCIPAL		SURETY	
Company:	(Corporate Seal)	Company:	(Corporate Seal)
Signature: ______________		Signature: ______________	
Name and Title:		Name and Title:	
Address:		Address:	

Appendix IV. Sample Payment Bond Form: AIA Document A312

Source: American Institute of Architects

THE AMERICAN INSTITUTE OF ARCHITECTS

AIA Document A312

Payment Bond

Any singular reference to Contractor, Surety, Owner or other party shall be considered plural where applicable.

CONTRACTOR (Name and Address):

SURETY (Name and Principal Place of Business):

OWNER (Name and Address):

CONSTRUCTION CONTRACT
Date:
Amount:
Description (Name and Location):

BOND
Date (Not earlier than Construction Contract Date):
Amount:
Modifications to this Bond: ☐ None ☐ See Page 6

CONTRACTOR AS PRINCIPAL
Company: (Corporate Seal)

Signature: ______________________
Name and Title:

SURETY
Company: (Corporate Seal)

Signature: ______________________
Name and Title:

(Any additional signatures appear on page 6)

(FOR INFORMATION ONLY—Name, Address and Telephone)
AGENT or BROKER:

OWNER'S REPRESENTATIVE (Architect, Engineer or other party):

1 The Contractor and the Surety, jointly and severally, bind themselves, their heirs, executors, administrators, successors and assigns to the Owner to pay for labor, materials and equipment furnished for use in the performance of the Construction Contract, which is incorporated herein by reference.

2 With respect to the Owner, this obligation shall be null and void if the Contractor:

2.1 Promptly makes payment, directly or indirectly, for all sums due Claimants, and

2.2 Defends, indemnifies and holds harmless the Owner from claims, demands, liens or suits by any person or entity whose claim, demand, lien or suit is for the payment for labor, materials or equipment furnished for use in the performance of the Construction Contract, provided the Owner has promptly notified the Contractor and the Surety (at the address described in Paragraph 12) of any claims, demands, liens or suits and tendered defense of such claims, demands, liens or suits to the Contractor and the Surety, and provided there is no Owner Default.

3 With respect to Claimants, this obligation shall be null and void if the Contractor promptly makes payment, directly or indirectly, for all sums due.

4 The Surety shall have no obligation to Claimants under this Bond until:

4.1 Claimants who are employed by or have a direct contract with the Contractor have given notice to the Surety (at the address described in Paragraph 12) and sent a copy, or notice thereof, to the Owner, stating that a claim is being made under this Bond and, with substantial accuracy, the amount of the claim.

4.2 Claimants who do not have a direct contract with the Contractor:

.1 Have furnished written notice to the Contractor and sent a copy, or notice thereof, to the Owner, within 90 days after having last performed labor or last furnished materials or equipment included in the claim stating, with substantial accuracy, the amount of the claim and the name of the party to whom the materials were furnished or supplied or for whom the labor was done or performed; and

.2 Have either received a rejection in whole or in part from the Contractor, or not received within 30 days of furnishing the above notice any communication from the Contractor by which the Contractor has indicated the claim will be paid directly or indirectly; and

.3 Not having been paid within the above 30 days, have sent a written notice to the Surety (at the address described in Paragraph 12) and sent a copy, or notice thereof, to the Owner, stating that a claim is being made under this Bond and enclosing a copy of the previous written notice furnished to the Contractor.

5 If a notice required by Paragraph 4 is given by the Owner to the Contractor or to the Surety, that is sufficient compliance.

6 When the Claimant has satisfied the conditions of Paragraph 4, the Surety shall promptly and at the Surety's expense take the following actions:

6.1 Send an answer to the Claimant, with a copy to the Owner, within 45 days after receipt of the claim, stating the amounts that are undisputed and the basis for challenging any amounts that are disputed.

6.2 Pay or arrange for payment of any undisputed amounts.

7 The Surety's total obligation shall not exceed the amount of this Bond, and the amount of this Bond shall be credited for any payments made in good faith by the Surety.

8 Amounts owed by the Owner to the Contractor under the Construction Contract shall be used for the performance of the Construction Contract and to satisfy claims, if any, under any Construction Performance Bond. By the Contractor furnishing and the Owner accepting this Bond, they agree that all funds earned by the Contractor in the performance of the Construction Contract are dedicated to satisfy obligations of the Contractor and the Surety under this Bond, subject to the Owner's priority to use the funds for the completion of the work.

9 The Surety shall not be liable to the Owner, Claimants or others for obligations of the Contractor that are unrelated to the Construction Contract. The Owner shall not be liable for payment of any costs or expenses of any Claimant under this Bond, and shall have under this Bond no obligations to make payments to, give notices on behalf of, or otherwise have obligations to Claimants under this Bond.

10 The Surety hereby waives notice of any change, including changes of time, to the Construction Contract or to related subcontracts, purchase orders and other obligations.

11 No suit or action shall be commenced by a Claimant under this Bond other than in a court of competent jurisdiction in the location in which the work or part of the work is located or after the expiration of one year from the date (1) on which the Claimant gave the notice required by Subparagraph 4.1 or Clause 4.2.3, or (2) on which the last labor or service was performed by anyone or the last materials or equipment were furnished by anyone under the Construction Contract, whichever of (1) or (2) first occurs. If the provisions of this Paragraph are void or prohibited by law, the minimum period of limitation available to sureties as a defense in the jurisdiction of the suit shall be applicable.

12 Notice to the Surety, the Owner or the Contractor shall be mailed or delivered to the address shown on the signature page. Actual receipt of notice by Surety, the Owner or the Contractor, however accomplished, shall be sufficient compliance as of the date received at the address shown on the signature page.

13 When this Bond has been furnished to comply with a statutory or other legal requirement in the location where the construction was to be performed, any provision in this Bond conflicting with said statutory or legal requirement shall be deemed deleted herefrom and provisions conforming to such statutory or other legal requirement shall be deemed incorporated herein. The intent is that this

Bond shall be construed as a statutory bond and not as a common law bond.

14 Upon request by any person or entity appearing to be a potential beneficiary of this Bond, the Contractor shall promptly furnish a copy of this Bond or shall permit a copy to be made.

15 DEFINITIONS

15.1 Claimant: An individual or entity having a direct contract with the Contractor or with a subcontractor of the Contractor to furnish labor, materials or equipment for use in the performance of the Contract. The intent of this Bond shall be to include without limitation in the terms "labor, materials or equipment" that part of water, gas, power, light, heat, oil, gasoline, telephone service or rental equipment used in the Construction Contract, architectural and engineering services required for performance of the work of the Contractor and the Contractor's subcontractors, and all other items for which a mechanic's lien may be asserted in the jurisdiction where the labor, materials or equipment were furnished.

15.2 Construction Contract: The agreement between the Owner and the Contractor identified on the signature page, including all Contract Documents and changes thereto.

15.3 Owner Default: Failure of the Owner, which has neither been remedied nor waived, to pay the Contractor as required by the Construction Contract or to perform and complete or comply with the other terms thereof.

MODIFICATIONS TO THIS BOND ARE AS FOLLOWS:

(Space is provided below for additional signatures of added parties, other than those appearing on the cover page.)

CONTRACTOR AS PRINCIPAL
Company: (Corporate Seal)

Signature: ______________________
Name and Title:
Address:

SURETY
Company: (Corporate Seal)

Signature: ______________________
Name and Title:
Address:

Appendix V. Sample Maintenance Bond Form: Fidelity and Deposit Company of Maryland

Source: Fidelity and Deposit Company of Maryland

MAINTENANCE BOND

KNOW ALL MEN BY THESE PRESENTS:

That ____

(hereinafter called Principal), as Principal, and FIDELITY AND DEPOSIT COMPANY OF MARYLAND, with its Home Office in the City of Baltimore, Maryland, (hereinafter called Surety), as Surety, are held and firmly bound unto ________________________________ (hereinafter called Owner), in the full and just sum of ______________________________ ____ Dollars ($______________), to the payment of which sum well and truly to be made, the Principal and Surety bind themselves, their and each of their heirs, executors, administrators, successors and assigns, jointly and severally, firmly by these presents.

WHEREAS, the Principal entered into a certain written agreement dated the ____ ______ day of ________________, A.D., 19 ___, with the Owner for ____________________________ ____ which agreement is or may be attached hereto for reference.

NOW, THEREFORE, THE CONDITION OF THIS OBLIGATION IS SUCH, That if the above bounden Principal shall well and truly make good, repair and replace at its own cost and expense, any defects in work done and/or material furnished in and about the performance of said contract, which defects may be discovered during the period of one year from and after the _______ day of ______ 19 ____, or in default thereof, shall save harmless and indemnify the Obligee from and against any and all loss and damage which it might sustain by reason of any such defects, then this bond to be void, otherwise of full force and effect.

Signed, sealed and dated this ______ day of __________, 19 ___.

____

Principal

BY: ______________________________

FIDELITY AND DEPOSIT COMPANY
OF MARYLAND

BY: ______________________________

Appendix VI. Specimen Clauses for Bid, Performance, and Payment Bonds

Source: Bonds & Insurance for Contractors 1989

SPECIMEN CLAUSES FOR BID INVITATIONS (Bonds & Insurance for Contractors 1989)

The following specimen provisions can be incorporated in bid invitations. Care should be taken to insure that the requirements fit the needs. If existing provisions are being removed and replaced, a clause such as the following can be included:

It is understood and agreed that the following provisions are hereby made part of the contract. In the event of conflict between any of the provisions of the contract and any of the following provisions, the following provisions shall prevail:

1. BID BONDS

a. *Each bid shall be accompanied by a Bid Bond in an amount not less than _______ percent of the amount of the bid, issued by a properly licensed surety company.*

b. *Each bid shall be accompanied by a Bid Bond in an amount not less than $_______, issued by a properly licensed surety company.*

Notes: (1) When it is desired to include the form of Bid Bond in the clause, the following phrase should be inserted after the words "Bid Bond": "on the form included herein."
(2) In some cases it is preferable to require a Bid Bond in a fixed amount, usually 5% or 10% of the architect's or engineer's estimate. To do this, alternative (b) should be used.

2. CONTRACT BONDS

a. PERFORMANCE BONDS

Within ten days of the award of contract, the contractor shall, at own expense, provide a Surety Bond in the amount of _______ percent of the contract price, covering the faithful performance of the contract and issued by a properly licensed surety company on a form acceptable to the owner.

b. SUPPLY BONDS

Within ten days of the award of contract, the bidder, shall, at own expense, provide a Surety Bond in the amount of _______ percent of the contract price, issued by a properly licensed surety company on a form acceptable to the owner covering the faithful performance of the bidder's contract to supply goods and/or services.

Note: In either case, if it is desired to include the form of bond in the specifications, the words "on the form contained herein" should be substituted for the words "on a form acceptable to the owner."

3. LABOR AND MATERIAL PAYMENT BONDS

Within ten days of the award of contract the contractor shall, at own expense, provide a Labor and Material Payment Bond in the amount of _______ percent of the contract price and issued by a properly licensed surety company on a form acceptable to the owner.

Note: When it is desired to include the form of bond in the clause, the words "on the form contained herein" should be substituted for the words "on a form acceptable to the owner."

4. NOTE: THE FOLLOWING CLAUSE CAN BE USED IN LIEU OF CLAUSES 2a and 3:

PERFORMANCE AND LABOR MATERIAL PAYMENT BONDS

Within ten days of the award of contract the contractor shall, at own expense, provide the following Surety Bonds, each in the amount of _______ percent of the contract price and issued by a properly licensed surety company on forms acceptable to the owner:

a. A Performance Bond covering the faithful performance of the contract

b. A Labor and Material Payment Bond.

Appendix VII. Sample Contractor Questionnaire Used by Surety

Source: National Association of Surety Bond Producers 1987

CONTRACTOR QUESTIONNAIRE

1. Name of Firm	
2. Address (city, state, zip)	**3. Telephone** ( )

4. Fiscal Year ________ 5. Contracting Specialty(ies)____________________________

6. Contact Person: ________________________ 7. Title: __________________

8. Year Business Started ________________ 9. Year Business Incorporated ______________

10. State of Incorporation ________________ 11. Area of Operation ___________________

12. Type of Business

__ Corporation __ Partnership __ Proprietorship Sub S Corporation

13. List the corporate officers, partners or proprietors of your firm: Please attach resumes.

	Name	Year of Birth	Position	Percent Owned	Name of Spouse
A.					
B.					
C.					
D.					

14. Will the above individuals and spouses personally indemnify the Surety?
If yes, concurrent personal financial statements will be needed. .... __ Yes __ No

If no, explain: ____

15. Is there a buy/sell agreement among the owners of he business? If yes, please attach a copy .... __ Yes __ No

16. Corporate Indemnity? __ Yes __ No

17. Cross/Corporate Indemnity? __ Yes __ No

18. Is this agreement funded by life insurance? __ Yes __ No

19. How many people does your firm employ? ______ 20. How many work crews? ______

21. Has your firm or any of its principals ever petitioned for bankruptcy, failed in business or defaulted so as to cause a loss to a Surety? __ Yes __ No

If yes, please explain: ______________________________

22. Is your firm or any of its owners or officers currently involved in any litigation? __ Yes __ No

If yes, please explain: ______________________________

23. What percentage of the firm's work is for:

Government Agencies ______ % Private _______ %

24. What percentage of the firm's work is normally sub-contracted? .... ________%

25. Which trades? ______________________________

26. Are bonds required of sub-contractors? .... __ Yes __ No

27. What is the largest amount of uncompleted work on hand a one time in the past?

Amount: ______________ Year: ______________

28. What is the largest job you expect to do during the next year? ....____________

29. What is the largest uncompleted work program expected during the next year? .... ____________

30. What is your expected annual volume next year? .... ____________

31. What trades do you normally undertake with your own forces? ____________________

32. SIC CODE: ______________

33. Do you lease equipment? __ Yes __ No If yes, name type of lease. _______________

34. What are the terms of the lease? ______________________________

35.

Name of your CPA	
Address (city, state, zip)	
Phone	**Contact Person**

36. On what bases are taxes paid?

__ Cash	__ Completed Job	__ Accrual	__ % of Completion

37. On what basis are financial statements prepared?

__ Cash	__ Completed Job	__ Accrual	__ % of Completion

38. On what level of assurance are financial statements prepared?

__ CPA Audit __ Review __ Compilation

39. How often are financial statements prepared?

__ Annually __ Semi-annually __ Quarterly __ Monthly

40. Do you have a full-time accountant on staff? .... __ Yes __ No

41. Years experience with your firm? ____________

42. Are job cost records kept? __ Yes __ No

43. How often reviewed? ______________ 44. How often updated? ____________

45. Do they show job detail? __

Yes __ No	46. Frequency? _______________

47.

Name of your Bank	
Address (city, state, zip)	
Phone	**Contact Person**

48. Amount of line of credit: ____________ 49. Expiration Date: ___________

50. What is interest rate? ______________	51. UCC Filing? __ Yes __ No
52. Is your firm union?	__ Yes __ No

53. What is firm's Dun & Bradstreet Number? ________________________________

54. D & B Rating: _______ 55. Pay Record: ____________ 56. Date of Rating: _________

Remarks: ____

57. Previous Bonding Companies:

	Name	**Reason for Change**
A.	______________________	______________________
B.	______________________	______________________
C.	______________________	______________________

58. List five of your largest contracts completed:

	Job Name	Contract Price	Gross Profit	Completion Date	Bonded?
A.	____________	$_________	$_________	______	__ Yes __ No
	Owner ______________		Design Professional ______________		
B.	____________	$_________	$_________	______	__ Yes __ No
	Owner ______________		Design Professional ______________		

C. ________________ $__________ $__________ ______ __ Yes __ No

Owner __________________ Design Professional ______________________

D. ________________ $__________ $__________ ______ __ Yes __ No

Owner __________________ Design Professional ______________________

E. ________________ $__________ $__________ ______ __ Yes __ No

Owner __________________ Design Professional ______________________

59. List five of your major suppliers:

	Name	**Address**	**Telephone**	**Contact**
A.				
B.				
C.				
D.				
E.				

60. List five sub-contractors (or contractors if you are a sub-contractor) that you do business with:

	Name	**Address**	**Telephone**	**Contact**	**Job**
A.					
B.					
C.					
D.					
E.					

61. List three Architects you do business with:

	Name	**Address**	**Telephone**	**Contact**	**Job**
A.					
B.					
C.					

62. List key personnel, foreman or supervisors:

	Name	**Position**	**Yr. of Birth**	**Yrs. Exper.**	**Prev. Employer**
A.					
B.					

C. ______ ______ ______ ______ ______

D. ______ ______ ______ ______ ______

E. ______ ______ ______ ______ ______

63. List any life insurance in effect on key personnel:

	Name	Beneficiary	Amount	Cash Value
A.	______	______	$______	$______
	Insurance Company: ______			
B.	______	______	$______	$______
	Insurance Company: ______			
B.	______	______	$______	$______
	Insurance Company: ______			

64. List other insurance coverage currently in effect:

	Limits In '000's			
Name	BI	PD	Carrier	Expiration Date
A. General Liability	______	______	______	______
B. Auto Liability	______	______	______	______
C. Umbrella:	______	______	______	______
D. Owner's Protection	______	______	______	______

65. List any subsidiaries and affiliates of the contracting firm:

	Firm Name	Ownership	Type Business	City and State
A.	______	______	______	______
B.	______	______	______	______
C.	______	______	______	______

Remarks: ______.

The above answers are true to the best of my knowledge and belief.

Completed by: ______

Title: ______ Date: ______

Appendix VIII. Sample Report of Independent Certified Public Accountants and Financial Statements for Example Contractor, Inc.

Source: AGCA Wisconsin Chapter 1990

EXAMPLE CONTRACTOR, INC.

Report of Independent Certified
Public Accountants
and
Financial Statements

Year Ended December 31, 19X1

CONTENTS

REPORT OF INDEPENDENT CERTIFIED PUBLIC ACCOUNTANTS

Board of Directors and Stockholders
EXAMPLE CONTRACTOR, INC.
Metropolis, U.S.A.

We have audited the accompanying balance sheet of Example Contractor, Inc. as of December 31, 19X1, and the related statements of earnings and retained earnings, and cash flows for the year then ended. These financial statements are the responsibility of the Company's management. Our responsibility is to express an opinion on these financial statements based on our audit.

We conducted our audit in accordance with generally accepted auditing standards. Those standards require that we plan and perform the audit to obtain reasonable assurance about whether the financial statements are free of material misstatement. An audit includes examining, on a test basis, evidence supporting the amounts and disclosures in the financial statements. An audit also includes assessing the accounting principles used and significant estimates made by management, as well as evaluating the overall financial statement presentation. We believe that our audit provides a reasonable basis for our opinion.

In our opinion, the financial statements referred to above present fairly, in all material respects, the financial position of Example Contractor, Inc., as of December 31, 19X1, and the results of their operations and their cash flows for the year thin ended in conformity with generally accepted accounting principles.

Certified Public Accountants
February 24, 19X2

EXAMPLE CONTRACTOR, INC.

BALANCE SHEET

December 31, 19X1

ASSETS

CURRENT ASSETS:	
Cash	$ 507,540
Marketable securities, at cost	289,000
Receivables (Note B)	1,721,209
Inventories (Note A-6)	57,803
Unbilled amounts on completed contracts (Note A-1)	80,467
Costs and estimated earnings in excess of billings on contracts in progress (Notes A-1 and C)	173,011
Prepaid expenses and other	34,055
TOTAL CURRENT ASSETS	2,863,085
PROPERTY AND EQUIPMENT, at cost (notes A-3 and D):	
Land	35,000
Buildings and improvements	103,000
Construction equipment and vehicles	370,998
Office furniture and equipment	51,283
	560,281
Less accumulated depreciation	(311,983)
	248,298
OTHER ASSETS	9,144
	$3,120,527

LIABILITIES AND STOCKHOLDERS' EQUITY

CURRENT LIABILITIES:	
Accrued payroll, payroll taxes, fringe benefits and amounts withheld from employees	$ 196,760
Accounts payable:	
Current	888,168
Retention	663,918
Accrued costs on completed contracts (Note A-1)	52,810
Billings in excess of costs and estimated earnings on contracts in progress (Notes A-1 and C)	233,633
Income taxes (Notes A-4 and E):	
Currently payable	16,777
Deferred	5,600
Current maturities of long-term debt (Note D)	36,530
TOTAL CURRENT LIABILITIES	2,094,196
LONG-TERM DEBT (Note D)	217,604
COMMITMENTS AND CONTINGENCY (Note F)	
STOCKHOLDERS' EQUITY	
Common stock, par value of $1.00 a share; authorized 500,000 shares, issued and outstanding 50,000	50,000
Retained earnings	758,727
	808,727
	$3,120,527

See Reports of Independent Certified Public Accountants and Notes to Financial Statements

EXAMPLE CONTRACTOR, INC.

STATEMENT OF EARNINGS AND RETAINED EARNINGS

Year Ended December 31, 19X1

EARNINGS FROM CONSTRUCTION (Notes A-1, B, and C):	
Revenues earned	$11,683,190
Cost of construction	11,002,186
Gross profit	681,004
SELLING, GENERAL AND ADMINISTRATIVE EXPENSES	533,296
EARNINGS FROM OPERATIONS	147,708
OTHER INCOME (EXPENSES):	
Interest income	27,451
Other income	10,421
Gain on sale of fixed assets	12,560
Interest expense	(37,651)
Other expense	(2,849)
	9,932
EARNINGS BEFORE INCOME TAXES	157,640
INCOME TAXES (Notes A-4 and E)	48,000
NET EARNINGS	109,640
RETAINED EARNINGS, beginning of year	649,087
RETAINED EARNINGS, end of year	$ 758,727

See Report of Independent Certified Public Accountants and Notes to Financial Statements

EXAMPLE CONTRACTOR, INC.
STATEMENT OF EARNINGS AND RETAINED EARNINGS
Year Ended December 31, 19X1

CASH FLOWS FROM OPERATING ACTIVITIES:		
Net earnings	$	109,640
Adjustments to reconcile net earnings to net cash provided by operating activities:		
Depreciation		111,226
Gain on sale of equipment		(12,560)
Deferred taxes		1,600
Change in:		
Marketable securities		(136,000)
Receivables		(181,871)
Underbillings and unbilled amounts		(108,022)
Prepaid expenses and other		(4,316)
Accounts payable and other liabilities		416,914
Overbillings and accrued costs		101,008
Income taxes		(90,286)
Total adjustments		97,693
Net cash provided by operating activities		207,333
CASH FLOWS FROM INVESTING ACTIVITIES:		
Proceeds from sale of construction equipment		40,000
Addition to equipment and vehicles		(35,000)
Increase in other assets		(1,119)
Net cash provided by investing activities		3,881
CASH FLOWS FROM FINANCING ACTIVITIES:		
Principal payments on long-term debt		(33,088)
Net cash used in financing activities		(33,088)
Net increase in cash		178,126
Cash, beginning of year		329,414
Cash, end of year		$ 507,540

EXAMPLE CONTRACTOR, INC.

STATEMENT OF EARNINGS AND RETAINED EARNINGS

Year Ended December 31, 19X1

SUPPLEMENTARY DISCLOSURE OF CASH FLOW INFORMATION

Cash paid during the year for:	
Interest	$ 28,795
Income taxes	$ 146,620

See Report of Independent Certified Public Accountants and Notes to Financial Statements.

EXAMPLE CONTRACTOR, INC.

STATEMENT OF EARNINGS AND RETAINED EARNINGS

Year Ended December 31, 19X1

NOTE A – SUMMARY OF SIGNIFICANT ACCOUNTING POLICIES

Example Contractor, Inc., is engaged as a general building contractor in the construction of commercial and light industrial buildings under long-term contracts primarily in the greater Metropolis area.

A summary of the significant accounting policies applied in the preparation of the accompanying financial statements follows:

1. Method of Accounting for Long-Term Construction Contracts

The accompanying financial statements have been prepared using the percentage-of-completion method of accounting and, therefore, take into account the cost and estimated earnings and revenue to date on contracts not yet completed.

The amount of revenue recognized at the financial statement date is that portion of the total contract price that cost incurred to date bears to anticipated final total cost, based on current estimates of cost-to-complete. The amount of revenue recognized is not related to the progress billings to customers.

Contract costs include all direct labor, materials unique to or installed in the project, subcontractor cost, and indirect cost allocations, including employee benefits and construction equipment expense.

As long-term contracts extend over one or more years, revisions in cost and earnings estimated during the course of the work are reflected in the accounting period in which the facts which require the revision become known.

At the time a loss on a contract becomes known, the entire amount of the estimated ultimate loss is recognized in the financial statements.

Contracts which are substantially complete are considered closed for financial statement purposes. Any accrued billings or costs related to these contracts are carried in the balance sheet as "Unbilled Amounts on Completed Contracts" and "Accrued Costs on completed Contracts," respectively.

EXAMPLE CONTRACTOR, INC.
STATEMENT OF EARNINGS AND RETAINED EARNINGS
Year Ended December 31, 19X0

NOTE A – SUMMARY OF SIGNIFICANT ACCOUNTING POLICIES – Continued

1. Method of Accounting for Long-Term Construction Contracts – Continued

The current asset, "Costs and Estimated Earnings in Excess of Billings on Contracts in Progress," represents revenues recognized in excess of amounts billed (underbillings) and the current liability, "Billings in Excess of Costs and Estimated Earnings on Contracts in Progress," represents billings in excess of revenues recognized (overbillings).

2. Operating Cycle

Assets and liabilities related to long-term contracts are included in current assets and current liabilities in the accompanying balance sheet, as they will be liquidated in the normal course of contract completion, although this may require more than one year.

3. Property, Equipment, and Depreciation

Property and equipment is carried at cost and is being depreciated principally on the straight-line method over the estimated useful lives of the assets.

4. Income Taxes

Deferred income taxes are provided for temporary differences in reporting income for financial statements and tax purposes arising from differences in the methods of accounting for depreciation. For tax purposes, contracts are reported on the "Percentage-of-Completion Method" as set forth in the 1986 Tax Act.

5. Allocation of Indirect Construction Cost

Indirect construction cost is allocated as follows:

Salaries and payroll benefits – dollars of labor incurred

Equipment – hours of use

The difference between actual expenditures for indirect construction cost and the amount allocated is charged to operations for the current year.

EXAMPLE CONTRACTOR, INC.
STATEMENT OF EARNINGS AND RETAINED EARNINGS
Year Ended December 31, 19X0

NOTE A – SUMMARY OF SIGNIFICANT ACCOUNTING POLICIES - Continued

6. Inventories

Inventories are stated at the lower of cost (first in – first out method) or market.

7. Cash Equivalents

For purposes of the statement of cash flows, the Company considers all highly liquid debt instruments purchased with a maturity of three months or less to be cash equivalents.

NOTE B – RECEIVABLES

Receivables December 31, 19X1 consist of the following:

Contract at receivables:	
Current	$ 688,245
Retention	1,000,608
	1,688,853
Other receivables	32,356
	$1,721,209

Contract receivables at December 31, 19X1 include $125,431 due from the majority stockholder on a contract in progress. This contract is for $690,000 of which the Company has reflected revenues of $454,597 and gross profits of $28,442 during the year ended December 31, 19X1. The Company collected $60,000 of this receivable during January 19X2. Collections on the remaining receivables from January 1, 19X2 through February 24, 19X2 were approximately $850,000.

Other receivables contain advances to the stockholders in the amount of $15,000 for the year ended December 31, 19X1. Current contract receivables at December 31, 19X1 are aged as follows:

	Current
0 – 30 days	$ 540,427
31 – 60 days	94,018
61 – 90 days	25,000
Over 90 days	28,800
	$ 688,245

EXAMPLE CONTRACTOR, INC.
STATEMENT OF EARNINGS AND RETAINED EARNINGS
Year Ended December 31, 19X1

NOTE C – CONTRACTS IN PROGRESS

Contract amounts, accumulated costs, estimated earnings, and the related billings to date on contracts in progress at December 31, 19X1 are as follows:

Total construction activity	$22,079,799	$11,683,190	$11,002,186	$681,004
Contracts completed during year	(13,299,589)	(6,479,484)	(6,155,432)	(324,052)
Over allocated indirect construction cost			3,199	(3,199)
Callbacks and adjustments		(8,244)	(14,984)	6,740
Activity during the year on contracts in progress		5,195,462	4,834,969	360,493
Activity in prior years on contracts in progress		567,566	526,417	41,149
Contracts in progress Dec. 31, 19X1	$ 8,780,210	5,763,028	$ 5,361,386	$401,642
Less progress billings to December 31, 19X1		5,823,650		
		$ (60,622)		

The above is included in the accompanying balance sheet under the following captions:

Costs and estimated earnings in excess of billings on contracts in progress	$ 173,011
Billings in excess of costs and estimated earnings on contracts in progress	(233,633)
	$ (60,622)

EXAMPLE CONTRACTOR, INC.
STATEMENT OF EARNINGS AND RETAINED EARNINGS
Year Ended December 31, 19X0
(Continued)

NOTE D – LONG-TERM DEBT AND LINE OF CREDIT

Long-term debt at December 31, 19X1 consists of the following:

Note payable to bank, collateralized by construction equipment vehicles with a carrying value of $192,600, due in monthly installments of $3,802 including interest of 9.75% through September 19X4	$142,969
Mortgage payable to bank, collateralized by land and building , with a carrying value of $104,600 due in monthly installments of $1,354 including interest at 11.75% through November 19X4	111,165
	254,134
Less current maturities	36,530
Long-term debt	$217,604

Aggregate maturities of long-term debt for the five years following December 31, 19X1 are as follows:

19X1	$ 36,530
19X2	44,525
19X3	37,660
19X4	5,381
19X5	6,051
Thereafter	123,987
	$254,134

The Company has available an unsecured line of credit with a bank for $1,000,000. At December 31, 19X1, the line of credit was unused. Under the terms of this agreement, interest is charged at the prime rate (9% at December 31, 19X1) plus 1%. This agreement is renewable annually on February 28.

EXAMPLE CONTRACTOR, INC.
STATEMENT OF EARNINGS AND RETAINED EARNINGS
Year Ended December 31, 19X1
(Continued)

NOTE E – INCOME TAXES

During 19X1, the Company was required to change its method of reporting long-term contracts for income tax purposes from the completed contract method to the "Percentage-of –Completion Method" as set forth in the 1986 Tax Act.

The provision for income taxes as of December 31, 19X1 consists of:

Current income taxes payable:	
Federal	$38,000
State	8,000
	46,000
Deferred income taxes	2,000
	$48,000

NOTE F – COMMITMENTS AND CONTINGENCIES

The Company is involved in various litigation with claims totaling approximately $310,000 relating to the failure of a subcontractor of the Company to pay its suppliers. In the opinion of management and counsel, the outcome of these claims will not have a material effect on the financial position of the Company.

EXAMPLE CONTRACTOR, INC.
STATEMENT OF EARNINGS AND RETAINED EARNINGS
Year Ended December 31, 19X1
(Continued)

NOTE G – BACKLOG

Following is a reconciliation of the approximate backlog of signed contracts as of December 31, 19X1:

Balance as of January 1, 19X1	$ 4,200,000
New contracts and change orders during the year	10,500,000
	14,700,000
Less contracts completed	11,700,000
Balance December 31, 19X1	$ 3,000,000

In addition, between January 1, 19X2 and February 24, 19X2, the Company entered into contracts approximating $3,400,000.

Appendix IX. Sample Dun & Bradstreet Report
Source: Dun and Bradstreet

Dun & Bradstreet

Business Information Report

SUBSCRIBER: 123-4567L
PREPARED FOR:

ANSWERING INQUIRY

DUNS: 00-007-7743
GORMAN MANUFACTURING CO. INC.
(SUBSIDIARY OF GORMAN HOLDING COMPANIES INC., LOS ANGELES, CA)

492 KOLLER ST
AND BRANCH(ES) OR DIVISION(S)
SAN FRANCISCO CA 94110
TEL: 415-555-0000

DATE PRINTED
OCT 30 199-

COMMERCIAL PRINTING
SIC NO.
2752

SUMMARY

RATING	3A3
STARTED	1965
SALES F	$13,007,229
WORTH F	$2,125,499
EMPLOYS	105(100 HERE)
HISTORY	CLEAR
FINANCING	SECURED
FINANCIAL CONDITION	FAIR

CHIEF EXECUTIVE: LESLIE SMITH, PRES

==

SPECIAL EVENTS

09/11/9- On Sept 9, 199-, the subject experienced a fire due to an earthquake. According to Leslie Smith, President, damages amounted to $35,000, which was fully covered by their insurance company. The business was closed for two days while employees settled personal matters due to the earthquake.

03/17/9- Subject moved from 400 KOLLER ST. to 492 KOLLER ST. on March 11, 199-.

==

* * * CUSTOMER SERVICE * * *

==

If you need any additional information, would like a credit recommendation, or have any questions regarding this report, please call our Customer Service Center at (800) 234-3867 from anywhere within the United States.

==

* * * SUMMARY ANALYSIS * * *

==

RATING SUMMARY . . . .

The "3A" portion of the Rating (Estimated Financial Strength) indicates that the company has a worth between $1 million and $10 million. The "3" on the right (Composite Credit Appraisal) indicates an overall "fair" credit appraisal. The "fair" credit appraisal was assigned because the company's overall payment record shows frequent slowness and because of D&B's "fair" assessment of the company's 12/31/9- fiscal financial statement.

Below is an overview of the company's D&B Rating(s) since 1-1-91:

RATING	DATE APPLIED
3A3	09/11/9-
--	01/01/91

==

* * * PAYMENT SUMMARY * * *

==

This Payment Summary section reflects payment information in D&B's file as of October 29, 199-.

The PAYDEX for this company is 67.

This PAYDEX score indicates that payments to suppliers average 18 days beyond terms, weighted by dollar amounts. When dollar amounts are not considered, approximately 75% of the company's payments are within terms.

Below is an overview of the company's dollar-weighted payments, segmented by its suppliers' primary industries:

	TOTAL RCV'D	TOTAL DOLLAR AMOUNTS	LARGEST HIGH CREDIT	W/IN TERMS	DAYS SLOW <31	31-60	61-90	91+
	#	$	$	%	%	%	%	%
Total in D&B's file	24	785,150	250,000					
Top 10 Industries:								
1 Whol Printing Paper	8	747,500	250,000	64	18	18	-	-
2 Repair Service	2	7,500	7,500	-	100	-	-	-
3 Mfg Industrial Mach	2	2,000	1,000	50	50	-	-	-
4 Misc Services	2	750	500	33	67	-	-	-
5 Whol Non Durables	1	15,000	15,000	50	50	-	-	-
6 Air Courier Service	1	7,500	7,500	100	-	-	-	-
7 Mfg Service Ind Mach	1	2,500	2,500	50	50	-	-	-
8 Transportation Svcs	1	1,000	1,000	50	-	-	50	-
9 Mfg Photo Equipment	1	500	500	100	-	-	-	-
10 Mfg Plastic Products	1	250	250	100	-	-	-	-
11 Other Industries	1	100	100	100	-	-	-	-
Other Payment Categories:								
Cash experiences	2	450	250					
Paying record unknown	1	100	100					
Unfavorable comments	0	0	0					
Placed for collection:								
with D&B	0	0						
other	0	N/A						

The highest "Now Owes" in D&B's file is $250,000
The highest "Past Due" in D&B's file is $90,000

The total dollar amount from the 24 trade experiences listed is 6.0% of this company's annual sales as presented in the Summary. D&B considers the trade experiences to be representative of this company's payment habits.

PAYMENTS (Amounts may be rounded to nearest figure in prescribed ranges)

- Antic - Anticipated (Payments received prior to date of invoice)
- Disc - Discounted (Payments received within trade discount period)
- Ppt - Prompt (Payments received within terms granted)

REPORTED	PAYING RECORD	HIGH CREDIT	NOW OWES	PAST DUE	SELLING TERMS	LAST SALE WITHIN
03/9-	Ppt-Slow 90	1000	500	-0-	N30	1 Mo
02/9-	Ppt	250	100			4-5 Mos
	Ppt-Slow 30	2500	2500	1000		1 Mo
	Slow 30	500	500			2-3 Mos
	(005)	200				
	Slow 30-60	70000	70000	6500		1 Mo

01/9-	Disc	2500	1000			1 Mo
	Disc-Ppt	25000	25000	0	2 to Prox	1 Mo
	Ppt-Slow 15	1000	500	250		1 Mo
	Ppt-Slow 30	15000	10000	5000		1 Mo
	Ppt-Slow 30	1000	0	0	N30	1 Mo
12/9-	Ppt	250000	250000	0		1 Mo
	Ppt	7500	250	0	N15	4-5 Mos
	(014)	250			Sales COD	1 Mo
	Ppt	500	0	0	N30	1 Mo
	Ppt	100	50	0	Regular terms	6-12 Mos
	Ppt-Slow 30	100000	100000	40000		1 Mo
	Ppt-Slow 30	70000	70000	50000	2 15 Prox	1 Mo
	Slow 30	7500	0	0		1 Mo
11/9-	Slow 30		0	0	N30	1 Mo
	Disc-Slow 30	30000	30000	7500		6-12 Mos
10/9-	Ppt	250	0	0		1 Mo
	Ppt-Slow 60	200000	200000	90000		6-12 Mos
09/9-	(024)	100	100		N30	1 Mo

* Payment experiences reflect how bills are met in relation to the terms granted. In some instances payment beyond terms can be the result of disputes over merchandise, skipped invoices, etc.

* Each experience shown represents a separate account reported by a supplier. Updated trade experiences replace those previously reported.

==

UPDATE

08/17/9- On August 17, 199- KEVIN J. HUNT Sec-Treas stated for the six months ended June 30, 199- profits were up compared to same period last year.

==

FINANCE 09/11/9-	Fiscal Dec 31, 199-	Fiscal Dec 31, 199-	Fiscal Dec 31, 199-
Curr Assets	4,643,821	4,825,611	5,425,125
Curr Liabs	3,595,821	3,625,000	4,135,718
Current Ratio	1.3	1.3	1.3
Working Capital	1,048,000	1,200,611	1,289,407
Other Assets	1,468,291	1,485,440	2,201,690
Worth	1,879,451	1,912,112	2,125,499
Sales	9,321,118	10,325,582	13,007,229
Net Income	24,211	32,661	213,387

Fiscal statement dated Dec. 31, 199-:

Cash	925,000	Accts Pay	2,125,114
Accts Rec	1,725,814	Notes Pay	450,000
Inventory	1,643,311	Bank Loan	1,100,000
Other Curr Assets	1,131,000	Other Curr Liabs	460,604
	--------		----------
Curr Assets	5,425,125	Curr Liabs	4,135,718
Fixt & Equip	1,667,918	L.T. Liab-Other	1,365,598
Other Assets	533,772	CAPITAL STOCK	50,000
		RETAINED EARNINGS	2,075,499
	--------		----------
Total Assets	7,626,815	Total	7,626,815

From JAN 1, 199- to DEC 31, 199- sales $13,007,229; cost of goods sold $9,229,554. Gross profit $3,777,675; operating expenses $3,751,661. Operating income $26,014; extraordinary gain $187,373. Net income $213,387.

Submitted SEPT 11, 199- by Leslie Smith, President. Prepared from statement(s) by Accountant: Ashurst & Ashurst, PC. Prepared from books without audit.

--0--

Accounts receivable shown net less $12,586 allowance. Other current assets consist of prepaid expenses $64,471 and $1,066,529 of a loan from an affiliated concern. Other assets consist of deposits. Bank loans are due to bank at the prime interest rate, are secured by accounts receivable and inventory, and mature in 3 years. Notes payable are due on printing equipment in monthly installments of $27,500. Other current liabilities are accrued expenses and taxes. Long term debt consist of the long term portion of the equipment note.

On SEPT 11, 199- Leslie Smith, president, submitted the above figure(s).

Leslie Smith submitted the following interim figures dated JUNE 30, 199-.

Cash	$ 1,011,812	Accts Pay	$ 1,932,118
Accts Rec	1,932,118	Owe Bank	1,100,000
Inventory	1,421,112	Notes Pay	350,000

Sales for 6 months were $7,325,001. Profits for 6 months were $103,782.

Projected annual sales are $14,000,000.

He also stated operating profits were below average due to heavy price competition in the industry, higher operating expenses, and decreased advertising budgets following the nationwide move towards cost containment.

PUBLIC FILINGS

The following data is for information purposes only and is not the official record. Certified copies can only be obtained from the official source.

* * * SUITS * * *

DOCKET NO:	12345	STATUS:	Pending
SUIT AMOUNT:	$1,000	DATE STATUS ATTAINED:	03/25/199-
PLAINTIFF:	MAZZUCA & ASSOC.	DATE FILED:	03/25/199-
DEFENDANT:	GORMAN MANUFACTURING CO. INC.	LATEST INFO RECEIVED:	03/31/199-
CAUSE:	Breach of contract		
WHERE FILED:	SAN FRANCISCO, CA		

* * * UCC FILING(S) * * *

COLLATERAL:	Accounts receivable - Inventory including proceeds and products		
FILING NO:	86188586	DATE FILED:	07/24/199-
TYPE:	Original	LATEST INFO RECEIVED:	10/04/199-
SEC. PARTY:	A.C. Paper, Palo Alto, CA	FILED WITH:	SECRETARY OF STATE/ UCC DIVISION, CA

BANKING

09/9- Account(s) averages high 6 figures. Account open over 10 years. Loans granted to low 7 figures on a secured basis. Now owing low 7 figures. Collateral consists of accounts receivable and inventory. Matures in 1 to 5 years. Borrowing account is satisfactory. Overall relations are satisfactory.

HISTORY
09/11/9- LESLIE SMITH, PRES KEVIN J. HUNT, SEC-TREAS
DIRECTOR(S): THE OFFICER(S)

BUSINESS TYPE: Corporation - Profit DATE INCORPORATED: 05/21/1965
AUTH SHARES - COMMON: 200 STATE OF INCORP: California
PAR VALUE - COMMON: No Par Value

Business started May 21, 1965 by Leslie Smith and Kevin J. Hunt. 100% of capital stock is owned by Parent Company.

SMITH born 1936. Married. Graduated from the University of California, Los Angeles, in June 1957 with a BS degree in Business Management. 1957-65 general manager for Raymor Printing Co., San Francisco, CA. 1965 formed subject with Kevin J. Hunt.

HUNT born 1935. Married. Graduated from Northwestern University, Evanston, IL. in June 1957. 1957-1965 controller for Raymor Printing Co., San Francisco, CA. 1965 formed subject with Leslie Smith.

RELATED COMPANIES: Through the financial interest of Gorman Holding Company Inc., subject's parent company, the Gorman Manufacturing Co. Inc., is related to two other companies:

1. Smith Lettershop Inc., San Diego, CA; commercial printing, started 1972.
2. Gorman Suppliers Inc., Los Angeles, CA; commercial print, started 1980.

Intercompany relations consist of loans.

OPERATION
09/11/9- Subsidiary of Gorman Holding Company Inc., Los Angeles, CA, started 1965 which operates as a holding company for its subsidiaries. Parent company owns 100% of capital stock. Parent company has 2 other subsidiaries. Intercompany relations: consist of loans and advances.

A consolidated financial statement of the parent company dated Dec. 31, 199- showed a worth of $4,125,112, with an overall fair financial condition.

Commercial Printing specializing in advertising posters, catalogs, circulars and coupons.
Net 30 days. Has 175 account(s). Sells to commercial concerns.
Territory: United States.
Nonseasonal.

EMPLOYEES: 105 which includes officer(s). 100 employed here.

FACILITIES: Rents 55,000 sq. ft. on first floor of one story cinder block building in good condition. Premises neat.

LOCATION: Central business section on well traveled street.

BRANCH(ES): Subject maintains a branch at 1073 Boyden Road, Los Angeles, CA.

Appendix X. Sample Robert Morris Associates Reports
Source: Robert Morris Associates

CONTRACTORS - CONCRETE WORK. SIC# 1771

Current Data Sorted By Revenue						Comparative Historical Data				
					# Postretirement Benefits					
					Type of Statement					
1	24	13		38	Unqualified	32	41	34	33	38
7	52	10	1	70	Reviewed	53	72	65	53	70
16	36	2	1	55	Compiled	45	52	52	49	55
7	4			11	Tax Returns	5	3	3	9	11
1	17	4	1	23	Other	18	19	24	21	23
51 (4/1-9/30/96)		146 (10/1/96-3/31/97)				4/1/92-3/31/93	4/1/93-3/31/94	4/1/94-3/31/95	4/1/95-3/31/96	4/1/96-3/31/97
0-1MM	**1-10MM**	**10-50MM**	**50 & Over**	**ALL**		**ALL**	**ALL**	**ALL**	**ALL**	**ALL**
32	133	29	3	197	**NUMBER OF STATEMENTS**	153	187	178	165	197
%	%	%	%	%	**ASSETS**	%	%	%	%	%
15.1	9.6	10.9		10.8	Cash & Equivalents	10.8	11.8	11.4	10.3	10.8
15.3	39.2	49.5		36.8	A/R - Progress Billings	35.4	39.4	39.5	40.5	36.8
2.6	3.4	4.6		3.5	A/R - Current Retention	3.7	4.4	3.8	3.2	3.5
2.6	2.0	1.6		2.1	Inventory	2.9	2.7	2.2	1.2	2.1
					Cost & Est. Earnings					
0.0	2.7	3.5		2.5	In Excess of Billings	1.4	1.6	1.3	1.8	2.5
1.9	3.2	3.0		3.0	All Other Current	3.6	2.9	2.8	2.3	3.0
37.4	60.1	73.0		58.6	Total Current	57.8	62.9	61.0	59.3	58.6
51.6	32.3	19.3		33.3	Fixed Assets (net)	34.2	28.4	30.4	34.1	33.3
2.6	0.0	1.2		1.1	Joint Ventures & Investments	1.4	1.3	0.9	0.8	1.1
1.3	1.6	0.6		1.4	Intangibles (net)	0.7	1.0	1.5	1.7	1.4
7.2	5.1	5.9		5.6	All Other Non-Current	5.9	6.4	6.3	4.1	5.6
100.0	100.0	100.0		100.0	Total	100.0	100.0	100.0	100.0	100.0
					LIABILITIES					
11.4	6.7	6.4		7.5	Notes Payable - Short Term	7.9	7.7	7.5	8.9	7.5
11.7	18.2	29.0		18.8	A/P - Trade	19.4	18.6	19.1	17.3	18.8
0.0	0.0	1.0		0.2	A/P - Retention	0.2	0.2	0.3	0.1	0.2
					Billings in Excess of Costs					
1.0	2.8	5.6		2.9	& Est. Earnings	1.5	1.6	2.1	1.8	2.9
0.6	0.4	1.0		0.6	Income Taxes Payable	0.6	1.1	0.8	0.7	0.6
9.1	6.1	4.6		6.3	Cur. Mat. - L/T/D	5.3	4.7	5.3	5.8	6.3
5.7	5.4	7.3		5.7	All Other Current	5.4	7.0	7.4	7.4	5.7
39.5	39.7	54.9		42.0	Total Current	40.3	41.0	42.5	42.0	42.0
23.3	15.3	6.2		15.2	Long Term Debt	14.2	11.4	12.4	13.7	15.2
0.3	1.4	1.4		1.2	Deferred Taxes	1.2	1.6	1.2	1.3	1.2
0.8	2.4	0.7		1.8	All Other Non-Current	1.0	2.8	1.5	1.6	1.8
36.0	41.2	36.8		39.8	Net Worth	43.3	43.2	42.3	41.5	39.8
100.0	100.0	100.0		100.0	Total Liabilities & Net Worth	100.0	100.0	100.0	100.0	100.0
					INCOME DATA					
100.0	100.0	100.0		100.0	Contract Revenues	100.0	100.0	100.0	100.0	100.0
40.6	23.9	11.2		24.6	Gross Profit	27.5	25.5	26.2	27.7	24.6
35.7	20.1	8.6		20.8	Operating Expenses	24.5	21.5	21.8	22.5	20.8
4.9	3.8	2.6		3.8	Operating Profit	3.0	4.0	4.4	5.2	3.8
1.0	0.2	-0.3		0.2	All Other Expenses (net)	0.3	0.7	0.5	0.9	0.2
3.9	3.6	2.9		3.6	Profit Before Taxes	2.7	3.3	3.8	4.4	3.6
					RATIOS					
2.2	2.4	1.7		2.2		2.3	2.4	2.1	2.1	2.2
1.2	1.6	1.3		1.5	Current	1.5	1.6	1.4	1.5	1.5
0.5	1.2	1.1		1.1		1.1	1.1	1.1	1.0	1.1
3.3	5.7	2.7		4.7		3.6	4.3	4.1	4.3	4.7
(27) 2.0	130) 2.7	1.7		(189) 2.3	Receivables/Payables	150) 2.1	182) 2.5	174) 2.4	163) 2.7	189) 2.3
0.6	1.6	1.4		1.4		1.4	1.6	1.6	1.7	1.4
0 UND	33 11.1	58 6.3		28 13.1		33 10.9	36 10.1	35 10.5	33 11.0	28 13.1
19 18.9	55 6.6	68 5.4		54 6.7	Revenues/Receivables	40 7.9	66 5.5	52 7.0	54 6.8	54 6.7
51 7.2	79 4.6	79 4.6		78 4.7		74 4.9	81 4.5	74 4.9	73 5.0	78 4.7
0 UND	14 26.3	26 14.1		14 26.9		15 24.9	15 24.3	14 27.0	12 29.9	14 26.9
19 19.5	24 15.2	43 8.5		26 14.0	Cost of Revenues/Payables	30 12.3	31 11.9	29 12.4	26 14.1	26 14.0
58 6.3	45 8.1	56 6.5		49 7.4		51 7.1	52 7.0	47 7.8	44 8.3	49 7.4
10.4	7.7	9.0		7.8		8.6	6.9	8.4	8.5	7.8
113.4	13.4	18.2		14.6	Revenues/Working Capital	15.3	12.3	16.2	14.5	14.6
-19.8	39.8	63.9		86.0		81.2	52.4	64.2	NM	86.0
11.8	11.1	11.8		11.9		10.7	9.0	12.4	12.9	11.9
(30) 3.6	111) 4.4	(24) 5.4		(168) 4.6	EBIT/Interest	138) 3.0	170) 3.6	165) 4.9	153) 4.0	168) 4.6
0.0	1.9	3.3		1.9		1.5	1.5	2.1	1.7	1.9
	5.8	4.7		5.4		5.0	4.7	4.2	5.9	5.4
	(55) 3.2	(15) 1.4		(79) 2.2	Net Profit + Depr., Dep.	(66) 1.9	102) 2.1	(81) 2.3	(54) 2.9	(79) 2.2
	1.4	0.7		0.9	Amort./Cur. Mat. L/T/D	1.0	1.1	1.3	1.6	0.9
0.7	0.3	0.3		0.3		0.3	0.3	0.4	0.4	0.3
1.7	0.7	0.4		0.7		0.7	0.6	0.7	0.7	0.7
2.7	2.0	0.9		1.9	Fixed Worth	1.3	1.3	1.3	1.8	1.9
0.6	0.7	1.1		0.7		0.6	0.7	0.7	0.7	0.7
1.5	1.5	1.9		1.5		1.2	1.5	1.4	1.5	1.5
4.5	3.6	3.6		3.6	Debt Worth	3.1	2.6	3.0	3.3	3.6
68.9	46.3	37.7		46.6		33.2	38.3	51.7	51.3	46.6
(28) 22.0	125) 25.6	22.0		(185) 25.0	% Profit Before Taxes/	143) 12.7	181) 13.6	173) 25.7	155) 27.5	185) 25.0
-1.6	9.8	8.2		8.7	Tangible Net Worth	1.5	2.6	9.3	11.2	8.7
31.0	17.4	14.3		17.4		14.3	15.7	18.8	19.2	17.4
9.7	8.2	6.4		8.1	% Profit Before Taxes/	5.1	5.7	9.1	9.9	8.1
-2.6	2.8	3.5		2.6	Total Assets	0.7	1.2	3.2	2.0	2.6
3.3	1.3	0.6		1.2		1.4	1.6	1.5	1.4	1.2
(28) 4.7	123) 2.6	(28) 1.1		(180) 2.7	% Depr., Dep. Amort./	142) 2.9	165) 2.5	168) 2.5	151) 2.7	180) 2.7
8.0	4.0	3.0		4.2	Revenues	5.3	4.5	3.8	4.6	4.2
5.2	2.3	1.1		2.1		2.9	2.1	2.9	2.2	2.1
(18) 11.0	(74) 3.4	(15) 2.0		(109) 3.4	% Officers', Directors',	(91) 4.8	105) 4.7	108) 4.8	102) 3.9	109) 3.4
18.5	7.0	2.7		7.3	Owners' Comp./Revenues	7.4	8.0	8.5	8.5	7.3
19644M	481962M	575590M	19238948M	20316144M	Contract Revenues ($)	693291M	1083401M	1099084M	2042636M	20316144M
9792M	188919M	220612M	9652821M	10072144M	Total Assets ($)	300434M	449962M	438373M	568644M	10072144M

Appendix XI. Listing of Surety Association of America Contract Types

Source: *Rate Manual: Surety Bonds*

The Surety Association of America 1987

Supply Contracts

Ammunition
Apparatus in public or private offices, stores, schools, colleges, etc.
Armament
Armor plate
Asbestos pipe covering
Automobiles, trucks or parts
Awnings
Blackboards and bulletin boards
Boilers
Cabinets
Clocks and master clocks
Coal
Concrete or cement-furnishing and delivering
Curtains, draperies, or scenery
Electric current-sale of
Electric Fixtures
Engines
Filing cases
Fire engines or fire hoses
Fuel
Furniture
Headstones and mortuary
Jail cells
Laboratory animals
Water softener
Laundry machinery
Linoleum, carpets, rugs-furnishing or laying
Lockers
Machinery
Munitions
Office furniture
Oil and oil burners
Partitions
Paving blocks
Pipe, concrete, clay or iron, furnishing only
Pipe covering, asbestos
Pipe organs
Placing fire or other insurance
Printing presses
Pumps
Safe deposit vaults or boxes
Showcases and counters
Steam, sale of
Stone, crushed
Track material, railroad ties plates, pads, and ballast
Uniforms
Venetian blinds
Window screens or shades

Class B Contracts

Airport buildings
Air conditioning systems
Aqueducts
Atomic energy plants
Buildings of all kinds including erection or repair
Breakwaters
Canals and canal lining
Carpentry
Coal stripping
Commercial buildings
Concrete work
Dams – all
Dikes, levees, and revetments
Ditches, including drainage
Docks and drydocks
Drilling contracts
Educational buildings
Electrical wiring
Embankments
Excavations
Filling stations
Filtering plants
Foundations
Fountains
Garbage disposal plants
Gasoline cracking plants
Gas compressor stations
Gas mains and laterals
Gas piping
Golf courses
Grain elevators
Gunite contracts
Heating systems
Hospital clinic and rest home buildings
Incinerators
Industrial buildings and plants
Jetties
Landscaping
Locks
Masonry
Missile installations
Nuclear reactors
Office buildings
Off-shore platforms and towers
Painting
Piers
Piling
Pipelines for water
Plastering
Plumbing
Power plants
Public improvements
Roadbeds for railroads
Sand blasting
Sculptures
Sea walls
Sewage disposal plants
Sewers and septic tanks
Shipyards
Spillways
Stone, cut or dressed
Subways
Swimming pools
Terminals for bus companies or others
Test boring
Tile and terrazzo
Transmission or distribution lines
Tunnels
Underwater cables
Ventilating systems
Water Works
Wells
Wharves

Class A Contracts

Airports-grading, clearing, drainage preparation, leveling, or surfacing of sites
Airport-runways
Aluminum siding
Athletic fields
Beacon or flood lights
Burial contracts
Ceilings-metal or acoustical tile
Coal storage
Construction forms
Curbing and guttering
Curtain walls-not structural
Ducts, underground-for power, light or telephone
Elevators and escalators
Filling up abandoned coal mines
Glazing
Greenhouses
High pressure power piping
Janitor service
Machinery made to special order
Map making
Mill work
Mural
Parking areas
Planting and cultivation of land
Playgrounds and parks
Research contracts
River bank protection
Road medians
Roofing
Ski lifts
Sprinkler systems
Stone, marble or granite-furnishing and delivering only, cut or dressed stone
Storage tanks, metal
Tennis courts
Water carriage of freight
Water proofing
Wind tunnel

Class A-1 Contracts

Arms
Ash conveyors
Automatic stokers
Automatic telephone exchange equipment
Automotive service contracts
Band concerts
Bird control
Boiler retubing and repair
Bookbinding
Cataloging
Coal handling machinery
Computers and data processing equipment
Conveyors
Data processing and computer work
Doors-all types
Dynamos
Exterminating contracts
Fire alarm systems
Fire escapes
Flag poles
Floats
Floors-asphalt tile, cork or rubber tile, parquet, wood, or wood block
Gas tanks
Furnishing food service
Generators
Grain doors, salvage, care and disposal of
Guard rails
Heating-temporary and/or portable
Incinerator operation
Insulation contracts
Kitchen equipment
Laboratory equipment and apparatus
Leasing of motor vehicles
Lightning rods
Lock gates-furnishing only or furnishing and installing
Mail Handling equipment
Metal windows and shutters
Mosquito control contracts
Movies
Office personnel
Organ repairs
Ornamental iron work
Parking meters
Photogrammetric work
Pipelines for oil or gas
Police alarm systems
Projectiles
Public address and music systems
Radio towers
Radiological equipment and apparatus
Recapping auto tires
Refrigerating plants, on land
Repair of automobiles and trucks
Resmelting old metal
Rip rap stone (furnishing)
Rolling stock
Scaffolding
Sidewalks
Signal systems on railroads
Signs (all)
Stack rooms
Street and subway lighting systems
Temporary personnel services
Thermostat equipment
Toll gates
Track laying
Traffic control systems on highways
Training manuals
Tree trimming and removal
Watchmen and signal services
Water towers
Weather stripping
Weed mowing
Window cleaning
Work and labor
X-ray inspection

Appendix XII. Underwriting Case Study Documents: Prometheus Construction Company, Inc.

Re: Prometheus Construction Co., Inc.
Anytown, USA

Gentlemen:

Enclosed please find the following underwriting data in reference to the above fine account.

1. September fiscal financial statements for the years of 1993 and 1994
2. Contractor's Information Record
3. Accountant's Information Blank
4. A current D & B report
5. A current bank Letter of Credit
6. Job Schedules
7. Personal financial Statements

Prometheus Construction, Inc., is an account which we have written since 1977. The firm is owned by J. Prometheus, who is one of the finest individuals I have had the pleasure to be associated with. J. Prometheus specializes in general contracting work on a national scale and has been quite successful. A review of the corporate financials, along with J. Prometheus' personal financials, should reveal a strong financial position which could easily support the desired work program. I think you'll agree that this is a successful business.

J. Prometheus has always surrounded himself with good people, i.e., highly qualified individuals (accountants, bankers, subcontractors, etc.). He has also always maintained an excellent relationship with his bankers, suppliers, subcontractors and owners. The enclosed letters should verify the above.

Reviewing the corporate financial should reveal an excellent financial condition with good cash balances, and a strong working capital and a net worth position. After reviewing the financials, I am sure you'll agree hat this is a very successful business which has been very profitable and has adequate financial resources.

Although the firm has a well balanced financial condition, J. Prometheus is willing to sign personally. However, Ms. Prometheus' indemnity is not available. Based on the financial resources of the firm, I do not think Ms. Prometheus' indemnity is needed.

Prometheus is interested in running an aggregate work program of $20mm. A single job limitation of $10mm should handle the typical needs of the firm. On occasion, we may ask that you bond $15mm projects, but those jobs will be few and far between.

Prometheus has done jobs in the $15mm range before and has handled work programs in the $18mm range. So, we are only talking about a little stretch from an experience standpoint. However, based on the financial resources and track record of this business, I feel very comfortable in recommending this case.

I would very much appreciate you reviewing the above case and getting back o me as soon as possible with a bond line as indicated above.

Sincerely,

J. Hermes

CONTRACTOR'S INFORMATION RECORD

1. Name of Firm Prometheus Construction Co., Inc.	
2. Address (city, state, zip) Anytown, USA	3. Telephone (555) 555-5555

4. Fiscal Year 9-30-94 5. Contracting Specialty(ies)________

6. Agency & Agent Hermes Insurance Agency

7. Year Business Started 1977 Year Business Incorporated 1977

8. State of Incorporation USA 9. Area of Operation USA

10. Type of Business

☐ Corporation ☐ Partnership ☐ Proprietorship ☒ Sub S Corporation

11. Is your firm union? .... ☒ Yes ☐ No

12. Are you involved in the design aspects of construction? .... ☒ Yes ☐ No

13. List any subsidiaries and affiliates of the contracting firm:

	Firm Name	Ownership	Type Business	City and State
A.				
B.				
C.				

14. Cross/Corporate Indemnity? If yes, concurrent financial statements will be needed ☐ Yes ☐ No

15. List the corporate officers, partners or proprietors of your firm: Please attach resumes.

	Name	Age	Position	Percent Owned	Name of Spouse
A.	J. Prometheus	65	President	100 %	E. Prometheus
B.	J. Prometheus Jr.	25	V. President	%	
C.	E. Prometheus	60	Treasurer	%	
D.				%	

16. Will the above individuals and spouses personally indemnify the Surety?
If yes, concurrent personal financial statements will be needed. .... ☐ Yes ☒ No

If no, explain: Mr. J. Prometheus signs personally

17. Is there a buy/sell agreement among the owners of he business? If yes, please attach a copy .... ☐ Yes ☒ No

18. Is this agreement funded by life insurance? ☐ Yes ☒ No

19. Other business in which any officer or partner is actively engaged:

Name of Business	**Partner of Officer Interested**	**Extent of Interest**
N/A		

20. What percentage of the firm's work is for:
Federal ______ % Public ______ % Private _______ %

21. What percentage of the firm's work is normally sub-contracted? .... 60 %

22. Which trades? Elec., Plumbing, HVAC, Equipment, Painting, Flooring, ACT, Glass, Glazing Roofing, Site work

23. Are bonds required of sub-contractors? .... ☐ Yes ☒ No

24. What trades do you normally undertake with your own forces? Concrete, masonry, studs & drywall, doors, hardware, str. steel, carpentry, specialties

26. What is the largest amount of uncompleted work on hand a one time in the past?
$15,000,000 Year 1991

27. What is the largest job you expect to do during the next year? .... $ 8,000,000

28. What is the largest uncompleted work program expected during the next year? $15,000,000

29. What is your expected annual volume next year? $10,000,000

30. How many people does your firm employ? 30

31. How many work crews? 4

32. What are your average annual equipment purchases? .... N/A

33. Do you lease equipment? ☐ Yes ☒ No If yes, name type of lease. As Needed

34. What are the terms of the lease? N/A

35.

Name of your CPA	
Hercules & Smith	
Address (city, state, zip) Anytown, USA	
Phone	**Contact Person** T. Hercules

36. Do you have a full-time accountant on staff? .... ☐ Yes ☒ No

If yes, please indicate professional designation ☐ Yes ☐ CMA ☐ PA ☐ Other BSBA Accounting

37. Years experience with your firm? 1

38. Years experience in the accounting profession? 3

39.

Name of Bank	Location	Maximum Line	Present Amount Owing	Nature of Security or Name of Endorser
A. Anytown Bank.	Anytown, USA	$250,000.		
B.		$		
C		$		

40. Previous Bonding Companies:

Name	Line of Surety Credit	Reason for Change	Agency
A. Action Bonding Co.	6,500,000/18,000,000	Bigger work program	Hermes Ins. Agency
B.			
C.			

41. List of largest contracts completed:

A. Restaurant $ 7,600,000 $ 320,000 1988 ☒ Yes ☐ No

Owner Jarrad Restaurant, Inc. Design Professional Danube Architectural

B. Nicholas Shopping Cntr $ 5,800,000 $ 150,000 1989 ☒ Yes ☐ No

Owner Nicholas Development Co. Design Professional Danube Architectural

C. Nicholas Office Bldg. $ 6,400,000 $ 500,000 1987 ☒ Yes ☐ No

Owner Nicholas Development Co. Design Professional Atlantis Archit.

D. Atlantis Waste Water Treatment Plant $ 7,100,000 $ 230,000 1986 ☒ Yes ☐ No

Owner Regional Sewer Dist. Design Professional So. Amer. Archit.

42. List three of your major suppliers:

	Name	Address	Telephone	Contact
A.	Bob's Supply House	Anytown, USA	311-7870	Bob Doe
B.	Barkley Lumber Yard	Anytown, USA	311-7210	H. Barkley
C.	Cartwright Supply House	Anytown, USA	311-3554	Hoss Cartwright

43. List three sub-contractors (or contractors if you are a sub-contractor) that you do business with:

	Name	Address	Telephone	Contact	Job
A.	Hercules Construction	Anytown, USA	311-8231	G. Hercules	Jarrad Restaurant
B.	Artemis Construction	Anytown, USA	311-8245	H. Artemis	Nicholas Office Bldg.
C.	Tigris, Inc.	Anytown, USA	311-8316	I. Tigris	Atlantic WWTP

44. List three Architects you do business with:

	Name	Address	Telephone	Contact	Job
A.	Danube Architectural	Anytown, USA	311-8256	A. Danube	Jarrad Restaurant
B.	Atlantis Architectural	Anytown, USA	311-7869	M. Atlantis	Nicholas Office Building
C.	S. Amer. Architectural	Anytown, USA	311-7059	R. American	Atlantis WWTP

45. List key personnel, foreman or supervisors:

	Name	Position	Yr. of Birth	Years Exper. Total	Years Exper. Your Firm	Prev. Employer
A.	E. Prometheus	Treasurer	1934	30	5	Action Const.
B.	M. Prometheus	Secretary	1936	13	13	A & P Const.
C.	R. Mack	Foreman	1938	24	13	Argus Const.

46. List any life insurance in effect on key personnel:

	Name	Beneficiary	Amount	Cash Value	Insurance Company
A.	A. Prometheus	Prometheus Const.	$ 100,000	$	Atlas Ins.
B.			$	$	
C.			$	$	

47. List other insurance coverage currently in effect:

Name	Limits in '000's Bodily Injury	Limits in '000's Property Damage	Carrier	Expiration Date
A. General Liability	$ 500,000	$ 500,000	Atlas Ins.	10-30-94
B. Auto Liability	$ 50	$ 100,000	Atlas Ins.	
C. Umbrella:	$ 1,000,000	$		
D. Owner's Protection	$	$		

48. Agency and Agent which handles insurance needs: J. Hermes Insurance

Liens, suits, judgments, defaults, and contingent liabilities:

49. Has your company or any officer or partner ever failed in business or compromised with creditors? .... ☐ Yes ☒ No

50. Has your company ever failed to complete a contract? .... ☐ Yes ☒ No

51. Has any partner or officer of your organization ever failed to complete a contract? .... ☐ Yes ☒ No

52. Are liens outstanding against any of your work? .... ☐ Yes ☒ No

53. Have you ever failed to qualify for a bond after an award? .... ☐ Yes ☒ No

54. Have you ever had any serious controversy with anyone on any of your work over the payment for labor or material? .... ☐ Yes ☒ No

55. Are there any judgments, suits, or claims pending against you? .... ☐ Yes ☒ No

56. Are there any judgments, suits, or claims pending against your partners? ☐ Yes ☒ No

57. Are you acting as corporate or personal surety for others? .... ☐ Yes ☒ No

58. Are you acting as endorser for others on their notes or accounts? .... ☐ Yes ☒ No

59. Does your company or any officer or partner owe any money to a bonding company? .... ☐ Yes ☒ No

60. Has your company or any officer or partner ever required any financial assistance or borrowed any money from a bonding company? .... ☐ Yes ☒ No

The above answers are true to the best of my knowledge and belief.

Completed by: J. Prometheus

Title: President Date: ____________

HERCULES & SMITH

Certified Public Accounts
1 State Street Plaza
Anytown, USA

INDEPENDENT AUDITOR'S REPORT

Members
American Institute of CPA's
Division For Securities
And Exchange Commission
Practice Section And
Private Companies Practice Section

Board of Directors and Stockholders
Prometheus Construction Co., Inc.
Anytown, USA

We have audited the accompanying balance sheet of Prometheus Construction Co., Inc. as of September 30, 1993, and the related statements of operations, stockholders' equity, and cash flows for the year then ended. These financial statements are the responsibility of the Company's management. Our responsibility is to express an opinion on these financial statements based on our audit.

We conducted our audit in accordance with generally accepted auditing standards. Those standards require that we plan and perform the audit to obtain reasonable assurance about whether the financial statements are free of material misstatement. An audit includes examining, on a test basis, evidence supporting the amounts and disclosures in the financial statements. An audit also includes assessing the accounting principles used and significant estimates made by management, as well as evaluating the overall financial statement presentation. We believe that our audit provides a reasonable basis for our opinion.

In our opinion, the financial statements referred to above present fairly, in all material respects, the financial position of Prometheus Construction Co., Inc. as of September 30, 1993, and the results of its operations and its cash flows for the year then ended in conformity with generally accepted accounting principles.

Our audit was made for purposes of forming an opinion on the basic financial statements, taken as a whole. The supplementary schedules on pages 10-13 are presented for purposes of additional analysis and are not a required part of the basic financial statements. Such information has been subjected to the auditing procedures applied in he audit of the basic financial statements, and in our opinion is fairly presented in all material respects in relation to the basic financial statements as a whole.

Anytown, USA
October 30, 1993

Hercules & Smith

Financial Statements

PROMETHEUS CONSTRUCTION CO., INC.

September, 30, 1993

BALANCE SHEET

PROMETHEUS CONSTRUCTION CO., INC.

September 30, 1993

ASSETS

CURRENT ASSETS	
Cash and cash equivalents	$ 724,893
Contracts receivable – Note B	3,764,797
Accrued interest receivable	92,467
Costs and estimated earnings in excess of billings on uncompleted contracts – Note C	106,150
Income tax refund receivable – Note A	57,549
Prepaid expenses and other	88,334
TOTAL CURRENT ASSETS	4,834,190
PROPERTY AND EQUIPMENT – Notes A and D	
Buildings and improvements 204,901	
Machinery and equipment	296,434
Furniture and fixtures	53,650
Vehicles and trailers	187,542
	742,527
	456,207
Less accumulated depreciation	286,320
OTHER ASSETS	
Notes receivable from stockholders (including accrued interest of $24,690)	215,915
Employee accounts receivable	47,815
	263,730
TOTAL ASSETS	$ 5,384,240

LIABILITIES AND STOCKHOLDERS' EQUITY

CURRENT LIABILITIES	
Accounts payable – Note E	$ 3,338,914
Accrued liabilities	38,515
Billings in excess of costs and estimated earnings on uncompleted contracts – Note C	374,652
Current maturities of long-term debt – Note D	25,393
TOTAL CURRENT LIABILITIES	3,777,474
LONG-TERM DEBT, less current portion – Note D	138,990
CONTINGENCIES – Note F	—
STOCKHOLDERS' EQUITY	
Common stock, no par value, 2,000 shares authorized, 1,660 shares issued and outstanding	16,600
Retained earnings	1,451,176
	1,467,776
TOTAL LIABILITIES & EQUITY	$ 5,384,240

See accompanying notes to financial statements

STATEMENT OF OPERATIONS

PROMETHEUS CONSTRUCTION CO., INC.

Year Ended September 30, 1993

Contract revenues earned	$ 15,840,529
Cost of revenues earned	15,492,609
Gross profit	347,920
General and administrative expense	597,912
Loss from operations	(249,992)
Other income (expense)	
Gain on sale of fixed assets	15,738
Interest income	204,962
Interest expense	(38,040)
Bad debt expense	(1,135)
	181,525
NET LOSS	$ (68,467)

See accompanying notes to financial statements

STATEMENT OF CHANGES IN STOCKHOLDERS' EQUITY

PROMETHEUS CONSTRUCTION CO., INC.

For the Year Ended September 30, 1993

	Common Stock		
	Number of Shares	Amount	Retained Earnings
BALANCE, SEPTEMBER 30, 1991	1,660	$16,600	$1,391,774
Distributions to stockholders	–	–	(144,000)
Net income	–	–	349,143
BALANCE, SEPTEMBER 30, 1992	1,660	16,600	1,596,917
Distribution to stockholders	–	–	(77,274)
Net loss	–	–	(68,467)
BALANCE, SEPTEMBER 30, 1993	1,660	$16,600	$1,451,176

See accompanying notes to financial statements

STATEMENT OF CASH FLOWS

PROMETHEUS CONSTRUCTION CO., INC.

Year Ended September 30, 1993

CASH FLOWS FROM OPERATING ACTIVITIES		
Net loss	$	(68,467)
Adjustments to reconcile net income loss to net cash used in operating activities		
Depreciation		82,443
Gain on sale of fixed assets		(15,738)
Changes in assets and liabilities:		
Contracts receivable		1,377,207
Costs and estimated earnings in excess of billings on uncompleted contracts		69,752
Income tax refund receivable		(57,549)
Prepaid expenses and other current assets		(135,333)
Accounts receivable from related parties		(139,086)
Accounts payable (1,707,910)		
Accrued income taxes		(59,249)
Accrued liabilities		(23,971)
Accrued dividends		(144,000)
Billings in excess of costs and estimated earnings on uncompleted contracts		106,771
TOTAL ADJUSTMENTS		(646,663)
Net cash used in operating activities		(715,130)
CASH FLOWS FROM INVESTING ACTIVITIES		
Purchase of fixed assets		(30,921)
Proceeds from sale of fixed assets		26,719
Proceeds from refunded deposit		18,310
Net cash provided from investing activities		14,738

Continued.

STATEMENTS OF CASH FLOWS – CONTINUED

PROMETHEUS CONSTRUCTION CO., INC.

Year Ended September 30, 1993

CASH FLOWS FROM FINANCING ACTIVITIES		
Payments on long-term debt	$	(47,556)
Distributions to stockholders		(77,274)
Net cash used in financing activities		(124,830)
NET DECREASE IN CASH		(855,222)
Cash and cash equivalents at beginning of year		1,550,115
Cash and cash equivalents at end of year	$	724,893

See accompanying notes to financial statements

NOTES TO FINANCIAL STATEMENTS

PROMETHEUS CONSTRUCTION CO., INC.

September 30, 1993

NOTE A – SIGNIFICANT ACCOUNTING POLICIES

Company's Activities: The Company is engaged in the construction of industrial and commercial buildings. The work is performed primarily under time-and-material, and fixed-price contracts modified by incentive and penalty provisions.

Revenue and Cost Recognition: Revenues from fixed-price and modified fixed-price contracts are recognized on the percentage-of-completion method, measured by the percentage of costs incurred to date to the estimated total costs for each contract. This method is used because management considers expended costs to be the best available measure of progress on these contracts. Revenues from time-and-material contracts are recognized currently as the work is performed.

Contract costs include all direct material and labor costs and those indirect costs related to contract performance, such as indirect labor, supplies, tools, and repair costs. General and administrative costs are charged to expense as incurred. Provisions for estimated losses on uncompleted contracts are made in the period in which such losses are determined. Changes in job performance, job conditions, and estimated profitability, including those arising from contract penalty provisions, and final contract settlements may result in revisions to costs and income and are recognized in the period in which the revisions are determined. Profit incentives are included in revenues when their realization is reasonably assured. An amount equal to contract costs attributable to claims is included in revenues when realization is probable and the amount can be reliably estimated.

Cash and Cash Equivalents: For purposes of the statement of cash flows, the Company considers all certificates of deposit to be cash equivalents. Interest paid during the year amounted to $38,040.

Property and Equipment: The Company records the purchase of property and equipment at cost, and maintenance and repairs are expensed as incurred. Depreciation, computed on the straight-line and accelerated methods, is based on the estimated useful lives of the assets.

NOTES TO FINANCIAL STATEMENTS – CONTINUED

PROMETHEUS CONSTRUCTION CO., INC.

September 30, 1993

NOTE A – SIGNIFICANT ACCOUNTING POLICIES – CONTINUED

Income Taxes: Effective October 1, 1991, management of the Company elected to be taxed as an S Corporation as defined in the Internal Revenue Code. Accordingly, no federal or state income taxes are paid by the Company since the net income or loss is reported by the stockholders in their individual income tax returns. The income tax refund receivable at September 30, 1993 represents an overpayment of taxes on income earned prior to September 30, 1992.

NOTE B – CONTRACT RECEIVABLES

Contract receivables	
Billed	
Completed contracts	$ 289,188
Contracts in progress	1,178,371
Retainage	2,205,967
	$ 3,673,526

NOTE C – COSTS AND ESTIMATED EARNINGS ON UNCOMPLETED CONTRACTS

Cost incurred on uncompleted contracts	$ 20,331,658
Estimated earnings	613,076
	20,944,734
Less billings to date	21,213,236
	$ (268,502)

Included in accompanying balance sheets under the following captions:

Costs and estimated earnings in excess of billings on uncompleted contracts	$ 106,150
Billings in excess of costs and estimated earnings on uncompleted contracts	(374,652)
	$ (268,502)

NOTES TO FINANCIAL STATEMENTS – CONTINUED

PROMETHEUS CONSTRUCTION CO., INC.

September 30,1993

NOTE D – LONG-TERM DEBT

Long-Term debt consists of the following:

10.5% to 13% notes payable to bank, due in monthly installments of $ 2,738 including interest, through February 1995, collateralized by certain trucks	$	32,842
11.8% mortgage note payable to bank due in monthly installments of $ 1,385 including interest through 2009, collateralized by land and building		119,649
5% note payable to finance company, due in monthly installments of $621, including interest, through May 1995, collateralized by certain equipment		11,892
		164,383
Less current portion		25,393
	$	138,990

As of September 30, 1993 long-term debt matures as follows:

1994 current portion	$	25,393
1995		21,185
1996		6,912
1997		3,681
1998 and thereafter		107,212
	$	164,383

NOTE E – ACCOUNTS PAYABLE

Accounts payable include amounts due to subcontractors totaling $ 1,830,068 which have been retained pending completion and customer acceptance of jobs.

NOTES TO FINANCIAL STAEMENTS – CONTINUED

PROMETHEUS CONSTRUCTION CO., INC.

September 30, 1993

NOTE F – CONTINGENCIES

The Company is a party to various lawsuits, claims, and other legal actions arising in the ordinary course of business. In the opinion of management, all such matters are without merit or are of such kind, or involve such amounts, that unfavorable disposition would not have a material effect on the financial position of the Company.

NOTE G – CONTRACT BACKLOG

The following schedule shows a reconciliation of backlog representing signed contracts in existence at September 30, 1993:

Balance at October 1, 1992	$ 9,702,283
Contracts adjustments	320,972
New contracts in 1993	10,518,486
	20,541,741
Less contract revenues earned in 1993	15,840,529
Balance at September 30, 1993	$ 4,701,212

SCHEDULE OF GENERAL AND ADMINISTRATIVE EXPENSE

PROMETHEUS CONSTRUCTION CO., INC.

Year Ended September 30, 1993

	Total	Louisville	Lexington
Salaries–officers	$ 116,191	$ 116,191	$ –
Salaries–others	153,089	114,411	38,678
Depreciation	15,219	15,219	–
Insurance	18,796	17,403	1,393
Office expense	27,268	24,912	2,356
Repairs and maintenance	14,880	13,030	1,850
Telephone	23,456	18,662	4,794
Advertising and promotion	4,772	4,772	–
Taxes and licenses	66,459	62,247	4,212
Dues and subscriptions	10,125	9,209	916
Travel and entertainment	8,112	8,112	–
Professional fees	61,735	61,735	–
Automobile expense	9,460	7,421	2,039
Utilities	6,514	5,445	1,069
Contributions	1,628	1,628	–
Computer services	16,693	16,693	–
Rent	10,815	–	10,815
Bonuses	32,700	32,700	–
	$ 597,912	$ 529,790	$ 68,122

SCHEDULE OF EARNINGS FROM CONTRACTS

PROMETHEUS CONSTRUCTION CO., INC.

Year Ended September 30, 1993

Contracts completed during the year ended September 30, 1993	$ 2,556,089	$ 2,336,880	$ 229,209
Contracts in progress at September 30, 1993	13,274,440	13,013,554	260,886
Depreciation-construction equipment	–	67,224	(67,224)
Call-backs and adjustments	–	74,951	(74,951)
	$15,840,529	$15,492,609	$ 374,920

SCHEDULE OF CONTRACTS COMPLETED
PROMETHEUS CONSTRUCTION COMPANY, INC.
Year Ended September 30, 1993

Contract		Contract Totals			Before October 1, 1993			During the Year Ended Sep. 30, 1993		
Number	Type	Revenues Earned	Cost of Revenues	Gross Profit	Revenues Earned	Cost of Revenues	Gross Profit	Revenues Earned	Cost of Revenues	Gross Profit (Loss)
RC-133	A	$ 1,835,982	$ 1,773,267	$ 112,715	$ 1,693,901	$ 1,576,916	$ 116,985	$ 192,081	$ 196,351	$ (4,270)
RC-139	A	486,570	445,910	40,660	414,333	387,373	26,290	72,237	58,537	13,700
RC-141	A	515,197	484,841	30,356	245,896	227,686	18,210	269,301	257,155	12,146
RC-142	A	550,624	522,362	28,262	291,456	276,871	14,585	259,168	245,491	13,677
RC-143	A	724,458	657,969	66,489	435,538	395,457	40,081	288,920	262,512	26,408
RC-145	A	664,055	564,019	100,036	176,383	153,387	22,996	487,672	410,632	77,040
RC-147	A	946,751	860,890	85,861	—	—	—	946,751	860,890	85,861
Small Contracts		49,659	45,312	4,647	—	—	—	49,959	45,312	4,647
		$ 5,823,596	$ 5,345,570	$ 469,026	$ 3,257,507	$ 3,017,690	$ 239,817	$ 2,566,089	$2,2336,880	$ 229,209

Contract Type
A—Fixed-price

SCHEDULE OF CONTRACTS-IN-PROGRESS
PROMETHEUS CONSTRUCTION COMPANY, INC.
September 30, 1993

Contract		Estimated Total Contract		Year Ended September 30, 1993			From Inception to September 30, 1993					
Number	Type	Revenues	Gross Profit	Revenues Earned	Cost	Gross Profit (Loss)	Revenue	Cost	Gross Profit	Billings to Date	Under (Over) Billed	Backlog
RC-137	A	$ 9,508,048	$215,741	$ 4,451,909	$ 4,451,196	$ 713	$ 9,495,611	$ 9,280,155	$215,456	$ 9,467,209	$ 28,402	$ 12,437
RC-138	A	6,616,122	23,431	3,878,419	3,992,827	(114,408)	6,505,011	6,481,972	23,039	6,555,927	(50,916)	111,111
RC-146	A	1,083,771	94,857	988,920	902,365	86,555	988,920	902,365	86,555	1,012,765	(23,845)	94,851
RC-148	A	4,117,869	446,816	1,723,068	1,536,104	186,964	1,723,068	1,536,104	186,964	2,022,959	(299,891)	2,394,801
RC-149	A	4,320,136	195,599	2,232,124	2,131,062	101,062	2,232,124	2,131,062	101,062	2,154,376	77,748	2,088,012
		$ 25,645,946	$976,444	$ 13,274,440	$ 13,013,554	$260,886	$ 20,944,734	$ 20,331,658	$613,076	$ 21,213,236	$ (286,502)	$ 4,701,212

Contract Type
A—Fixed-price

Costs and estimated earnings in excess of billings on uncompleted projects:	$ 106,150
Billings in excess of costs and estimated earnings on uncompleted projects	(374.652)
	$ (268,502)

Contracts signed subsequent to September 30,1993

$19,400,000 (Unaudited)

HERCULES & SMITH

Certified Public Accounts
1 State Street Plaza
Anytown, USA

INDEPENDENT AUDITOR'S REPORT

Members
American Institute of CPA's
Division For Securities
And Exchange Commission
Practice Section And
Private Companies Practice Section

Board of Directors and Stockholders
Prometheus Construction Co., Inc.
Anytown, USA

We have audited the accompanying balance sheets of Prometheus Construction Co., Inc. (an S Corporation) as of September 30, 1994 and 1993, and the related statements of operations, stockholders' equity, and cash flows for the years then ended. These financial statements are the responsibility of the Company's management. Our responsibility is to express an opinion on these financial statements based on our audit.

We conducted our audit in accordance with generally accepted auditing standards. Those standards require that we plan and perform the audit to obtain reasonable assurance about whether the financial statements are free of material misstatement. An audit includes examining, on a test basis, evidence supporting the amounts and disclosures in the financial statements. An audit also includes assessing the accounting principles used and significant estimates made by management, as well as evaluating the overall financial statement presentation. We believe that our audit provides a reasonable basis for our opinion.

In our opinion, the financial statements referred to above present fairly, in all material respects, the financial position of Prometheus Construction Co., Inc. as of September 30, 1994 and 1993, and the results of its operations and its cash flows for the years then ended in conformity with generally accepted accounting principles.

Our audit was made for purposes of forming an opinion on the basic financial statements, taken as a whole. The supplementary schedules on pages 10-13 are presented for purposes of additional analysis and are not a required part of the basic financial statements. Such information has been subjected to the auditing procedures applied in he audit of the basic financial statements, and in our opinion is fairly presented in all material respects in relation to the basic financial statements as a whole.

Anytown, USA
October 30, 1994

Hercules & Smith

BALANCE SHEET

PROMETHEUS CONSTRUCTION CO., INC.

	September 30 1994	1993
ASSETS		
CURRENT ASSETS		
Cash and cash equivalents	$ 604,933	$ 724,893
Contracts receivable–Note B	3,955,984	3,673,526
Accrued interest receivable	9,093	92,467
Other receivables	84,921	—
Costs and estimated earnings in excess of billings on uncompleted contracts–Note C	432,527	106,150
Income tax refund receivable–Note A	—	57,549
Prepaid expenses and other	101,802	88,234
TOTAL CURRENT ASSETS	5,189,260	4,742,819
PROPERTY AND EQUIPMENT–Notes A and D		
Buildings and improvements	204,901	204,901
Machinery and equipment	320,865	296,434
Furniture and fixtures	53,315	53,650
Vehicles and trailers	217,949	187,542
	797,030	742,527
Less accumulated depreciation	459,701	456,207
	337,329	286,320
OTHER ASSETS		
Notes receivable from stockholders (including accrued interest of $38,027 in 1994 and $24,690 in 1993	320,523	307,186
Accounts receivable from related parties	40,263	47,815
	360,786	355,001
TOTAL ASSETS	$5,887,375	$5,384,240

	September 30	
	1994	1993
LIABILITIES AND STOCKHOLDERS' EQUITY		
CURRENT LIABILITIES		
Accounts payable–Note E	$ 3,811,318	$ 3,338,914
Accrued liabilities	85,741	38,515
Billings in excess of costs and estimated earnings on uncompleted contracts–Note C	94,134	374,652
Current maturities of long-term debt–Note D	89,461	25,393
Accrued distributions	116,547	77,274
TOTAL CURRENT LIABILITIES	4,197,201	3,777,474
LONG-TERM DEBT, less current portion–Note D	117,347	138,990
CONTINGENCIES–Note F	—	—
STOCKHOLDERS' EQUITY		
Common stock, no par value, 2,000 shares authorized, 1,660 shares issued and outstanding	16,600	16,600
Retained earnings	1,556,227	1,451,176
	1,572,827	1,467,776
TOTAL LIABILITIES & EQUITY	$ 5,887,375	$5,384,240

See accompanying notes to financial statements

STATEMENT OF OPERATIONS

PROMETHEUS CONSTRUCTION CO., INC.

	Year Ended September 30	
	1994	1993
Contract revenues earned	$ 18,899,903	$ 15,840,529
Cost of revenues earned	18,145,934	15,492,609
Gross profit	753,969	347,920
General and administrative expense	588,250	597,912
Loss from operations	165,719	(249,992)
Other income (expense)		
Gain on sale of fixed assets	27,426	15,738
Interest income	161,066	204,962
Other Income	22,001	—
Interest expense	(24,759)	(38,040)
Bad debt expense	129,855)	(1,135)
	55,879	181,525
NET GAIN (LOSS)	$ 221,598	$ (68,467)

See accompanying notes to financial statements

STATEMENT OF CASH FLOWS

PROMETHEUS CONSTRUCTION CO., INC.

	Year Ended September 30	
	1994	1993
CASH FLOWS FROM OPERATING ACTIVITIES		
Net income (loss)	$ 221,598	$ (68,467)
Adjustments to reconcile net income loss to net cash used in operating activities		
Depreciation	85,752	82,443
Gain on sale of fixed assets	(27,426)	(15,738)
Changes in assets and liabilities:		
Contracts receivable	(282,458)	1,377,207
Costs and estimated earnings in excess of billings on uncompleted contracts	(326,377)	69,752
Income tax refund receivable	57,549	(57,549)
Prepaid expenses and other current assets	(7,463)	(135,333)
Accounts receivable from related parties	(13,337)	(139,086)
Accounts payable	549,678	(1,707,910)
Accrued income taxes	–	(59,249)
Accrued liabilities	47,226	(23,971)
Accrued dividends	39,273	(144,000)
Billings in excess of costs and estimated earnings on uncompleted contracts	(280,518)	106,771
TOTAL ADJUSTMENTS	(158,101)	(646,663)
Net cash used in operating activities	63,497	(715,130)
CASH FLOWS FROM INVESTING ACTIVITIES		
Purchase of fixed assets	(154,985)	(30,921)
Proceeds from sale of fixed assets	45,650	26,719
Proceeds from refunded deposit	–	18,310
Net cash provided from investing activities	(109,335)	14,738

Continued.

STATEMENTS OF CASH FLOWS – CONTINUED

PROMETHEUS CONSTRUCTION CO., INC.

	Year Ended September 30	
	1994	1993
CASH FLOWS FROM FINANCING ACTIVITIES		
Proceeds from long-term borrowings	$ 70,000	$ –
Payments on long-term debt	(27,575)	(47,556)
Distributions to stockholders	(116,547)	(77,274)
Net cash used in financing activities	(74,122)	(124,830)
NET DECREASE IN CASH	(119,960)	(855,222)
Cash and cash equivalents at beginning of year	724,893	1,550,115
Cash and cash equivalents at end of year	$ 604,933	$ 724,893

See accompanying notes to financial statements

STATEMENT OF CHANGES IN STOCKHOLDERS' EQUITY

PROMETHEUS CONSTRUCTION CO., INC.

For the Year Ended September 30, 1993

	Common Stock		
	Number of Shares	Amount	Retained Earnings
BALANCE, SEPT. 30, 1992	1,660	$16,600	$1,391,774
Distributions to stockholders	—	—	(77,274)
Net income	—	—	68,467
BALANCE, SEPT. 30, 1993	1,660	16,600	1,451,176
Distribution to stockholders	—	—	(116,574)
Net loss	—	—	221,598
BALANCE, SEPT. 30, 1994	1,660	$16,660	$1,556,227

See accompanying notes to financial statements

NOTES TO FINANCIAL STATEMENTS
PROMETHEUS CONSTRUCTION CO., INC.
September 30, 1993

NOTE A – SIGNIFICANT ACCOUNTING POLICIES

Company's Activities: The Company is engaged in the construction of industrial and commercial buildings. The work is performed primarily under time-and-material, and fixed-price contracts modified by incentive and penalty provisions.

Revenue and Cost Recognition: Revenues from fixed-price and modified fixed-price contracts are recognized on the percentage-of-completion method, measured by the percentage of costs incurred to date to the estimated total costs for each contract. This method is used because management considers expended costs to be the best available measure of progress on these contracts. Revenues from time-and-material contracts are recognized currently as the work is performed.

Contract costs include all direct material and labor costs and those indirect costs related to contract performance, such as indirect labor, supplies, tools, and repair costs. General and administrative costs are charged to expense as incurred. Provisions for estimated losses on uncompleted contracts are made in the period in which such losses are determined. Changes in job performance, job conditions, and estimated profitability, including those arising from contract penalty provisions, and final contract settlements may result in revisions to costs and income and are recognized in the period in which the revisions are determined. Profit incentives are included in revenues when their realization is reasonably assured. An amount equal to contract costs attributable to claims is included in revenues when realization is probable and the amount can be reliably estimated.

Cash and Cash Equivalents: For purposes of the statement of cash flows, the Company considers all certificates of deposit to be cash equivalents. Interest paid during the year amounted to $24,176, in 1994 and $38,040 in 1993.

Property and Equipment: The Company records the purchase of property and equipment at cost, and maintenance and repairs are expensed as incurred. Depreciation, computed on the straight-line and accelerated methods, is based on the estimated useful lives of the assets.

Income Taxes: Effective October 1, 1991, management of the Company elected to be taxed as an S Corporation as defined in the Internal Revenue Code. Accordingly, no federal or state income taxes are paid by the Company since the net income or loss is reported by the stockholders in their individual income tax returns.

NOTE A – SIGNIFICANT ACCOUNTING POLICIES – CONTINUED

Income Taxes (Continued): The income tax refund receivable at September 30, 1993 represents an overpayment of taxes on income earned prior to September 30, 1992.

Reclassification: Certain reclassifications of amounts in the 1993 balance sheet have been made to conform with the 1994 presentation.

	September 30	
	1994	1993
NOTE B – CONTRACT RECEIVABLES		
Contract receivables		
Billed		
Completed contracts	$ 559,864	$ 289,188
Contracts in progress	2,281,106	1,178,371
Retainage	1,115,014	2,205,967
	$ 3,955,984	$ 3,373,526

NOTE C – COSTS AND ESTIMATED EARNINGS ON UNCOMPLETED CONTRACTS

	1994	1993
Cost incurred on uncompleted contracts	$ 13,302,376	$ 20,331,658
Estimated earnings	588,749	613,076
	13,891,125	20,944,734
Less billings to date	13,552,732	21,213,236
	$ 338,393	$ (268,502)

Included in accompanying balance sheets under the following captions:

	1994	1993
Costs and estimated earnings in excess of billings on uncompleted contracts	$ 432,527	$ 106,150
Billings in excess of costs and estimated earnings on uncompleted contracts	(94,134)	(374,652)
	$ 338,393	$ (268,502)

NOTES TO FINANCIAL STATEMENTS – CONTINUED

PROMETHEUS CONSTRUCTION CO., INC.

September 30, 1994 and 1993

NOTE D – LONG-TERM DEBT

Long – Term debt consists of the following:

	September 30 1994	September 30 1993
10.5% to13% notes payable to bank, due in monthly installments of $ 2,738 including interest, through February 1996, collateralized by certain trucks	$ 15,775	$ 32,842
11.8% mortgage note payable to bank due in monthly installments of $ 1,385 including interest through 2009, collateralized by land and building	116,758	119,649
5% note payable to finance company, due in monthly installments of $621, including interest, through May 1995, collateralized by certain equipment	70,000	11,892
	206,808	164,383
Less current portion	89,461	25,393
	$ 117,347	$ 138,990

As of September 30, 1993 long-term debt matures as follows:

1991 current portion	$ 89,461
1992	7,056
1993	3,821
1994	4,276
1995 and thereafter	102,194
	$ 206,808

NOTES TO FINANCIAL STATEMENTS – CONTINUED

PROMETHEUS CONSTRUCTION CO., INC.

September 30, 1994

NOTE E – ACCOUNTS PAYABLE

Accounts payable include amounts due to subcontractors which have been retained pending completion and customer acceptance of jobs totaling $1,260,812 and $1,830,068 at September 30, 1994 and 1993, respectively.

NOTE F – CONTINGENCIES

The Company is a party to various lawsuits, claims, and other legal actions arising in the ordinary course of business. In the opinion of management, all such matters are without merit or are of such kind, or involve such amounts, that unfavorable disposition would not have a material effect on the financial position of the Company.

NOTE G – CONTRACT BACKLOG

The following schedule shows a reconciliation of backlog representing signed contracts in existence at September 30, 1994:

Balance at October 1, 1993	$ 4,701,212
Contracts adjustments	108,691
New contracts in 1994	19,839,877
	24,649,780
Less contract revenues earned in 1994	18,899,903
Balance at September 30, 1994	$ 5,749,877

SCHEDULE OF GENERAL AND ADMINISTRATIVE EXPENSE

PROMETHEUS CONSTRUCTION CO., INC.

Year Ended September 30, 1994

	1994	1993
Salaries – officers	$ 116,740	$ 116,191
Salaries – others	112,144	158,089
Depreciation	13,400	15,219
Insurance	15,296	18,796
Office expense	20,840	27,268
Repairs and maintenance	16,372	14,880
Telephone	16,635	23,456
Advertising and promotion	3,818	4,772
Taxes and licenses	33,162	66,459
Dues and subscriptions	9,443	10,125
Travel and entertainment	2,349	8,112
Professional fees	106,302	61,735
Automobile expense	5,388	9,460
Utilities	5,309	6,514
Contributions	1,263	1,628
Computer services	13,789	16,693
Bonuses	96,000	32,700
Rent	–	10,815
	$ 588,250	$ 597,912

SCHEDULE OF EARNINGS FROM CONTRACTS

PROMETHEUS CONSTRUCTION CO., INC.

Year Ended September 30, 1994

	1994			1993
	Contract Revenues Earned	Cost of Revenues Earned	Gross Profit (Loss)	Gross Profit (Loss)
Contracts completed during the year ended Sept. 30, 1994	$ 5,008,778	$ 4,710,890	$ 297,888	$ 229,209
Contracts in progress at Sept. 30,1993	13,891,125	13,302,376	588,749	260,886
Depreciation–construction equipment	–	72,352	(72,352)	(67,224)
Call-backs and Adjustments	–	60,316	(60,316)	(74,951)
	$18,899,903	$18,145,934	$ 753,969	$ 374,920

SCHEDULE OF CONTRACTS COMPLETED
PROMETHEUS CONSTRUCTION COMPANY, INC.
Year Ended September 30, 1994

Contract		Contract Totals			Before October 1, 1994			During the Year Ended Sep. 30, 1994		
Number	Type	Revenues Earned	Cost of Revenues	Gross Profit	Revenues Earned	Cost of Revenues	Gross Profit	Revenues Earned	Cost of Revenues	Gross Profit (Loss)
RC-137	A	$ 9,508,048	$ 9,283,885	$ 224,163	$ 9,495,611	$ 9,280,155	$ 215,456	$ 12,437	$ 3,730	$ 8,707
RC-138	A	6,589,936	6,566,791	23,145	6,505,011	6,481,972	23,039	84,925	84,819	106
RC-146	A	1,086,513	996,931	89,582	988,920	902,365	86,555	97,593	94,566	3,027
RC-148	A	4,182,233	3,733,882	448,351	1,723,068	1,536,104	186,964	2,459,165	2,197,778	261,387
RC-149	A	4,387,907	4,277,262	110,645	2,232,124	2,131,062	101,062	2,155,783	2,146,200	9,583
RC-155	A	87,630	82,887	4,743	—	—	—	87,630	82,887	4,743
Small Contracts		111,245	100,910	10,335	—	—	—	111,245	100,910	10,355
		$ 25,953,512	$ 25,042,548	$ 910,964	$ 20,944,734	$ 20,331,658	$ 613,076	$ 5.008,778	$ 4,710,890	$ 297,888

Contract Type
A—Fixed-price

SCHEDULE OF CONTRACTS-IN-PROGRESS
PROMETHEUS CONSTRUCTION COMPANY, INC.
September 30, 1994

Contract		Estimated Total Contract		Year Ended September 30, 1994			From Inception to September 30, 1994					
Number	Type	Revenues	Gross Profit	Revenues Earned	Cost	Gross Profit (Loss)	Revenue	Cost	Gross Profit	Billings to Date	Under (Over) Billed	Backlog
RC-150	A	$ 403,709	$ 61,013	$ 24,868	$ 21,110	$ 3,758	$ 24,868	$ 21,110	$ 3,758	$ 12,000	$ 12,868	$ 378,841
RC-151	A	4,392,665	183,550	3,282,037	3,144,496	137,451	3,282,037	3,144,496	137,541	3,256,394	25,643	1,110,628
RC-152	A	5,367,520	244,218	4,128,321	3,940,092	188,229	4,128,321	3,940,092	188,229	3,938,992	189,329	1,239,199
RC-153	A	5,170,647	222,053	3,331,841	3,188,267	143,574	3,331,841	3,188,267	143,574	3,425,975	(94,134)	1,838,806
RC-154	A	4,306,461	158,923	3,124,058	3,008,411	115,647	3,124,058	3,008,411	115,647	2,919,371	204,687	1,182,403
		$ 19,641,002	$869,927	$ 13,891,125	$ 13,302,376	$588,749	$ 13,891,125	$ 13,302,376	$ 588,749	$ 13,522,732	$ 338,393	$ 5,749,877

Contract Type
A—Fixed-price

Costs and estimated earnings in excess of billings on uncompleted projects:	$ 423,527
Billings in excess of costs and estimated earnings on uncompleted projects	(94.134)
	$ 338,393

Contracts signed subsequent to September 30,1994 **$15,400,000 (Unaudited)**

Appendix XIII. Effective Claim Management: Clem Winter, C. Winter Enterprises, Inc.

EFFECTIVE CLAIM MANAGEMENT: INVESTIGATION—MANAGEMENT—SALVAGE

Contract Surety Losses

With a contract surety loss, it's the unknown one fears, not the known! The goal of the surety with notice of a contractor's inability to continue its performance is to bring about a termination of its liability in a manner that will 1) discharge its obligations without exposing it to liability beyond the penal sum of its bonds, and 2) minimize its financial loss and consider and secure its salvage opportunities.

Principals in desperate financial conditions do desperate things, seldom to their own benefit, and NEVER to the benefit of the surety. There is no question about it, upon notice of a claim today's surety has a duty to conduct a reasonable investigation. Most states have adopted a policy known as "The Fair Claim Practice Act" that requires companies to investigate.

On a contract surety loss, one must determine the exposure (per job and overall) with the best information available. Reinsurers' studies have found the exposure range is 30% to 40% of the uncompleted work (remaining contract balance) and the exposure range would include the accounts payable. Thus with $1 million in remaining balance to earn on contract, it's suggested an exposure of $300,000 to $400,000. This is, however, a rule of thumb. The surety's decision on what to do when faced with a default must be based upon a solid foundation of facts and figures; a foundation available only through comprehensive and detailed fact gathering. Surety bond claim personnel in charge of that effort must carefully marshal and then analyze a great volume of critical information, which is a most difficult and demanding task. The sources of this information are divided into two classes: 1) people and 2) paper. Using experience and common sense, the claim person must reach a conclusion or set in place a course of action.

Generally, the claim personnel will use a report format to quantify the surety's exposure and evaluate the loss potential. In this context, a report is defined as the setting out, outlining, reviewing, and analyzing of a course of action and/or recommendations based upon given information forming a clear and factual picture of the claim situation. This can only be achieved after an investigation has been completed, and the following attempts to outline this process.

Investigation

With the goal a concise and clear report outlining the options and best course of action for the surety, the claim personnel must scrutinize contractor documents, job sites, and personnel. When investigating a contract surety loss, there are a few areas that time and again must be studied, understood, and discussed before being summarized in a report. Together, these pieces of information can be found on the following "Contract Surety Claim Checklist."

Contract Surety Claim Checklist

1) **Contract—per project**
   - A) Total contract amount—obligee—location
   - B) Change orders approved—pending—date—amounts
   - C) Last paid estimate & amount—balance to draw
   - D) Liquidated damage per day—pending and approved extensions
   - E) Completion date
   - F) Retainage earned to date
   - G) Percentage completed to date—amount
   - H) Percentage completed to date—time

2) **Cost-to-complete**
   - A) Cost of material (less material on job site)
   - B) Cost of labor—principal's
   - C) Cost of equipment—principal's
   - D) Rental of equipment—principal's
   - E) Trucking
   - F) Subcontractor's estimate to complete
   - G) Insurance cost—direct
   - H) Leases
   - I) Outstanding payroll
   - J) Field overhead
   - K) Living expenses and travel
   - L) Royalties

   2-A Subcontractors
   - A) Subcontractor
   - B) Line item of work (general's price)
   - C) Contract amount
   - D) Change orders—approved—pending and unapproved
   - E) Date of last paid estimate and amount—outstanding estimates
   - F) Bonded subcontractors—sureties
   - G) Retainage earned to date

   2-B Truckers
   - a)Trucker and date of contract
   - b) Amount per hour, per ton
   - c) Copy of contract

   2-C Outstanding purchase orders
   - a) Date
   - b) Purchase order number—supplier
   - c) Amount

   2-D Overhead
   - a) Shop personnel—estimated cost
   - b) Office personnel—estimated cost
   - c) Repairs on equipment to complete
   - d) Office equipment—rentals
   - e) Payable union dues—last payment
   - f) Office rent, electricity, gas, telephone—cost estimate
   - g) Miscellaneous expenses—expense reports

3) **Accounts payable**

A) Supplier
B) Amount—invoice number—statement amount—date
C) Unsecured notes—payer—amount—date
D) General ledger balance—accounts payable—date
E) Amounts due subcontractors—retainage due
F) The new surety bond form AIA 312 requires payment of undisputed claims within 45 days.

3-A Taxes

a) Federal withholding taxes—filed—paid—quarter
b) State unemployment taxes—filed—paid—quarter
c) Federal unemployment taxes—filed—paid
d) Current payable taxes
c) Contractor's first loan is non-payment on withholding taxes

3-B Bank accounts

a) Outstanding checks
b) Bank balance
c) Date of last reconciliation
d) Copy of bank statement
e) All issued checks mailed

3-C Outstanding Notes

a) Date of note
b) Amount of note
c) Secured by
d) Due date
e) Amount of payment
f) Payment
g) Interest date
h) Notes due banks—amount—due date—outstanding balance

4) **Insurance**

A) Workmen's compensation policy limit expiration date amount of deposit
B) Public liability policy limit—expiration date—payment gate—amount
C) Equipment floater policy limit—expiration date—payment—amount
D) Copies of Policies' Coverage
E) Amount due agent of broker

5) **Equipment**

A) List of all equipment serial numbers
B) Appraised value of each
C) Mortgage holder—amount of note—date—payment amount
D) Current balance on note
E) Insurance schedule of equipment and replacement cost

6) **Assets**

A) Property

a) Equipment
b) Automobiles
c) Real state—description—outstanding mortgage
d) Life insurance
e) Tax refunds

B) Indemnitors—names and addresses

a) Assets

b) Current financial statement(s)

7) **Miscellaneous (copies needed)**

A) Copy of bidding estimate

B) Copy of bid sheet on contract—obligee's tabulation—bidders' names, addresses, phone numbers

C) Copies of <u>all</u> estimates from owner and subcontractors

D) Copies of <u>all</u> subcontractors' and truckers' contracts

E) Copies of <u>all</u> leases on land and/or equipment

F) Copies of <u>all</u> notes and mortgages

G) Copy of contract, bond, indemnity agreement, and bond application

H) Copies of certificates of insurance on subcontractors and truckers

I) Copies of pertinent correspondence and letters of acceptance

J) Copies of financial statements

K) Copy of punch list from owners—architects

L) Copy of notification of indemnitors—date

8) **Additional Information Would Include:**

A) Unbonded projects—current status of each completed and uncompleted

B) Unbonded payables

C) Type of work

D) Principal's financial condition—(changes)

E) Indemnitor's financial condition—(changes)

F) Collateral available

G) Surety attorney and firm's assignment

H) Surety consultant and their assignment

I) Determine estimated overhead cost—it's the major factor in continuing with a contractor, followed by personnel, which may have to be carried through winter, shutdown, or rain.

First Priorities

Strengths and weaknesses are interesting in the surety business. The obligee's major strength is knowledge of the construction business. Their major weakness is their inexperience and lack of knowledge of what or how to handle a bond claim default. A seasoned bond claim person's biggest plus is their knowledge of the surety's rights and obligations. The bond claim person simply must have a set of real life experiences on which to draw.

To use an insurance expression, "They had an accident and they're going to bleed to death if they don't get some help." In the bonding industry, "help" is spelled M-O-N-E-Y! The only option available to a bonded principal with cash-flow problems is the bonding company—not the bank, not the government, not the obligee, and generally not the family. The surety is the only game in town for the bonded contractor in trouble.

As soon as possible, the surety's maximum performance exposure must be identified and controlled. The surety's payment exposure in most cases is known or can be obtained, but it's the performance exposure that costs money! When a contract claim arises, the surety must act rather than react. However, the first people to know there is a problem are, in order: the contractor, subcontractors, materialmen, architect, owner, bank, agent, and finally, the surety. Yet who must act first? The surety!

Any experienced and seasoned bond claim person's biggest plus is their knowledge of the surety obligations and their past experience to draw upon. They must know the facts, consider and understand the options, and then set in place a plan to limit the surety's exposure and to fulfill its bonded obligations. As a seasoned contract bond claim person, the first reaction might be to take no action at all until all problems are understood, including shutting all jobs down. However, the first priority is generally to obtain a written default notice. Without this document, a surety cannot act. Nonetheless, a surety has a duty to investigate notices from non-payment creditors as set out in AIA Bond 312 and follow up in performance with obligee on performance obligations. As soon as a surety learns of any trouble, the claim personnel will generally obtain 1) a formal letter of default, 2) a request for assistance from the contractor, or 3) a written notice from unpaid creditors. Exhibits A and B provide samples of such notification letters.

Surety Loss Reporting

It is customary for the surety to prepare a factual and narrative report on any contract bond loss when the potential exposure exceeds an agreed-upon fixed amount, as set by its management or the reinsurers. The form on the following page has been generally accepted as the standard reporting format for Contract Surety Claims. The standard narrative section of the report would review:

A) Contractor background
B) Agent
C) Overview of each project
D) Problems: project—others
E) Indemnity—parties
F) Assets available
G) Firms retained to assist
H) Suggested course of action
I) Options on unbonded projects.

Other items that may be reviewed in the narrative section of the report might include:

1) Taxes
2) Bank accounts
3) Outstanding notes and loans
4) Insurance
5) Equipment
6) Miscellaneous

Review and Analysis

When a review and analysis is set out in a report, the bottom line figure should be the total exposure, per project, and overall. Reserves are usually set in the same amount. An exception could be: 1) maximum cash flow needs, 2) anticipated claims to be receive, 3) sale of assets. Assets of the principal and/or indemnitor are not considered in the standard industry's reserving practice, but, if identified and secured, they might be considered in the overall loss or exposure.

Construction Contract Claim Report					
1	Principal:	Grand Total All Projects	Claim Number	Claim Number	Claim Number
2	Bond Number				
3	Obligee				
4	Job Number				
5	Project Description				
6	Performance Bond Penalty				
7	Payment Bond Penalty				
8	Combination Bond Penalty				
9	Bond Effective Date				
10	Percentage of Funds Earned				
11	Contractual Completion Date				
12	Anticipated Completion Date				
13	Liquidated Damages Per Day				
14	Original Contract Price				
15	Approved Change Orders (+/-)				
16	Revised Contract Price				
17	Amount Paid to Date				
18	Liquidated Damages to Date				
19	Balance in Contract				
20	Retainage				
21	Unearned Portion				
22	Estimate to Complete				
23	Sub-Contract				
24	Materials				
25	Labor				
26	Overhead				
27	Future Liquidated Damages				
28	Total Cost to Complete (Items 22 through 26)				
29	Unpaid Bills				
30	Total Completion Cost + Bills (27 + 28)				
31	Estimated Surplus/Deficit (18 – 29)				
32	Surety's Payments to Date				
33	Surety's Received to Date				
34	Surety's Net Loss to Date				
35	Total Estimated Loss (30 + 33)				
36	Cash Flow Anticipated				
37	Other Contingency				
38	Total Reserve Required				
39	Portion Reinsured				

Projects: 1.__________ 2.__________ 3.__________

All sureties have or should have a set of well-defined options. The important course of actions to consider are:

1) Funding/Financing
2) Take-Over and Relet
3) Relet and Walk Away
4) Protective Bid
5) Buy Bond Back
6) Do Nothing

1) Funding/Financing

A vexing question for claim personnel is often, "to finance or not to finance?" A decision to finance the contractor to complete the work and pay the creditors involves a much more critical judgment than a decision to use any of the other techniques. Treatment of this option first is not to suggest that it should be the first to consider using. To the surety, one might suggest that it should be the first option to consider eliminating!

Let's take a look at some of the pros and cons of financing considered before the final decision is made. Always remember that financing is in no way mandatory even though all signals appear to be go.

Arguments in favor:

1) Job momentum and continuity
2) Contractor's image
3) Additional bonds
4) Salvage made easier
5)Avoid "up price" of completion by another contractor
6) Status of work—substantial completion

Arguments against:

1) No reduction in bond penalty
2) Payment of non-bonded obligations
3) Uncertainty as to amount of loss
4) Status of work—barely commenced
5) Problem of tail-end let-downs
6) Time budget strain
7) Item joint ventures
8) Extensive unbonded work

If funding or financing (i.e., offering an extension of credit) is being considered for the principal, usually the reinsurer's prior approval is needed. When financing is being considered as the ultimate course of action, the information set forth in the reports should be provided to the reinsurers along with the request for their consent. The extension of credit without the reinsurer's consent can jeopardize treaty coverage.

Knowledge of the contractor's unbonded work would be needed when a funding arrangement is being considered; specifically, the number of unbonded projects, estimated total completion cost, and the total amount of unbonded payables. If funding or financial assistance is an option, the following methods should be reviewed and understood:

A) Guaranteed bank loan
B) Additional bonds for the principal in claim
C) Early payment of funds to or for the contractor during investigation—"look see money"
D) Funds to satisfy payment bond claims—"backdoor" financing

The other topics to be clearly understood:

1) All assets should be turned over to the surety-includes those of the principal and indemnitors.
2) The surety has the sole right to terminate any funding arrangement.
3) Will principal be in business and ongoing after completion of projects.
4) Status of all unbonded projects.
5) Status of bank loans.
6) Outstanding payroll taxes.
7) Ongoing (new) work.
8) Ongoing overhead projections.

2) Take-Over and Relet

A take-over and relet is the most commonly used surety claim option. It is usually the only technique that can be used to relet Federal Government contracts, and with some other obligees. The key to using this method is a take-over agreement acceptable to the surety and a carefully negotiated and drafted relet contract with the completing contractor.

The take-over and relet approach is much more cumbersome because the surety steps into the shoes of its defaulted principal. The surety receives payments from the obligee under the original pay estimate breakdown and pays the completing contractor on the basis of their pay estimate earnings. Generally, the surety appoints the completing contractor to act as its representative at the job site, but it is practically impossible to avoid becoming involved in some disputes using this course of action. Remember: All agreements should have an off-ramp, if the surety finds that the plan should be aborted for some reason. Also, for corrective work, the surety should obtain separate prices and negotiate any up front agreement.

3) Relet and Walk Away

Relet and walk away is the surety's best approach in most cases because it is less burdensome but usually is hard to obtain an agreeable understanding with all parties. For projects under 50% complete, reletting is a top consideration.

Reletting allows for the surety to tender a new contractor and new bonds to the obligee to compete for an agreed-upon contract price. Whatever the negotiated upset price is, the surety can pay that amount as soon as the completing contractor signs a contract with the obligee and furnishes the required bonds. At this point, the surety has set its completion loss and cost. Negotiations of the terms of the completion contract and price may include some consideration for a discount of price over contract funds for up front payment by the surety versus spreading additional payments by the surety over the length of the project. In addition, the surety usually will have to pay the defaulted contractor's payment bond claims.

4) Protective Bid

This approach is designed to limit the surety's loss when the surety is unable to get a satisfactory take-over agreement, or when the obligee is required by statute or inexperienced counsel to rebid the project in a default situation. It's also an approach to consider for use as a means of fixing the loss in situations where there is a dispute between the owner and the contractor, and maybe the surety decides it should not side with the owner. The surety negotiates an agreement with the bidding contractor to submit a mutually acceptable bid supported by the required bonds. The surety then has established its completion loss, whether or not the negotiated bid is the low bid.

5) Buy Bond Back

This allows for a reasonable cash settlement based upon the various claims, counterclaims, and defenses between the parties. Reasons for choosing this option include:

A) Evidence that the obligee has materially breached the contract.
B) Evidence that the obligee has materially changed the nature or quantity of the work that the original contract ceases to exist.
C) Evidence that the obligee has prematurely paid the principal to the prejudice of the surety.
D) When the loss obviously exceeds the amount of the bond.
E) When the obligee for some reason would rather settle for cash immediately. This approach should be considered on subcontract bonds where the obligee is a prime contractor.

6) Do Nothing

The use of this seemingly simple approach is the most difficult decision a surety has to make. It should be considered when:

A) Indemnitors are induced to take-over and complete the work and pay the bills to minimize their loss and experience.
B) Adequate financing can be quickly arranged by the principal with the aid of the other indemnitors or friends without the surety's assistance or support.
C) A bank is in so deep it can't let go and decides to support the principal with additional financing to recoup a severe loss to avoid default.
D) The obligee and principal actually enter into and commence performance of an entirely different contract than the one the surety bonded.
E) The owner and/or the contractor are different entities than those shown on the bond.
F) The surety receives a trouble notice from someone and its investigation reveals the principal is financially sound and able to handle its own affairs.
G) On Federal Government contracts where the surety is unable to get a satisfactory take-over agreement in the face of a serious tax or other set-off probabilities and it is impossible or impractical to line up a protective bid.

Claim Management

In addition to understanding the options available to the surety, there are other factors that help the claim personnel and surety management make decision. Additional items to consider include:

1) Summary of accounting figures
2) Computation to determine exposure
3) Financial statements (principal and indemnitors)
4) Recap listing of mortgages and loans—balance due

5) List of assets and equipment—current appraised value
6) Equipment—condition—(pictures)—they are worth a 1000 words
7) Property (buildings and land)—(pictures)—they are worth a 1000 words
8) Agreements—legal—problem provisions
9) Jobs—problem areas and engineer's technical comments
10) Job superintendents—level of skill and experience

Overhead (not direct job cost) should be thoroughly reviewed—it may be a major factor. To figure this, the following should be considered:

1) Principal's and family member's salaries (in-laws and outlaws)
2) Funds to bring loans current—usually in arrears
3) Office staff and its cost
4) Premiums unpaid on insurance (same company as surety) (agent—uncollected premium with no recourse)
5) Payroll taxes and union dues outstanding

Additional concern is the cost of outside assistance. Outside assistance includes: 1) construction management assistance, 2) attorney's fees and expenses, 3) consulting, engineering and accounting fees and expenses. The surety claim person must be the quarterback and in charge. Legal, engineering, and financial experts cost the surety money and should be used only when necessary.

Fidelity and surety Loss Adjustment Expense runs 10% to 12% of earned premium, or $215 million in 1990. An estimate for outside assistance would therefore be $172 million. Might I submit that this is about the same as the salvage received. Yet here's the second biggest mystery in the world: how can consultants estimate a job's cost and hours required to complete, but cannot project or provide an upset figure to complete the assignment? It must be added to any funding/financing cost figure for completion. "Suretyship is a field in which everyone is an expert and the less experience one has had, the more experienced they are!"

A caveat: a contract surety loss never—I repeat—never gets better! Exposure must be controlled, and experience is the only plus the surety has on its side. Attorney and consultants are retained to advise and counsel. It's not their role to mange or direct claims; that's the primary's job. Outside firms should provide timely monthly or quarterly reports that set out 1) current status, 2) any problems, 3) any adjustments, 4) changes in completion dates and why, and 5) professional opinion as to options and present course of action. Also, detailed statements should be submitted (date, work performed, time spent, etc.).

The following provisions are suggested for a funding agreement including the basis for terminating said agreement:

1) Cut overhead by percentage figure and date certain.
2) Target date for completion (each project).
3) Review date on job progress (set bench marks for completion).
4) An increase in cost to complete over set percentage would be grounds for a relet.
5) Sell property and equipment.
    a) Schedule starting dates.

b) Determine selling price.
c) Set final date for sale on pre-determined selling price.
d) Consider options to set in place for action after final date.
e) Identify party with final right to approve or reject.

6) Review claims for extras and set date to determine bedrock settlement price versus ongoing cost. Thereafter set a date to reduce.
7) Consider bonus to employees—received percentage of settlement over bedrock figure.
8) My visits with outside firms has revealed their biggest concern is the client doesn't or can't identify the assignment. *Be specific*.
9) Option to consider with consultants: their estimated cost to complete. Will they bond it?

Salvage

At this stage, the surety should secure all available assets belonging to 1) the principal and 2) the indemnitors. The surety should take no prisoners: why should it trust the principal if it wants to withhold assets? In addition, the surety should review again the options from best to worst. Fear has no place in the decision-making process.

1) Principal
2) Indemnitor
3) Banks
4) Surety
   a) Some form of relet
   b) Funding/financing
   c) Protective bid/buy back bond
   d) Do nothing

The surety must determine:

1) Where does it want to be?
2) Why does it want to be there?
3) How to get there
4) Impact on other parties
5) Decision as to course of action

Conclusion

Full knowledge of the facts, figures, and options, coupled with the application of mature and seasoned judgment will lead the surety to the proper solution as to a course of action. The surety must:

1) Investigate
2) Consider options including off ramps and termination provisions
3) Determine exposure and establish reserve
5) Outline a course of action
6) Secure all assets
7) Manage claim and exposure
8) Set target date to sell assets
9) Reserve the right to terminate all agreements
10) Manage and direct outside firm by outlining in detail their assignment.

The successful surety will manage to accomplish all of the above in a short amount of time. The longer it takes to determine the facts, the longer it takes to determine a course of action, and the more money the surety stands to lose. It is not a race to the finish, but in a default situation, a seasoned surety claim person will act with alacrity. With experience and an informed process, the surety can minimize its losses.

(Contractor's letterhead) Exhibit A

(Date)

(Surety's Name, Address)

_________________________________.
_________________________________.
_________________________________.

Re: Principal: ____
Projects(s): ____________________________________
Bond No.: ____________ Amount: ______________
Obligee: ____________________________________

Dear _____________________: (Surety's claim person)

This is to formally notify __________ (Surety) ___________ as our Surety on the referenced contracts that we are unable to complete these contracts and pay the obligations in connection therewith. In view of the situation, we hereby voluntarily declare __________________ (Contractor) ____________________ to be in default of these contracts, hereby waiving all our rights, title and interest to all contract funds, balances and retainages in connection therewith whether earned or unearned and hereby assign all such monies due or to become due on these contracts to ________________ (Surety) ________________, our Surety.

We further reaffirm all of the terms and covenants contained in the bond applications and general indemnity agreements previously executed by us as condition to issuance by ________________(Surety)____________ of the referenced bonds. __________ ______(Surety)____________ as our Surety should proceed to take whatever measures it deems necessary to protect its interest with respect to these defaults and contracts, and we hereby agree to allow free access to all books and records to you and your representatives.

Very truly yours,

____
(Contractor)

Witness:

_________________________ By: ___________________________
(name of individual indemnitor)

____________ (Surety) ___________ Company Exhibit B
--- Blank Street
Anywhere, U.S.A.

Certified Mail – Return Receipt Requested

______ (Principal) _________Construction Company
Anytown, U.S.A.

Gentlemen:

Re: Bonds Issued ________ (Surety) __________Company
Principal: __________________ Construction Company

You have requested financial assistance from _ (Surety) ________Company for the purpose of completing the above projects and paying obligations to various subcontractors and suppliers of labor, material and services used in the performance of the work on such projects incurred by ______(Principal) ________ Construction Company.

Before _______(Surety)________Company can respond to your request, an investigation must be conducted by its representatives. You must not construe any statement or action of _______(Surety)________Company's representatives during such investigation to be an agreement or promise to render financial assistance to ___ ______(Principal)__________or to engage in any course of conduct for the benefit of __________(Principal)__________Construction Co.

________(Principal)_______Construction Company represents that it has no other source of funding its financial needs available to it and that it is unable to perform the obligation secured by the bonds without financial and other assistance from _____ ______(Surety)__________Company.

________(Principal)________Construction Company gives permission to ______(Surety)________Company and its representatives to conduct such an investigation as ________(Surety's)________representatives deem necessary including, but not limited to: A review of _________(Principal's)_______ Books, records and files; interviews with all officers and employees of ________(Principal) _______; interviews with __________(Principal's)________attorneys, accountants and other professionals retained by _________(Principal)__________; interviews with ______(Principal's)_______banker and other creditors; and interviews with

the owner's representatives on each project of (Principal) ______ whether secured by bond or not.
Surety Company

You, _______(Principal)_______Construction Company, and others executed a General Indemnity Agreement in favor of _______(Surety)_______Company. You and ______(Principal) _Construction Company, upon acceptance of this letter, affirm the obligations undertaken by you and _(Principal)________, Construction Company pursuant to such agreement and agree to execute all documents necessary to secure ________(Surety)________Company from all loss and expense it may incur in behalf of ___(Principal)________Construction Company. You also agree to execute any other documents necessary for you or _______(Principal)________Construction Company to perform and implement the obligations undertaken under the General Indemnity Agreement.

Yours truly,

______________________________.
(Surety Company representative)

cc: Individual Indemnitors (Certified Mail – Return Receipt Requested)

1.
2.
3.

Accepted.

____
_____(Principal)__________Construction Company

______ ______
A. B.

Appendix XIV. Common Disputes Arising From Surety Contract Bonds

Contract bond claims can result in disputes. Listed below are the causes of typical disputes that arise in practice between the surety and owner (obligee):

1. The owner does not follow payment terms (e.g., overpayment, premature payments, or payment for substandard work).
2. The owner's actions contradict its obligation to mitigate loss within reason, through such actions as:
   a. Hiring a more expensive contractor to complete the project.
   b. Deciding he or she desires a higher quality than that which was originally contracted.
3. Owner requests recovery of additional architect/engineer charges for revisions, inspections, and administrative expenses associated with a contract default.
4. Owner challenges surety's loss resolution.
5. The timeliness of the surety's response to the situation, especially if the contractor has filed bankruptcy at or near the same time as the default.
6. Contractor disputes the owner's declaration that there is default.
7. Either the owner or contractor supplied the surety with incorrect information that affected crucial underwriting decisions.
8. Contractor assumes additional responsibility on a project without informing surety (i.e., significant change in scope or material alteration.
9. Specialty contractors and/or material suppliers' fail to follow statutory notice requirements ("Sub Waits Too Long" 1987).
10. Owner fails to make payments and becomes insolvent.
11. Owner fails to inspect and monitor contractor's progress.
12. Owner fails to obtain signed power of attorney from company underwriting bonds.

In some cases contractor default and/or disputes are traceable to poor underwriting procedures. Owners must realize that surety bonds do not, in themselves, insure that a contractor is qualified to perform the contract (Russell 1990). As explained previously, *surety bonds are not insurance. Nor do they provide the same benefits as insurance.* Poor underwriting procedures can result from underwriting designed to generate cash flow from collected bond premiums, as was the case in the early 1980s; the surety personnel's construction inexperience; or an incorrectly focused analysis. Hence, the low bid that manages to secure the required contract bonds is not necessarily the "lowest responsible bidder." The term "lowest responsible bidder" refers to a contractor who is competent and capable of performing the work and whose bid substantially complies with the material requirements that the bid invitation contains. *Thus, an owner should not rely on the ability of the contractor to secure surety bonds as the sole gauge of a contractor's competence. For these reasons and others, many owners provide their own qualification evaluations.*

Additionally, the structure of the surety company may be responsible for some problems with underwriting. Underwriting departments typically function independently within the greater surety company and do not often consult with the claims department. This can deprive them of important information and assessment tools that can be valuable in the contractor evaluation process. Fierce competition between surety companies for business from bond-seeking contractors can also result in complacent attitudes or the lack sound judgment by underwriters. Some industry experts fear such is the current case at the end of the 1990s, as close to 600 surety companies or divisions are writing bonds. How competition has and will affect the surety industry is dealt with in more detail in the final chapter of the book.

Appendix XV. Sources of Information About Suretyship

National Association of Minority Contractors (NAMC)
1333 F Street, NW, Suite 500
Washington, D.C. 20004
Phone (202) 347-8259
Fax: (202) 628-1876
www.namc.org/

National Association of Surety Bond Producers (NASBP)
5225 Wisconsin Avenue, NW
Suite 600
Washington, D.C. 20015-2014
Phone: (202) 686-3700
Fax: (202) 686-3656
www.nasbp.org/

Surety Association of America (SAA)
1101 Connecticut Ave, NW, Suite 800
Washington D.C. 20036
Phone: (202) 463-0600
Fax: (202) 463-0606
www.surety.org/

Surety Information Office (SIO)
5225 Wisconsin Avenue NW, Suite 600
Washington, DC 20015-2014
Phone: (202) 686-7463
Fax: (202) 686-3656
www.sio.org

Surety Bond Program Area Office Contacts for Small Business Association (SBA)
www.sba.gov/OSG/

Area Office 1
Phone: (215) 580-2717
Fax: (215) 580-2759
Geographic Territory: MA, RI, ME, NH, VT, CT, NY, NJ, PR, VI, PA, DC, MD, VA, WV, DE, OH, IL, IN, MN, MI, WI

Area Office 2
Phone: (404) 347-0210 ext. 155
Fax: (404) 347-6448
Geographic Territory: GA, FL, NC, SC, AL, MS, KY, TN, TX, OK, AR, NM, LA

Area Office 3
Phone: (303) 844-5231 Ext. 251
Fax: (303) 844-6490
Geographic Territory: KS, IA, MO, NE, CO, SD, UT, WY, ND, MT

Area Office 4
Phone: (206) 553-0961
Fax: (206) 553-6259
Geographic Territory: WA, AK, OR, ID, CA, AZ, HI, NV, GU

REFERENCES

American Institute of Certified Public Accountants. *Contractors-Audit and Accounting Guide*. New York, NY: AICPA, 1981.

Associated General Contractors of America. Working Document of the Wisconsin Chapter, Madison, WI: AGCA, 1990.

"Bad Bonds Bring 'Fresh Start' Firm To Stand Still." *Morning News Tribune*, Seattle, WA, Oct. 15, 1990.

Bauman, J. "Mitigating Surety Construction Losses." *Best's Review Property/Casualty Insurance Edition* 10(1989):58.

Best's Insurance Reports, Property-Casualty. A.M. Best Company, Oldwick, NJ: A.M. Best Company, 1997.

Bond Default Manual. American Bar Association, Chicago, IL: ABA, 1987.

Bond and Insurance Coverages for Highway Contractors. National Cooperative Highway Research Program Report 341, Transportation Research Board, National Research Council, Washington, D.C., 1991.

"Bonds & Insurance for Contractors." Pasadena, CA: Corroon & Black, Construction Industry Division, 1989.

"Bonding, Insurance Falling Short." *Engineering News-Record* 13(1988): 38-43.

"Bonding Sources Grow." (1991). *Engineering News-Record* 4(1991): 11-12.

"Brown and Root Strikes Back." *Engineering News-Record* 24(1982): 11.

Canadian Construction Documents Committee. *A Guide to Construction Surety Bonds*. Ottawa, Canada, Canadian Construction Documents Committee, 1994.

"Changing of the Guard Remains a Problem." *Engineering News-Record* 20(1990): 15-16.

"Bond-Financed Self-Insurance for Public Entities." *Journal of Chartered Property and Casualty Underwriters* March (1998): 21-33.

Clough, R. H. *Construction Contracting, Fourth Edition*. New York: John Wiley and Sons, Inc., 1981.

"Construction Trends for 1991 and Beyond." Fails Management Institute (FMI), memo dated April 10, 1991, Denver, CO.

"Contract Change Let Surety Off Hook." *Engineering News-Record* 22(1988): 40.

Cushman, R. F. and Stamm, C. H. *Handling Fidelity and Surety Claims.* New York: John Wiley and Sons, 1984.

Dockray, G. "Do Sureties Care About Life Insurance and Estate Planning?" *Suretyscope: Journal of the National Association of Surety Bond Producers* 3(1979): 18-21.

"DOD Finds Liability Fears Could Hurt Cleanup Effort." *Engineering News-Record* 2(1991): 7.

"FBI probes Fife Contract Bond." *Morning News Tribune*, Seattle, WA, Oct. 15, 1990.

Federal Register. U.S. Dept. of the Treasury, Fiscal Services, 127(1988): 198.

"Firms Seek Legislation To Ease Bonding Problem." *Engineering News-Record* 13(1990): 20.

"Florida Bonding Case Has Sureties Worried." *Engineering News-Record* 10(1989): 16.

"GAO Suggests Bias in Surety Market." *The Insurance Regulator* 26(1995): 1.

"Gaps or Going Bare? What the Insurance and Bonding Industry Is Doing To Help Design-Build." Panel held at the Design-Build Institute of America and the American Institute of Architect's Professional Design-Build Conference, Chicago, Illinois, October 14-16, 1998.

Garvin, J. H. "Contract Bond Underwriting—Success by Default or Failure by Design." *Journal of Chartered Property and Casualty Underwriters* 32(1979): 93-103.

Gastel, Ruth. "Environmental Pollution: Insurance News." *Insurance Issues Update* June, 1998.

Gilbert, Evelyn. "Environmental Pros Lack Coverage." *National Underwriters Property & Casualty* 40(1995): 29.

Glossary Fidelity Surety. Iselin, NJ: The Surety Association of America, 1997.

"Government Reins in Individual Sureties." *Engineering News-Record* 24(1989): 18.

"Guarantees With Nothing Behind Them." (1989). *Forbes* 2(1989): 41-42.

Halpin, D.W. *Financial and Cost Concepts for Construction Management.* New York: John Wiley and Sons Inc., 1985.

Halpin, D.W. and Woodhead, R. W. *Construction Management.* New York: John Wiley and Sons Inc., 1985.

Hammond, D. *Granting Credit to Contractors.* New York: Touche Ross and Co., 1982.

Hannum, John L. "SBA Program Increases Access To Surety Bonds." Washington, DC: Surety Information Office, 1997.

Hirsh, Jeff. "Surety: Strong Profits Continue To Attract a Plethora of Players and Capacity." *Engineering News-Record* 6(1998): S-8.

"Individual Sureties May Make a Comeback." *Engineering News-Record* 24(1990): 21-22.

Jewler, R. The United States Bankruptcy Code. Washington, DC: Congressional Research Service, 1989.

Johnston, B.M. "The Surety Evaluation Process in the U.S. Construction Industry." Thesis presented at Massachusetts Institute of Technology, Cambridge, Massachusetts, in partial fulfillment of the requirements for the degree of Master of Science in Civil Engineering, 1981.

Klinger, Marilyn. "The Times They Are a Changin': An Evolution of Surety Claims Handling." *Suretyscope: Journal of the National Association of Surety Bond Producers* 11(1997): 1-6.

Levy, S. M. *Japan's Big Six: Inside Japan's Construction Industry.* New York: McGraw-Hill, Inc., 1993.

Mc Kenna, E. J. *Surety Underwriting Manual.* Indianapolis, IN: The Rough Notes Co, Inc., 1972.

McGreevy, Susan Linden. "Here's Bonding Made Easier: SBA's New 'Plan B' Offers Fewer Hassles." *Contractor* 9(1991): 56.

"Miller Act Frustrates Subs." *Engineering News-Record* 7(1997): 11.

"Minority-Owned Firms Get Fewer Bonds." *Engineering News-Record* 15(1995): 18.

Mrozek, D. "The Surety's Responsibility for Withholding Taxes." Mid-Winter Meeting, Fidelity and Surety Law Committee, Jan. 1978.

Opfer, N. D. and Koehler, K. "Improving Construction Bonding Capacity." Proceedings, Thirty-Fourth Annual Conference of the American Association of Cost Engineers, Boston, MA, E.4.1-E.4.8, 1991.

"Opponents Find Common Ground." *Engineering News-Record* 26(1990): 11.

Perez, M. "Unsure Future in Surety Line." *Best's Review: Property/Casualty Insurance Edition* 2(1988): 139-140.

Peroutka, Raymond J. Jr. "Accounting Issues in Surety Underwriting & Claims." Presentation given to National Association of Surety Bond Producers, August, 1991.

Petro, J.J. "Suretyship in 2750 B.C.: The Second Oldest Profession." *Suretyscope: Journal of the National Association of Surety Bond Producers* 11(1987): 1-5.

Pietroforte, Roberto. *Building International Construction Alliances: Successful Partnering for Construction Firms.* London: E & FN Spon, 1997.

Port-Hull, Candace. "Design/Build Insurance: Filling in the Gaps." *Civil Engineering* 9(1997): 56-59.

Pryce, Deborah. "Pryce Applauds Brownfields Pilot Program." Press release, July 15, 1998.

"Profitable Companies May Miss Trouble Signs." *Engineering News-Record* 27(1990): 10-11.

Provost, Lloyd. "USA: Despite Worrying Times, the Surety Industry Will Remain Profitable." *Reuters Textline Review*, March 6, 1991, p. 23.

Rate Manual: Surety Bonds. Iselin, NJ: The Surety Association of America, 1987.

Rejda, G. *Principles of Insurance, Third Edition.* Glenview, IL: Scott Foresman Publishers, 1988.

Remmen, A. *The Contract Bond Book.* Cincinnati, OH: The National Underwriters Co., 1977.

Rice, H.L. "Construction Acquisition Trends." *Construction Business Review* Nov./Dec., 1991, pp. 68-73.

RMA Annual Statement Studies. Philadelphia, PA: Robert Morris and Associates, 1997.

Russell, J.S. "Construction Contract Bonds." *Journal of Management in Engineering* 3(1991b): 299-313.

Russell, J.S. "Contractor Failure: Analysis." *Journal of Performance of Constructed Facilities* 3(1991c): 163-180.

Russell, J.S. *Constructor Prequalification: Choosing the Best Constructor and Avoiding Constructor Failure.* New York: ASCE Press, 1996.

Russell, J.S. "Insurance Industry: Overview." *Journal of Management in Engineering* 1(1991a): 98-118.

Russell, J.S. "Surety Industry: Overview." *Journal of Management in Engineering* 3(1990): 323-341.

Russell, J.S. "Underwriting Process for Construction Contract Bonds." *Journal of Management in Engineering* 1(1992): 63-80.

Schleifer, T. C. *A Contractor's Survival Guide Managing With Confidence.* McLean, VA: Aslan Press, 1987.

Schriener, J. "Letters of Credit Return." *Engineering News-Record* 13(1992): 26.

"Scrap companies win Superfund case." *New Steel* 7(1998): 23.

"Senate Panel Passes Superfund Liability Bill." *Engineering News-Record* 15(1990): 7.

Severson, G.D.; Russell, J.S.; and Jaselskis, E.J. "Predicting Construction Contractor Surety Bond Claims Based on Constructor Financial Data." *Journal of Construction Engineering and Management* 2(1994): 405-20.

Shiatte, K.W. "A Preliminary Analysis on Self-Insurance at NYSDOT." Unpublished final report, Division of Construction, New York State Department of Transportation, 1989.

"Small Shops Face Tightening Credit." *Contractor* 11(1991): 1.

South, A.J. "Risk Analysis and Allocation." *Management of International Construction Projects, Proceedings of Institution of Civil Engineers, London, November 14-15, 1994.* London: Thomas Telford, 1985, pp. 15-22.

Stevens, J. D. and Al-Dulaijan, S. U. "Contractor Financing, Public Works in Saudi Arabia." *Journal of Construction Engineering and Management* 1(1989): 1-13.

"Subs Find That Bonding Is a Growing Problem." *Engineering News-Record* 10(1988): 11-12.

"Subs To Offer Own Bonding." *Engineering News-Record* 17(1989): 15-16.

"Sub Waited Too Long To File Defect Claim." *Engineering News-Record* 8(1987): 60.

"Sureties In the Black in '88.'" *Engineering News-Record* 20(1989): 10.

"Surety Asking More of Contractors." *Engineering News-Record* 9(1990): 13-14.

"Surety Bond Crunch Hiking Superfund Costs." *Engineering News-Record* 8(1990): 8-9.

"Surety Bonds and Minority Contractors." Hearing Before the Subcommittee on Commerce, Consumer Protection, and Competitiveness of the Committee on Energy and Commerce, House of Representatives, One Hundredth Congress, second session, June 30, 1988, Washington, DC.

"Surety Companies: What They Are and How To Find Out About Them." Bethesda, MD: National Association of Surety Bond Producers.

"Surety Controls Proposed." *Engineering News-Record* 22(1988): 16.

"Surety firm implicated in other schemes." (1990). *Morning News Tribune*, Seattle, WA, Oct. 15, 1990.

"Surety Firm Competition is Saving Contractors Money." *Engineering News-Record* 19(1996): 21.

"Surety Seminar Summary." Notes on a seminar regarding problems facing the surety industry. Indianapolis, IN; November 13, 1987.

"Surety Underwriting Considerations." *Contractor* 5(1989): 22-26.

"Survey Examines 6 Issues in Succession Planning for Contracting Firms." *Contractors Business Manager Reports* August, 1998, p. 1.

Sweeney, Patrick M. "Heightened Competition Impeded Surety and Fidelity Lines in 1996." *Best Week: Property/Casualty Supplement*, August 25, 1997.

Sweeney, Patrick M. "No Surety or Fidelity Thing; Property and Casualty Insurance." *Best's Review—Property-Casualty Insurance Edition* 5(1997): 70.

Tiong, Robet L. K. "Strategies in Risk Management of On-Demand Guarantees." *Journal of Construction Engineering and Management* 2(1992): 229-43.

Trodden, S. "Negotiating A Smooth Course Through the Guarantee Business." *Euromoney Trade Finance* April, 1986, pp. 44-46.

United States Bankruptcy Code (1988). Chapter 11 §362(a), Washington, DC.

"Use of Surety Bonds in Federal Construction Should Be Improved." Washington, DC: General Accounting Office, 1975.

"Users Information." Iselin, NJ: The Surety Association of America, 1991.

Walford, C. *The Insurance Cyclopaedia, Volume 3.* London: Charles and Edwin Layton, 1875.

Weitz, S. M. "The Limited Liability Company: An Alternative To S Corporations and Limited Partnerships." *Construction Business Review* Jan.-Feb., 1992, p. 37.

Wentz, G. R. *Fidelity and Surety Bonding.* St. Paul, MN: Lee Company, 1939.

Winston, Sherie. "Miller Act Under New Scrutiny." *Engineering News-Record* 12(1998): 21.

Winter, C. H. "Communication and Cooperation-Agents, Underwriters and Bond Claim Personnel." *Suretyscope: Journal of the National Association of Surety Bond Producers* 6(1984): 17-22.

Wright, Andrew, and Roe, Andrew. "Brownfield Cleanups Debug Development." *Focus on Environment* 17(1997): 22.

INDEX

ABOUT THE AUTHOR

Jeffrey S. Russell is a professor at the University of Wisconsin-Madison and serves as chair of the Construction Engineering and Management Program. Dr. Russell's research interests center on developing decision-support tools for the managerial aspects of construction including constructor prequalification, surety bonds for construction contracts, constructor failure, constructability, maintainability, quality control and quality assurance, innovative project delivery systems, and construction automation and robotics. He has taught such topics as surety contract bonding, construction project management, constructability analysis, construction equipment and heavy construction methods, and construction automation.

Dr. Russell has been a recipient of four national American Society of Civil Engineers (ASCE) awards: the Collingwood Prize, the Edmund Friedman Young Engineer Award, the Walter L. Huber Civil Engineering Research Prize (for modeling of construction contractor evaluation and failure), and the Thomas Fitch Roland Prize. He has been active on the publications committee for ASCE's *Journal of Management in Engineering*, and currently serves as the journal's editor-in-chief. He has co-authored numerous articles for ASCE journals, and he compiled a how-to manual entitled *Constructor Prequalification: Choosing the Best Constructor and Avoiding Constructor Failure* (1996). In addition, he served as general chair for the 1997 ASCE Construction Congress held in Minneapolis, Minnesota.

At the local level, Dr. Russell has chaired the Wisconsin ASCE's construction committee and was faculty advisor to the university's ASCE student chapter in 1996 when the students hosted the national concrete canoe competition. He has five times been selected by the ASCE student chapter as the outstanding professor in Civil and Environmental Engineering. He has also received the Polygon College of Engineering Award.

Dr. Russell received his B.S.C.E. degree from the University of Cincinnati and his M.S.C.E. and Ph.D. degrees from Purdue University. In addition, Dr. Russell plays an active role in his church and devotes himself fully to his family.